Dance Music Manual

Cirencester College, GL7 1XA
Telephone: 01285 640994

348227

Dance Music Manual
Tools, Toys and Techniques

Second Edition

Rick Snoman

AMSTERDAM • BOSTON • HEIDELBERG • LONDON • NEW YORK • OXFORD
PARIS • SAN DIEGO • SAN FRANCISCO • SINGAPORE • SYDNEY • TOKYO

Focal Press is an imprint of Elsevier

Focal Press is an imprint of Elsevier
The Boulevard, Langford Lane, Kidlington, Oxford OX5 1GB, UK
30 Corporate Drive, Suite 400, Burlington, MA 01803, USA

First edition 2009
Reprinted 2010 (twice), 2011 (three times), 2012

Notice
No responsibility is assumed by the publisher for any injury and/or damage to persons
or property as a matter of products liability, negligence or otherwise, or from any use
or operation of any methods, products, instructions or ideas contained in the material
herein. Because of rapid advances in the medical sciences, in particular, independent
verification of diagnoses and drug dosages should be made.

British Library Cataloguing in Publication Data
A catalogue record for this book is available from the British Library

Library of Congress Cataloging-in-Publication Data
A catalog record for this book is available from the Library of Congress

ISBN: 978-0-2405-2107-7

For information on all Focal Press publications
visit our website at www.elsevierdirect.com

Printed and bound in China

12 13 14 15   12 11 10 9 8 7

Working together to grow
libraries in developing countries

www.elsevier.com | www.bookaid.org | www.sabre.org

ELSEVIER   BOOK AID
International   Sabre Foundation

This book is dedicated to my children: Neve and Logan.

Contents

Acknowledgements

I would like to personally thank the following for their invaluable help, contributions and/or encouragement in writing this book:

Catharine Steers at Elsevier (for being so patient)
Colin and Janice Lewington
Darren Gash at the SAE Institute of London
Dave 'Cannockwolf' Byrne
DJ 'Superstar' Cristo
Mark Penicud
John Mitchell
Mick 'Blackstormtrooper' Byrne,
Helen and Gabby Byrne
Mike at the Whippin' post
Richard James
Steve Marcus
Everyone on the Dance Music Production Forum

All music featured on the CD – ©Phiadra
(R. Snoman & J. Froggatt)
Vocals on Chill Out supplied by Tahlia Lewington
Vocals on Hip Hop supplied by MC Darkstar
Vocals on Trance and Garage supplied by Kate Lesing
Cover Design: Daryl Tebbut

ix

Preface

If a book is worth reading then it's worth buying ...

Welcome to the *Dance Music Manual – Second Edition*. After the release of the first edition way back in May 2004, I received numerous emails with suggestions for a second edition of the book and I've employed as many of them as possible, as well as updating some of the information to reflect the continually updated technology that is relevant to dance musicians. I'd like to personally thank everyone who took the time to contact me with their suggestions.

As with the first edition, the purpose of the *Dance Music Manual* is to guide you through the technology and techniques behind creating professional dance and club-based music. While there have been numerous publications written on this important subject, the majority have been written by authors who have little or no experience of the scene nor the music, but simply rely on 'educated guesswork'. With this book, I hope to change the many misconceptions that abound and offer a real-world insight into the techniques on how professional dance music is written, produced and marketed.

I've been actively involved in the dance music scene since the late 1980s and, to date, I've produced and released numerous white labels and remixes. I've held seminars across the country on remixing and producing club-based dance music, and authored numerous articles and reviews.

This book is a culmination of the knowledge I've attained over the years and I believe it is the first publication of its kind to actively discuss the real-world applications behind producing and remixing dance music for the twenty-first century.

The Dance Music Manual has been organized so as to appeal to professionals and novices alike, and to make it easier to digest it has been subdivided into three parts.

The first part discusses the latest technology used in dance music production, from the basics of synthesis and sampling to music theory, effects, compression, microphone techniques and the principles behind the all-important sound design. If you're new to the technology and theory behind much of today's dance music, then this is the place to start.

The second part covers the techniques for producing musical styles including, among others, trance, drum 'n' bass, trip-hop, rap and house. This not only discusses the general programming principles behind drum loops, basses and leads for the genres, but also the programming and effects used to create the sounds. If you already have a good understanding of sampling rates, bits, synthesis programming and music theory, then you can dip into these sections and start practicing dance music straight away.

The third part is concerned with the ideology behind mixing, mastering, remixing, pressing and publication of your latest masterpiece. This includes the theory and practical applications behind mixing and mastering, along with a realistic look at how record companies work and behave; how to copyright your material, press your own records and the costs involved.

At the end of the book you'll also find a chapter in which an international DJ has submitted his view on dance music and DJing in general.

Of course, I cannot stress enough that this book will *not* turn you into a superstar overnight and it would be presumptuous to suggest that it would even guarantee you a successful dance record. Dance music has always evolved from musicians pushing the technology further. Rather, it is my hope that it will give you an insight into how the music is produced and from there it's up to you to push in a new direction.

Creativity can never be encapsulated in words, pictures or software and it's our individual creative instincts and twists on a theme that produces the dance floor hits of tomorrow.

Experimentation always pays high dividends.

Finally, I'd also like to take this opportunity to thank you for buying *The Dance Music Manual*. By purchasing this book, you are rewarding me for all the time and effort I've put into producing it and that deserves some gratitude. I hope that, by the end, you feel it was worth your investment.

> "I would like to remind record companies that they have a cultural responsibility to give the buying public great music. Milking a trend to death is not contributing to culture and is ultimately not profitable."

Dance music has always relied on sampling. From its first incarnation of mixing two records together on a pair of record decks to make a third mashup to the evolution and consequent increased power of the sampler and audio workstation. It would be fair to say that without sampling, dance music would be a very different beast and may not have even existed at all.

For legal reasons I cannot suggest that any artist choosing to read this book dig through their record collections for musical ideas to sample, but now more than ever, it has become a cornerstone of the production of dance-based music. The importance of sampling should not be underestimated and when used creatively, it can open new boundaries and spawn entirely new genres of music. Perhaps the best example of this is the Amen Break.

Back in 1969, a song was released by the Winston's called Colour Him Father. The B side to the record contained a track named Amen Brother. The middle eight of this particular song contained a drum break just less than 6 s long which was later named the 'Amen Break'. It contained nothing more than a drummer freelancing on his kit, but it became one of the largest factors in the evolution of dance music.

Although used well before the emergence of dance music by Hip Hop artists, it was the release of the SP1200 sampler by Emu that kick started the sampling revolution. And soon to follow was the first 'commercial' dance record to feature the Amen break Mantronix "King of the Beats" in 1990, followed by UK Apachi and Shy FX "Original Nuttah" in 1994. Over the years, the Amen Break became more and more widely used and appeared in tracks such as:

- 2 Live Crew: *Feel Alright Yall*
- 4 Hero: *Escape That*
- Amon Tobin: *Nightlife*
- Aphex Twin: *Boy/Girl Song*
- Atari Teenage Riot: *Burn Berlin Burn*
- Brand Nubian: *The Godz Must Be Crazy*
- Deee-Lite: *Come on In, the Dreams are Fine*
- Dillinja: *The Angels Fell*
- Eric B & Rakim: *Casualties of War*

- Freestylers: *Breaker beats*
- Funky Technicians: *Airtight*
- Heavy D: *MC Heavy D!*
- Heavy D: *Let it Flow*
- Heavy D: *Flexin'*
- Heavyweight: *Oh Gosh*
- J. Majik: *Arabian Nights*
- J. Majik: *Your Sound*
- Lemon D: *This is Los Angeles*
- Level Vibes: *Beauty & the Beast*
- Lifer's Group: *Jack U. Back* (*So You Wanna Be a Gangsta*)
- Ltj Bukem: *Music*
- Maestro Fresh Wes: *Bring it On* (*Remix*)
- Movement Ex: *KK Punani*
- Nice & Smooth: *Dope Not Hype*
- Salt-N-Pepa: *Desire*
- Scarface: *Born Killer*
- Schoolly D: *How a Black Man Feels*
- Goldie: *Chico: Death of a Rock Star*
- Roni Size: *Brown Paper Bag* (*Nobukazu Takemura Remix*)
- Oasis : *Do Y'Know What I Mean*
- Frankie Bones: *Janets Revenge*

Perhaps more importantly, though, the evolution of both breakbeat and jungle – alongside their many offshoots – are accredited to these 6 s of a drum break. Indeed, an entire culture arose from a 1969 break as samplers became more experimental and artists began cutting and rearranging the loop to produce new rhythms.

If this break had suffered the same copyright laws that music does today it is entirely possible that breakbeat, jungle, techstep, artcore, 2-step and drum 'n' bass – among many others – would never have come to light. It takes a large audience to appreciate new material before it becomes publicized.

As record companies and artists continue to tighten the laws on copyright, our musical palette is becoming more and more limited, and experimental music born from utilizing past samples is being replaced with 'popcorn' music in order to ensure that the monies returned are enough to cover the original fee for using a sample.

Whereas scientists are free to build upon past work without having to pay their peers and film directors are free to copy the past, disgracefully music no longer exhibits that same flexibility. With the current copyright laws, musicians can no longer appropriate from the past without a vast amount of paperwork, a good solicitor, a large wallet and an understanding record company.

Record companies anxiously await the next 'big thing' and voice concerns over the lack of new musical ideas and genres. Yet, in the same breath they are – perhaps unintentionally – industriously locking down culture and placing countless limitations on our very creativity. The musical freedom that our predecessors experienced and built upon has all but vanished.

Without the freedom to borrow and develop on the past, creativity is stifled and with that our culture can only slow to a grinding pace. To quote Laurence Leesig: '*A society free to borrow and build upon the past is culturally richer than a controlled one*'.

PART 1
Technology and Theory

CHAPTER 1
The Science of Synthesis

'Today's recording techniques would have been regarded as science fiction forty years ago.' ...

Today's dance- and club-based music relies just as heavily on the technology as it does on the musicality; therefore, to be proficient at creating this genre of music it is first necessary to fully comprehend the technology behind its creation. Indeed, before we can even begin to look at how to produce the music, a thorough understanding of both the science and the technology behind the music is paramount. You wouldn't attempt to repair a car without some knowledge of what you were tweaking, and the same applies for dance- and club-based music.

Therefore, we should start at the very beginning and where better to start than the instrument that encapsulated the genre – the analogue synthesizer. Without a doubt, the analogue synthesizers were responsible for the evolution of the music, and whilst the early synthesizers are becoming increasingly difficult to source today, nearly all synthesizers in production, whether hardware or software, follow the same path first laid down by their predecessors. However, to make sense of the various knobs and buttons that adorn a typical synthesizer and observe the effects that each has on a sound, we need to start by examining some basic acoustic science.

ACOUSTIC SCIENCE

When any object vibrates, air molecules surrounding it begin to vibrate sympathetically in all directions creating a series of sound waves. These sound waves then create vibrations in the ear drum that the brain perceives as sound.

The movement of sound waves is analogous to the way that waves spread when a stone is thrown into a pool of water. The moment the stone hits the water, the reaction is immediately visible as a series of small waves spread outwards in every direction. This is almost identical to the way in which sound behaves, with each wave of water being similar to the vibrations of air particles.

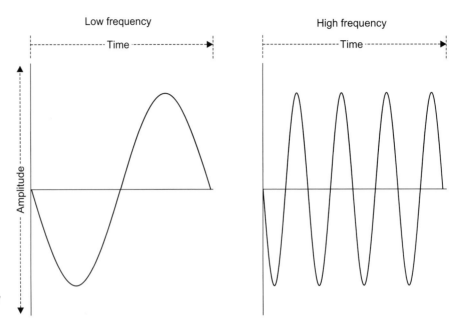

Low frequency

High frequency

FIGURE 1.1
Difference between low and high frequencies

For instance, when a tuning fork is struck, the forks first move towards one another compressing the air molecules before moving in the opposite direction. In this movement from 'compression' to 'rarefaction' there is a moment where there are less air molecules filling the space between the forks. When this occurs, the surrounding air molecules crowd into this space and are then compressed when the forks return on their next cycle. As the fork continues to vibrate, the previously compressed air molecules are pushed further outwards by the next cycle of the fork and a series of alternating compressions and rarefactions pass through the air.

The numbers of rarefactions and compressions, or 'cycles', that are completed every second is referred to as the operating frequency and is measured in Hertz (Hz). Any vibrating object that completes, say, 300 cycles/s has a frequency of 300 Hz while an object that completes 3000 cycles/s has a frequency of 3 kHz.

The frequency of a vibrating object determines its perceived pitch, with faster frequencies producing sounds at a higher pitch than slower frequencies. From this we can determine that the faster an object vibrates, or 'oscillates', the shorter the cycle between compression and rarefaction. An example of this is shown in Figure 1.1.

Any object that vibrates must repeatedly pass through the same position as it moves back and forth through its cycle. Any particular point during this movement is referred to as the 'phase' of the cycle and is measured in degrees, similar to the measurement of a geometric circle. As shown in Figure 1.2, each cycle starts at position zero, passes back through this position, known as the 'zero crossing', and returns to zero.

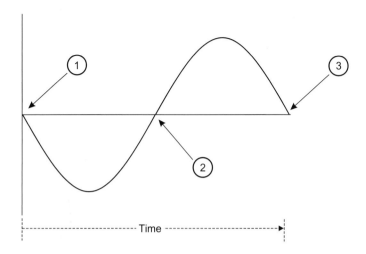

FIGURE 1.2
The zero crossing in a
waveform

Consequently, if two objects vibrate at different speeds and the resulting waveforms are mixed together, both waveforms will start at the same zero point but
the higher frequency waveform will overtake the phase of the lower frequency.
Provided that these waveforms continue to oscillate, they will eventually catch
up with one other and then repeat the process all over again. This produces an
effect known as 'beating'.

The speed at which waveforms 'beat' together depends on the difference in frequency between them. It's important to note that if two waves have the same
frequency and are 180° out of phase with one another there, one waveform
reaches its peak while the second is at its trough, and no sound is produced.
This effect, where two waves cancel one another out and no sound is produced,
is known as 'phase cancellation' and is shown in Figure 1.3.

As long as waveforms are not 180° out of phase with one another, the interference
between the two can be used to create more complex waveforms than the simple
sine wave. In fact, every waveform is made up of a series of sine waves, each slightly
out of phase with one another. The more complex the waveform this produces,
the more complex the resulting sound. This is because as an increasing number of
waves are combined a greater number of harmonics are introduced. This can be
better understood by examining how an everyday piano produces its sound.

The strings in a piano are adjusted so that each oscillates at an exact frequency.
When a key is struck, a hammer strikes the corresponding string forcing it
to oscillate. This produces the fundamental pitch of the note and also, if the
vibrations from this string are the same as any of the other strings natural
vibration rates, sets these into motion too. These are called 'sympathetic vibrations' and are important to understand because most musical instruments are
based around this principle. The piano is tuned so that the strings that vibrate
sympathetically with the originally struck string create a series of waves that are
slightly out of phase with one another producing a complex sound.

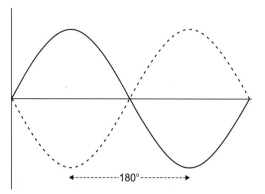

FIGURE 1.3
Two waves out of phase

Any frequencies that are an integer multiple of the lowest frequency (i.e. the fundamental) will be in harmony with one another, a phenomenon that was first realized by Pythagoras, from which he derived the following three rules:

- If a note's frequency is multiplied or divided by two, the same note is created but in a different octave.
- If a note's frequency is multiplied or divided by three, the strongest harmonic relation is created. This is the basis of the western musical scale. If we look at the first rule, the ratio 2:3 is known as a perfect fifth and is used as the basis of the scale.
- If a note's frequency is multiplied or divided by five, this also creates a strong harmonic relation. Again, if we look at the first rule, the ratio 5:4 gives the same harmonic relation but this interval is known as the major third.

A single sine wave produces a single tone known as the fundamental frequency, which in effect determines the pitch of the note. When further sine waves that are out of phase from the original are introduced, if they are integer multiples of the fundamental frequency they are known as 'harmonics' and make the sound appear more complex, otherwise if they are not integer multiples of the fundamental they are called 'partials', which also contribute to the complexity of the sound. Through the introduction and relationship of these harmonics and partials an infinite number of sounds can be created.

As Figure 1.4 shows, the harmonic content or 'timbre' of a sound determines the shape of the resulting waveform. It should be noted that the diagrams are simple representations, since the waveforms generated by an instrument are incredibly complex which makes it impossible to accurately reproduce it on paper.

In an attempt to overcome this, Joseph Fourier, a French scientist, discovered that no matter how complex any sound is it could be broken down into its frequency components and, using a given set of harmonics, it was possible to reproduce it in a simple form.

To use his words, 'Every periodic wave can be seen as the sum of sine waves with certain lengths and amplitudes, the wave lengths of which have harmonic relations'. This is based around the principle that the content of any sound is determined by the relationship between the level of the fundamental frequency and its harmonics and their evolution over a period of time. From this theory, known as the Fourier theorem, the waveforms that are common to most synthesizers are derived.

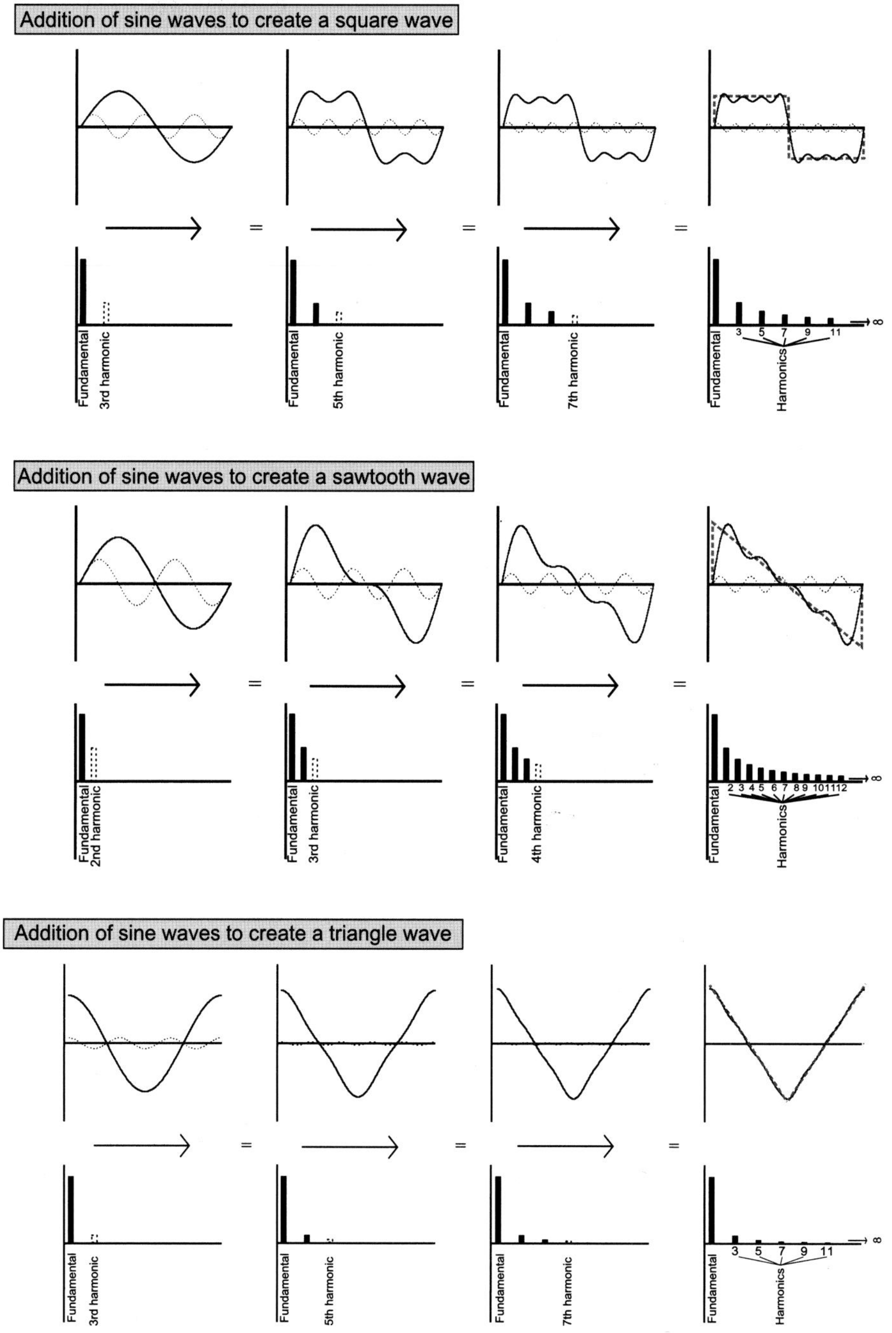

Addition of sine waves to create a square wave

Addition of sine waves to create a sawtooth wave

Addition of sine waves to create a triangle wave

Note: In the above diagram, the 3rd and 4th images in the series on the top row appear to have none, or virtually no wave being added. This is because the odd harmonics decrease in level exponentially. For example, the 3rd harmonic is 3^2 of the level (1/9), the 5th harmonic is 5^2 of the level (1/25), and so forth.

Legend
—— Current wave form
·········· Next wave to be added
------ Eventual wave form

FIGURE 1.4
How multiple sound waves create harmonics

So far we've looked at how both the pitch and the timbre are determined. The final characteristic to consider is volume. Changes in volume are caused by the amount of air molecules an oscillating object displaces. The more air an object displaces, the louder the perceived sound. This volume, also called 'amplitude', is measured by the degree of motion of the air molecules within the sound waves, corresponding to the extent of rarefaction and compression that accompanies a wave. The problem, however, is that many simple vibrating objects produce a sound that is inaudible to the human ear because so little air is displaced; therefore, for the sound wave to be heard most musical instruments must amplify the sound that's created. To do this, acoustic instruments use the principle of forced vibration that utilizes either a sounding board, as in a piano or similar stringed instruments, or a hollow tube, as in the case of wind instruments.

When a piano string is struck, its vibrations not only set other strings in motion but also vibrate a board located underneath the strings. Because this sounding board does not share the same frequency as the vibrating wires, the reaction is not sympathetic and the board is forced to resonate. This resonance moves a larger number of air particles than the original sound alone, in effect amplifying the sound. Similarly, when a tuning fork is struck and placed on a tabletop, the table's frequency is forced to match that of the tuning fork and the sound is amplified.

Of course, neither of these methods of amplification offers any physical control over the amplitude. If the level of amplification can be adjusted, then the ratio between the original and the changed amplitude is called the 'gain'.

It should be noted, however, that loudness itself is difficult to quantify because it's entirely subjective to the listener. Generally speaking, the human ear can detect frequencies from as low as 20 Hz up to 20 kHz; however, this depends on a number of factors. Indeed, whilst most of us are capable of hearing (or more accurately feeling) frequencies as low as 20 Hz, the perception of higher frequencies changes with age. Most teenagers are capable of hearing frequencies as high as 18 kHz while the middle-aged tend not to hear frequencies above 14 kHz. A person's level of hearing may also have been damaged, for example, by overexposure to loud noise or music. Whether it is possible for us to perceive sounds higher than 18 kHz with the presence of other sounds is a subject of debate that has yet to be proven. However, it is important to remember that sounds that are between 3 and 5 kHz appear perceivably louder than frequencies that are out of this range.

SUBTRACTIVE SYNTHESIZER

Having looked into the theory of sound, we can look at how this relates to a synthesizer. Subtractive synthesizer is the basis of many forms of synthesizers

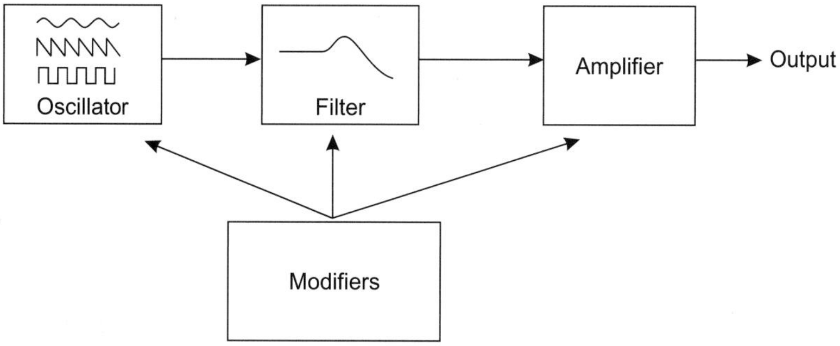

FIGURE 1.5
Layout of a basic
synthesizer

and is commonly related to analogue synthesizer. It is achieved by combining a number of sounds or 'oscillators' together to create a timbre that is very rich in harmonics.

This rich sound can then be sculpted using a series of 'modifiers'. The number of modifiers available on a synthesizer is entirely dependent on the model, but all synthesizers offer a way of filtering out certain harmonics and of shaping the overall volume of the timbre.

The next part of this chapter looks at how a real analogue synthesizer operates, although any synthesizer that emulates analogue synthesizer (i.e. digital signal processing (DSP) analogue) will operate in essentially the same way, with the only difference being that the original analogue synthesizer voltages do not apply to their DSP equivalents.

An analogue synthesizer can be said to consist of three components (Figure 1.5):

- An oscillator to make the initial sound.
- A filter to remove frequencies within the sound.
- An amplifier to define the overall level of the sound.

Each of these components and their role in synthesizer are discussed in the sections below.

VOLTAGE-CONTROLLED OSCILLATOR (VCO)

When a key on a keyboard is pressed, a signal is sent to the oscillator to activate it, followed by a specific control voltage (CV) to determine the pitch. The CV that is sent is unique to the key that is pressed, allowing the oscillator to determine the pitch it should reproduce. For this approach to work correctly, the circuitry in the keyboard and the oscillator must be incredibly precise in order to prevent the tuning from drifting, so the synthesizer must be serviced regularly. In addition, changes in external temperature and fluctuations in the power supply may also cause the oscillator's tuning to drift.

This instability gives analogue synthesizers their charm and is the reason why many purists will invest small fortunes in second-hand models rather than use the latest DSP-based analogue emulations. Although, that said, if too much detuning is present, it will be immediately evident and could become a major problem! There is still an ongoing argument over whether it's possible for DSP oscillators to faithfully reproduce analogue-based synthesizers, but the argument in favour of DSP synthesizers is that they offer more waveforms and do not drift too widely, and therefore prove more reliable in the long run.

In most early subtractive synthesizers the oscillator generated only three types of waveforms: square, sawtooth and triangle waveforms. Today this number has increased and many synthesizers now offer additional sine, noise, tri-saw, pulse and numerous variable wave shapes as well.

Although these additional waveforms produce different sounds, they are all based around the three basic wave shapes and are often introduced into synthesizers to prevent mixing of numerous basic waveforms together, a task that would reduce the number of oscillators.

For example, a tri-saw wave is commonly a sample of three sawtooth waves blended together to produce a sound that is rich in harmonics, with the advantage that the whole sound is contained in one oscillator. Without this waveform it would take three oscillators to recreate this sound, which could be beyond the capabilities of the synthesizer. Even if the synthesizer could utilize three oscillators to produce this one sound, the number of available oscillators would be reduced. Subsequently, while there are numerous oscillator waves available, knowledge of only the following six types is required.

The Sine Wave

A sine wave is the simplest wave shape and is based on the mathematical sine function (Figure 1.6). A sine wave consists of the fundamental frequency alone and does not contain harmonics. This means that they are not suitable for sole use in a subtractive sense, because if the fundamental is removed no sound is produced (and there are no harmonics upon which the modifiers could act). Consequently, the sine wave is used independently to create sub-basses

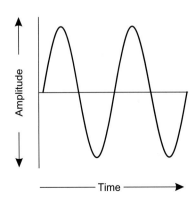

FIGURE 1.6
A sine wave

or whistling timbres or is mixed with other waveforms to add extra body or bottom end to a sound.

The Square Wave

A square wave is the simplest waveform for an electrical circuit to generate because it exists in only two states: high and low (Figure 1.7). This wave produces only odd harmonics resulting in a mellow, hollow sound. This makes it particularly suitable for emulating wind instruments, adding width to strings and pads, or for the creation of deep, wide bass sounds.

The Pulse Wave

Although pulse waves are often confused with square waves, there is a significant difference between the two (Figure 1.8). Unlike a square wave, a pulse wave allows the width of the high and low states to be adjusted, thereby varying the harmonic content of the sound. Today it is unusual to see both square and pulse waves featured in a synthesizer. Rather the square wave offers an additional control allowing you to vary the width of the pulses. The benefit of this is that reductions in the width allow you to produce thin reed-like timbres along with the wide, hollow sounds created by a square wave.

The Sawtooth Wave

A sawtooth wave produces even and odd harmonics in series and therefore produces a bright sound that is an excellent starting point for brassy, raspy sounds (Figure 1.9). It's also suitable for creating the gritty, bright sounds needed for leads and raspy basses. Because of its harmonic richness, it is often employed in sounds that will be filter swept.

The Triangle Wave

The triangle wave shape features two linear slopes and is not as harmonically rich as a sawtooth wave since it only contains odd

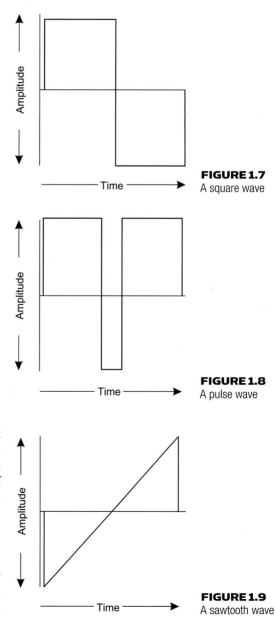

FIGURE 1.7
A square wave

FIGURE 1.8
A pulse wave

FIGURE 1.9
A sawtooth wave

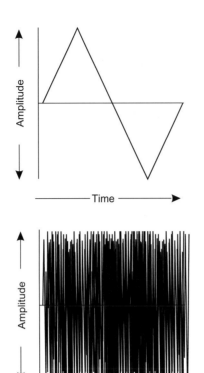

FIGURE 1.10
A triangle wave

FIGURE 1.11
A noise wave

harmonics (partials) (Figure 1.10). Ideally, this type of waveform is mixed with a sine, square or pulse wave to add a sparkling or bright effect to a sound and is often employed on pads to give them a glittery feel.

The Noise Wave

Noise waveforms are unlike the other five waveforms because they create a random mixture of all frequencies rather than actual tones (Figure 1.11). Noise waveforms can be 'pink' or 'white' depending on the energy of the mixed frequencies they contain. White noise contains equal amounts of energy at every frequency and is comparable to radio static, while pink noise contains equal amounts of energy in every musical octave and therefore we perceive it to produce a heavier, deeper hiss.

Noise is useful for generating percussive sounds and was commonly used in early drum machines to create snares and hand-claps. Although this remains its main use, it can also be used for simulating wind or sea effects, for producing breath effects in wind instrument timbres or for producing the typical trance leads.

CREATING MORE COMPLEX WAVEFORMS

Whether oscillators are created by analogue or DSP circuitry, listening to individual oscillators in isolation can be a mind-numbing experience. To create interesting sounds, a number of oscillators should be mixed together and used with the available modulation options.

This is achieved by first mixing different oscillator waveforms together and then detuning them all or just those that share the same waveforms so that they are out of phase from one another, resulting in a beating effect. Detuning is accomplished using the detune parameter on the synthesizer, usually by odd rather than even numbers. This is because detuning by an even number introduces further harmonic content that may mirror the harmonics already provided by the oscillators, causing the already present harmonics to be summed together.

It should be noted here that there is a limit to the level that oscillators can be detuned from one another. As previously discussed, oscillators should be detuned so that they beat, but if the speed of these beats is increased by any more than 20 Hz the oscillators separate, resulting in two noticeably different sounds. This can sometimes be used to good effect if the two oscillators are to be mixed with a timbre from another synthesizer because the additional timbre can help to fuse the two separate oscillators. As a general rule of thumb, it is unusual to detune an oscillator by more than an octave.

Additional frequencies can also be added into a signal using ring modulation and sync controls. Oscillator sync, usually found within the oscillator section of a synthesizer, allows a number of oscillators' cycles to be synced to one another. Usually all oscillators are synced to the first oscillators' cycle; hence, no matter where in the cycle any other oscillator is, when the first starts its cycle again the others are forced to begin again too.

For example, if two oscillators are used, with both set to a sawtooth wave and detuned by −5 cents (one-hundredth of a tone), every time the first oscillator restarts its cycle so too will the second, regardless of the position in its own cycle. This tends to produce a timbre with no harmonics and can be ideal for creating big, bold leads. Furthermore, if the first oscillator is unchanged and pitch bend is applied to the second to speed up or slow its cycle, screaming lead sounds typical of the Chemical Brothers are created as a consequence of the second oscillator fighting against the syncing with the first.

After the signals have left the oscillators, they enter the mixer section where the volume of each oscillator can be adjusted and features such as ring modulation can be applied to introduce further harmonics. (The ring modulation feature can sometimes be found within the oscillator section but is more commonly located in the mixer section, directly after the oscillators). Ring modulation works by providing a signal that is the sum and difference compound of two signals (while also removing the original tones). Essentially, this means that both signals from a two-oscillator synthesizer enter the ring modulator and come out from the other end as one combined signal with no evidence of the original timbre remaining.

As an example, if one oscillator produces a signal frequency of 440 Hz (A4 on a keyboard) and the second produces a frequency of 660 Hz (E5 on a keyboard), the frequency of the first oscillator is subtracted from the second.

$$660\,Hz - 440\,Hz = 220\,Hz\,(A3)$$

Then the first oscillator's frequency is added to that of the second.

$$660\,Hz + 440\,Hz = 1100\,Hz\,(C\#6)$$

Based on this example, the difference of 220 Hz provides the fundamental frequency while the sum of the two signals, 1100 Hz, results in a fifth harmonic overtone. When working with synthesizer, though, this calculation is rarely performed. This result is commonly achieved by ring modulating the oscillators together at any frequency and then tuning the oscillator. Ring modulation is typically used in the production of metallic-type effect (ring modulators were used to create the Dalek voice from *Dr Who*) and bell-like sounds. If ring modulation is used to create actual pitched sounds, a large number of in-harmonic overtones are introduced into the signal creating dissonant, unpitched results.

The option to add noise may also be included in the oscillator's mix section to introduce additional harmonics, making the signal leaving the oscillator/mix section full of frequencies that can then be shaped further using the options available.

VOLTAGE-CONTROLLED FILTERS

Following the oscillator's mixer section are the filters for sculpting the previously created signal. In the synthesizer world, if the oscillator's signal is thought of as a piece of wood that is yet to be carved, the filters are the hammer and chisels that are used to shape it. Filters are used to chip away pieces of the original signal until a rough image of the required sound remains.

This makes filters the most vital element of any subtractive synthesizer because if the available filters are of poor quality, few sound sculpting options will be available and it will be impossible to create the sound you require. Indeed, the choice of filters combined with the oscillator's waveforms is often the reason why specific synthesizers must be used to recreate certain 'classic' dance timbres.

The most common filter used in basic subtractive synthesizers is a low-pass filter. This is used to remove frequencies above a defined cut-off point. The effect is progressive, meaning that more frequencies are removed from a sound, the further the control is reduced, starting with the higher harmonics and gradually moving to the lowest. If this filter cut-off point is reduced far enough, all harmonics above the fundamental can be removed, leaving just the fundamental frequency. While it may appear senseless to create a bright sound with oscillators only to remove them later with a filter, there are several reasons why you may wish to do this.

- Using a variable filter on a bright sound allows you to determine the colour of the sound much more precisely than if you tried to create the same effect using oscillators alone.
- This method enables you to employ real-time movement of a sound.

This latter movement is an essential aspect of sound design because we naturally expect dynamic movement of sound throughout the length of the note. Using our previous example of a piano string being struck, the initial sound is very bright, becoming duller as it dies away. This effect can be simulated by opening the filter as the note starts and then gradually sweeping the cut-off frequency down to create the effect of the note dying away.

Notably, when using this effect, frequencies that lie above the cut-off point are not attenuated at right angles to the cut-off frequency; therefore, the rate at which they die away will depend on the transition period. This is why different

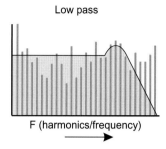

Low pass

F (harmonics/frequency)

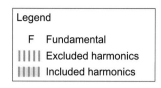

Legend

F   Fundamental

||||| Excluded harmonics

||||| Included harmonics

FIGURE 1.12
Action of the low-pass filter

filters that essentially perform the same function can make beautiful sweeps, whilst others can produce quite uneventful results (Figure 1.12).

When a cut-off point is designated, small quantities of the harmonics that lie above this point are not removed completely and are instead attenuated by a certain degree. The degree of attenuation is dependent on the transition band of the filter being used. The gradient of this transition is important because it defines the sound of any one particular filter. If the slope is steep, the filter is said to be 'sharp' and if the slope is more gradual the filter is said to be 'soft'. To fully understand the action of this transition, some prior knowledge of the electronics involved in analogue synthesizer is required.

When the first analogue synthesizers appeared in the 1960s, different voltages were used to control both the oscillators and the filters. Any harmonics produced by the oscillators could be removed gradually by physically manipulating the electrical current. This was achieved using a resistor (to reduce the voltage) and a capacitor (to store a voltage), a system that is often referred to as a resistor–capacitor (RC) circuit. Because a single RC circuit produces a 6 dB transition, the attenuation increases by 6 dB every time a frequency is doubled.

One RC element creates a 6 dB per octave 1-pole filter that is very similar to the gentle slope created by a mixing desks EQ. Consequently, manufacturers soon implemented additional RC elements into their designs to create 2-pole filters, which attenuated 12 dB per octave, and 4-pole filters, to provide 24 dB per octave attenuation. Because 4-pole filters attenuate 24 dB per octave, making substantial changes to the sound, they tend to sound more synthesized than sounds created by a 2-pole filter; so it's important to decide which transition period is best suited to the sound. For example, if a 24 dB filter is used to sweep a pad, it will result in strong attenuation throughout the sweep, while a 12 dB will create a more natural flowing movement (Figure 1.13).

If there is more than one of these available, some synthesizers allow them to be connected in series or parallel, which gives more control over the timbre from the oscillators. This means that two 12 dB filters could be summed together to produce a 24 dB transition, or one 24 dB filter could be used in isolation for aggressive tonal adjustments with the following 12 dB filter used to perform a real-time filter sweep.

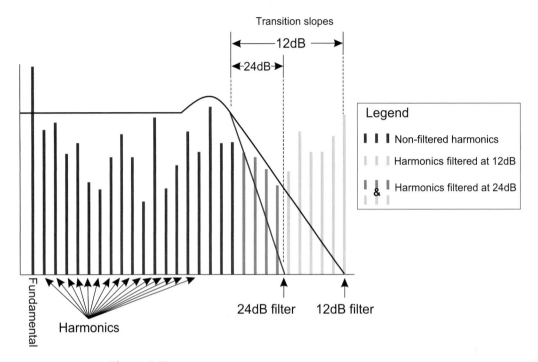

Figure 1.13
The difference between 12 dB and 24 dB slopes

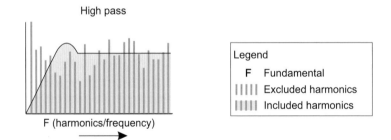

FIGURE 1.14
Action of a high-pass
filter

Although low-pass filters are the most commonly used type, there are numerous variations including high pass, band pass, and notch and comb. These utilize the same transition periods as the low-pass filter but each has a widely different effect on the sound (Figure 1.14).

A high-pass filter has the opposite effect to a low-pass filter, first removing the low frequencies from the sound and gradually moving towards the highest. This is less useful than the low-pass filter because it effectively removes the fundamental frequency of the sound, leaving only the fizzy harmonic overtones. Because of this, high-pass filters are rarely used in the creation of instruments

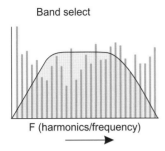

Band select

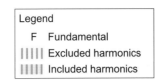

Legend
F   Fundamental
||||| Excluded harmonics
||||| Included harmonics

F (harmonics/frequency)

FIGURE 1.15
Action of the band-pass filter

and are predominantly used to create effervescent sound effects or bright timbres that can be laid over the top of another low-pass sound to increase the harmonic content.

The typical euphoric trance leads are a good example of this, as they are often created from a tone with the fundamental overlaid with numerous other tones that have been created using a high-pass filter. This prevents the timbre from becoming too muddy as a consequence of stacking together fundamental frequencies. In both remixing and dance music, it's commonplace to run a high-pass filter over an entire mix to eliminate the lower frequencies, creating an effect similar to a transistor radio or a telephone. By reducing the cut-off control, gradually or immediately, the track morphs from a thin sound to a fatter one, which can produce a dramatic effect in the right context.

If high- and low-pass filters are connected in series, then it's possible to create a band-pass, or band-select, filter. These permit a set of frequencies to pass unaltered through the filter while the frequencies either side of the two filters are attenuated. The frequencies that pass through unaltered are known as the 'bandwidth' or the 'band pass' of the filter, and clearly, if the low pass is set to attenuate a range of frequencies that are above the current high-pass setting, no frequencies will pass through and no sound is produced.

Band-pass filters, like high-pass filters, are often used to create timbres consisting of fizzy harmonics (Figure 1.15). They can also be used to determine the frequency content of a waveform, as by sweeping through the frequencies each individual harmonic can be heard. Because this type of filter frequently removes the fundamental, it is often used as the basis of sound effects or lo-fi and trip-hop timbres or to create very thin sounds that will form the basis of sound effects.

Although band-pass filters can be used to thin a sound, they should not be confused with band-reject filters, which can be used for a similar purpose. Band-reject filters, often referred to as notch filters, attenuate a selected range of frequencies effectively creating a notch in the sound – hence the name – and usually leave the fundamental unaffected. This type of filter is handy for

Notch

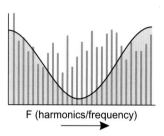

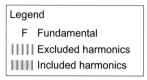

Legend

F   Fundamental

||||| Excluded harmonics

||||| Included harmonics

FIGURE 1.16
Action of the notch
filter

F (harmonics/frequency)

Comb

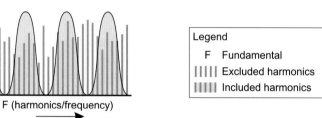

Legend

F   Fundamental

||||| Excluded harmonics

||||| Included harmonics

F (harmonics/frequency)

FIGURE 1.16
Action of the notch
filter

FIGURE 1.17
Action of the comb
filter

scooping out frequencies, thinning out a sound while leaving the fundamental intact, making them useful for creating timbres that contain a discernable pitch but do not have a high level of harmonic content (Figure 1.16).

One final form of filter is the comb filter. With these, some of the samples entering the filter are delayed in time and the output is then fed back into the filter to be reprocessed to produce the results, effectively creating a comb appearance, hence the name. Using this method, sounds can be tuned to amplify or reduce specific harmonics based on the length of the delay and the sample rate, making it useful for creating complex sounding timbres that cannot be accomplished any other way. Because of the way they operate, however, it is rare to find these featured on a synthesizer and are usually available only as a third-party effect.

As an example, if a 1 kHz signal is put through the filter with a 1 ms delay, the signal will result in phase because 1 ms is coincident with the inputted signal, equalling one. However, if a 500 Hz signal with a 1 ms delay were used instead, it would be half of the period length and so would be shifted out of phase by 180°, resulting in a zero. It's this constructive and deconstructive period that creates the continual bump then dip in harmonics, resulting in a comb-like appearance when represented graphically, as in Figure 1.17. This method applies to all frequencies, with integer multiples of 1 kHz producing ones and odd multiples of 500 Hz (1.5, 2.5, 3.5 kHz etc.) producing zeros. The effect of

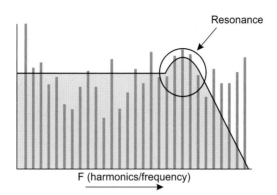

Resonance

F (harmonics/frequency)

FIGURE 1.18
The effect of resonance

using this filter can at best be described as highly resonant, and forms the basis of flanger effects; therefore, its use is commonly limited to sound design rather than the more basic sound sculpting.

One final element of sound manipulation in a synthesizer's filter section is the resonance control. Also referred to as peak, this refers to the amount of the output of the filter that is fed back directly into the input, emphasizing any frequencies that are situated around the cut-off frequency. This has a similar effect to employing a band-pass filter at the cut-off point, effectively creating a peak. Although this also affects the filter's transition period, it is more noticeable at the actual cut-off frequency than anywhere else. Indeed, as you sweep through the cut-off range the resonance follows the curve, continually peaking at the cut-off point. In terms of the final sound, increasing the resonance makes the filter sound more dramatic and is particularly effective when used in conjunction with low-pass filter sweeps (Figure 1.18).

On many analogue and DSP-analogue-modelled synthesizers, if the resonance is turned up high enough it will feed back on itself. As more and more of the signal is fed back, the signal is exaggerated until the filter breaks into self-oscillation. This produces a sine wave with a frequency equal to that of the set cut-off point and is often a purer sine wave than that produced by the oscillators. Because of this, self-oscillating filters are commonly used to create deep, powerful sub-basses that are particularly suited to the drum 'n' bass and rap genres.

Notably, some filters may also feature a saturation parameter which essentially overdrives the filters. If applied heavily, this can be used to create distortion effects, but more often it's used to thicken out timbres and add even more harmonics and partials to the signal to create rich sounding leads or basses.

The keyboard's pitch can also be closely related to the action of the filters, using a method known as pitch tracking, keyboard scaling or more frequently 'key follow'. On many synthesizers the depth of this parameter is adjustable,

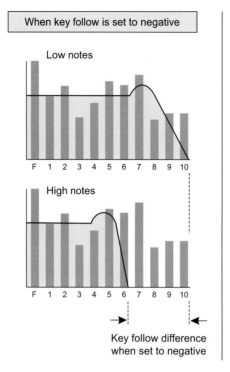
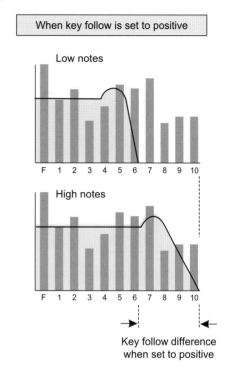

FIGURE 1.19
The effect of filter key follow

allowing you to determine how much or how little the filter should follow the pitch.

When this parameter is set to its neutral state (neither negative nor positive), as a note is played on the keyboard the cut-off frequency tracks the pitch and each note is subjected to the same level of filtering. If this is used on a low-pass filter, for example, the filter setting remains fixed, so as progressively higher notes are played fewer and fewer harmonics will be present in the sound, making the timbre of the higher notes mellower than that of the lower notes. If the key follow parameter is set to positive, the higher notes will have a higher cut-off frequency and the high notes will remain bright (Figure 1.19). If, on the other hand, the key follow parameter is set to negative, the higher notes will lower the cut-off frequency, making the high notes even mellower than when key follow is set to its neutral state. Key follow is useful for recreating real instruments such as brass, where the higher notes are often mellower than the lower notes, and is also useful on complex bass lines that jump over an octave, adding further variation to a rhythm.

VOLTAGE-CONTROLLED AMPLIFIER (VCA)

Once the filters have sculpted a sound, the signal then moves into the final stage of synthesizer: the amplifier. When a key is pressed, rather than the volume

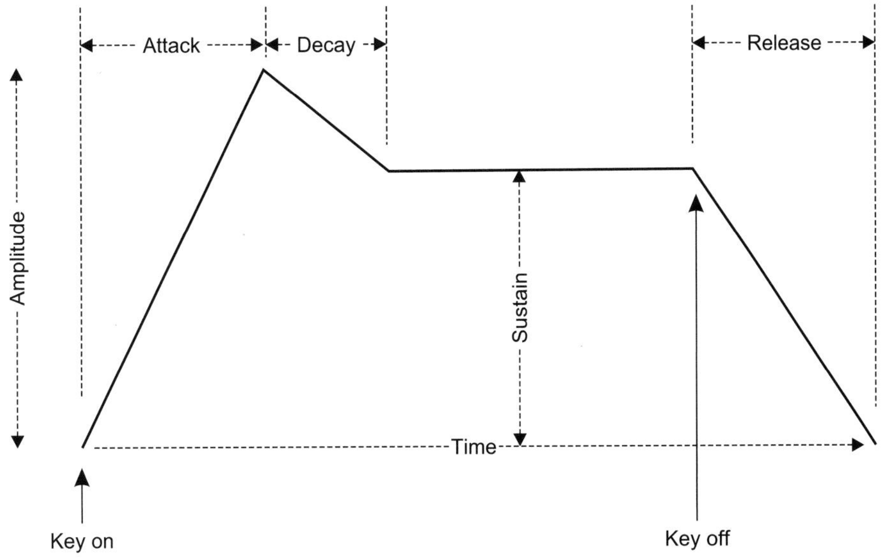

FIGURE 1.20
The ADSR envelope

rising immediately to its maximum and falling to zero when released, an 'envelope generator' is employed to emulate the nuances of real instruments.

Few, if any, acoustic instruments start and stop immediately. It takes a finite amount of time for the sound to reach its amplitude and then decay away to silence again; thus, the 'envelope generator' – a feature of all synthesizers – can be used to shape the volume with respect to time. This allows you to control whether a sound starts instantly the moment a key is pressed or builds up gradually and how the sound dies away (quickly or slowly) when the key is released. These controls usually comprise four sections called attack, decay, sustain, and release (ADSR), each of which determines the shaping that occurs at certain points during the length of a note. An example of this is shown in Figure 1.20.

- Attack: The attack control determines how the note starts from the point when the key is pressed and the period of time it takes for the sound to go from silence to full volume. If the period set is quite long, the sound will 'fade in', as if you are slowly turning up a volume knob. If the period set is short, the sound will start the instant a key is pressed. Most instruments utilize a very short attack time.
- Decay: Immediately after a note has begun it may initially decay in volume. For instance, a piano note starts with a very loud, percussive part but then drops quickly to a lower volume while the note sustains as the key is held down. The time the note takes to fade from the initial peak at the attack stage to the sustain level is known as the 'decay time'.
- Sustain: The sustain period occurs after the initial attack and decay periods and determines the volume of the note while the key is held down. This means that if the sustain level is set to maximum, any decay period

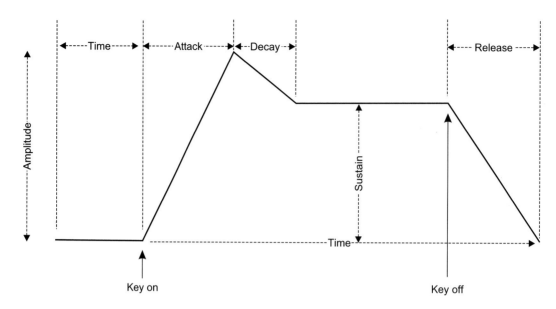

FIGURE 1.21
The TADSR envelope

will be ineffective, because at the attack stage the volume is at maximum and so there is no level to decay down to. Conversely, if the sustain level were set to zero, the sound peaks following the attack period and will fade to nothing even if you continue to hold down the key. In this instance, the decay time determines how quickly the sound decays down to silence.

- Release: The release period is the time it takes for the sound to fade from the sustain level to silence after the key has been released. If this is set to zero, the sound will stop the instant the key is released, while if a high value is set the note will continue to sound, fading away as the key is released.

Although ADSR envelopes are the most common, there are some subtle variations such as attack–release (AR), time–attack–delay–sustain–release (TADSR), and attack–delay–sustain–time–release (ADSTR). Because there are no decay or sustain elements contained in most drum timbres, AR envelopes are often used on drum synthesizers. They can also appear on more economical synthesizers simply because the AR parameters are regarded as having the most significant effect on a sound, making them a basic requirement. Both TADSR and ADSTR envelopes are usually found on more expensive synthesizers. With the additional period, T (time), in TADSR, for instance, it is possible to set the amount of time that passes before the attack stage is reached (Figure 1.21).

It's also important to note that not all envelopes offer linear transitions, meaning that the attack, decay and release stages will not necessarily consist

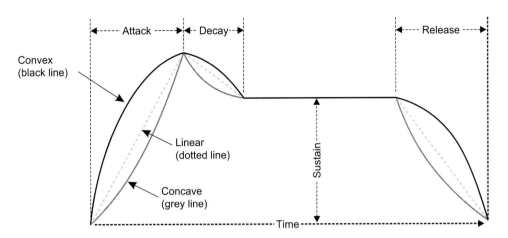

FIGURE 1.22
Linear and exponential envelopes

entirely of a straight line as it is shown in Figure 1.22. On some synthesizers these stages may be concave or convex, while other synthesizers may allow you to state whether the envelope stages should be linear, concave, or convex. The differences between the linear and the exponential envelopes are shown in Figure 1.22.

MODIFIERS

Most synthesizers also offer additional tools for manipulating sound in the form of modulation sources and destinations. Using these tools, the response or movement of one parameter can be used to modify another totally independent parameter, hence the name 'modifiers'.

The number of modifiers available, along with the destinations they can affect, is entirely dependent on the synthesizer. Many synthesizers feature a number of envelope generators that allow the action of other parameters alongside the amplifier to be controlled.

For example, in many synthesizers, an envelope may be used to modify the filter's action and by doing so you can make tonal changes to the note while it plays. A typical example of this is the squelchy bass sound used in most dance music. By having a zero attack, short decay and zero sustain level on the envelope generator, a sound that starts with the filter wide open before quickly sweeping down to fully closed is produced. This movement is archetypal to most forms of dance music but does not necessarily have to be produced by envelopes. Instead, some synthesizers offer one-shot low-frequency oscillators (LFOs) which can be used in the envelope's place. For instance, by using a triangle waveform LFO to modulate the amp, there is a slow rise in volume before a slowdrop again.

LOW-FREQUENCY OSCILLATOR

LFOs produce output frequencies in much the same way as VCOs. The difference is that a VCO produces an audible frequency (within the 20 Hz–20 kHz range) while an LFO produces a signal with a relatively low frequency that is inaudible to the human ear (in the range 1–10 Hz).

The waveforms an LFO can utilize depend entirely upon the synthesizer in question, but they commonly employ sine, saw, triangle, square, and sample and hold waveforms. The sample and hold waveform is usually constructed with a randomly generated noise waveform that momentarily freezes every few samples before beginning again.

LFOs should not be underestimated because they can be used to modulate other parameters, known as 'destination', to introduce additional movement into a sound. For instance, if an LFO is set to a relatively high frequency, say 5 Hz, to modulate the pitch of a VCO, the pitch of the oscillator will rise and fall according to the speed and shape of the LFO waveform and an effect similar to that of vibrato is generated. If a sine wave is used for the LFO, then it will essentially create an effect similar to that of a wailing police siren. Alternatively, if this same LFO is used to modulate the filter cut-off, then the filter will open and close at a speed determined by the LFO, while if it were used to modulate an oscillator's volume, it would rise and fall in volume recreating a tremolo effect.

This means that an LFO must have an amount control (sometimes known as depth) for varying how much the LFO's waveform augments the destination, a rate control to control the speed of the LFO's waveform cycles, and a fade-in control in some. The fade-in control adjusts how quickly the LFO begins to affect the waveform after a key has been pressed. An example of this is shown in Figure 1.23.

The LFO on more capable synthesizers may also have access to its own envelope. This gives control of the LFO's performance over a specified time period, allowing it not only to fade in after a key has been pressed but also to decay, sustain, and fade away gradually. It is worth noting, however, that the destinations an LFO can modulate are entirely dependent on the synthesizer being used. Some synthesizers may only allow LFOs to modulate the oscillator's pitch and the filter, while others may offer multiple destinations and more LFOs. Obviously, the more LFOs and destinations that are available, the more creative options you will have at your disposal.

If required, further modulation can be applied with an attached controller keyboard or the synthesizer itself in the form of two modulation wheels. The first, pitch bend, is hard-wired and provides a convenient method of applying a modulating CV to the oscillator(s). By pushing the wheel away from you, you can bend the pitch (i.e. frequency) of the oscillator up. Similarly, you can bend the pitch down by pulling the wheel towards you. This wheel is normally spring loaded to return to the centre position, where no bend is applied,

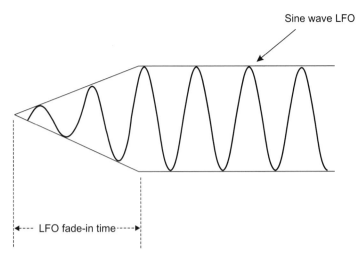

FIGURE 1.23
LFO fade-in

if you let go of it, and is commonly used in synthesizer solos to give additional expression. The second wheel, modulation, is freely assignable and offers a convenient method of controlling any on-board parameters, such as the level of the LFO signal sent to the oscillator, filter or VCA or to control the filter cut-off directly. Again, whether this wheel is assignable will depend on the manufacturer of the synthesizer.

On some synthesizers the wheels are hard coded to only allow oscillator modulation (for a vibrato effect), while some others do not have a separate modulation wheel and instead the pitch bend lever can be pushed forward to produce LFO modulation.

PRACTICAL APPLICATIONS

While there are other forms of synthesis – which will be discussed later in this chapter – most synthesizers used in the production of dance music are of an analogue/subtractive nature; therefore, it is vital that the user grasps the concepts behind all the elements of subtractive synthesis and how they can work together to produce a final timbre. With this in mind, it is sensible to experiment with a short example to aid in the understanding of the components.

Using the synthesizer of your choice, clear all the current settings so that you start from nothing. On many synthesizers, this is known as 'initializing a patch', so it may be a button labelled 'init', 'init patch' or similar.

Begin by pressing and holding C3 on your synthesizer, or alternatively controlling the synthesizer via MIDI programme in a continual note. If not, place something heavy on C3. The whole purpose of this exercise is to hear how the sound develops as you begin to modify the controls of the synthesizer, so the note needs to play continually.

Select sawtooth waves for two oscillators; if there is a third oscillator that you cannot turn off, choose a triangle for this third oscillator. Next, detune one sawtooth from the other until the timbre begins to thicken. This is a tutorial to grasp the concept of synthesis, so keep detuning until you hear the oscillators separate from one another and then move back until they become one again and the timbre is thickened out. Generally speaking, detuning of 3 cents should be ample but do not be afraid to experiment – this is a learning process. If you are using a triangle wave, detune this against the two saws and listen to the results. Once you have a timbre you feel you can work with, move onto the next step.

Find the VCA envelope and start experimenting. You will need to release C3 and then press it again so you can hear the effect that the envelope is having on the timbre. Experiment with these envelopes until you have a good grasp on how they can adjust the shape of a timbre; once you're happy you have an understanding, apply a fast attack with a short decay, medium sustain and a long release. As before, for this next step you will need to keep C3 depressed.

Find the filter section, and experiment with the filter settings. Start by using a high-pass filter with the resonance set around midway and slowly turn the filter cut-off control. Note how the filter sweeps through the sound, removing the lower frequencies first, slowly progressing to the higher frequencies. Also experiment with the resonance by rotating it to move upwards and downwards and note how this affects the timbre. Do the same with the notch and band pass etc. (if the synthesizer has these available) before finally moving to the low pass. Set the low-pass filter quite low, along with a low-resonance setting – you should now have a static buzzing timbre.

The timbre is quite monotonous, so use the filter envelope to inject some life into the sound. This envelope works on exactly the same principles as the VCA, with the exception that it will control the filter's movement. Set the filter's envelope to a long attack and decay, but use a short release and no sustain and set the filter envelope to maximum positive modulation. If the synthesizer has a filter key follow, use this as it will track the pitch of the note being played and adjust itself. Now try depressing C3 to hear how the filter envelope controls the filter, essentially sweeping through the frequencies as the note plays.

Finally, to add some more excitement to the timbre, find the LFO section. Generally, the LFO will have a rotary control to adjust the rate (speed), a selector switch to choose the LFO waveform, a depth control and a modulation destination. Choose a triangle wave for the LFO waveform, Hold down C3 on the synthesizer's keyboard, turn the LFO depth control up to maximum and set the LFO destination to pitch. As before, hold down the C3 key and slowly rotate the LFO rate (speed) to hear the results. If you have access to a second LFO, try modulating the filter cut-off with a square wave LFO, set the LFO depth to maximum and experiment with the LFO rate again.

If you would like to experiment more with synthesis to help get to grips with the principles, jump to Chapter 4 for further information on programming specific synthesizer timbres. Note, however, that different synthesizers will produce timbres differently and some are more suited to reproducing particular timbres than others.

OTHER SYNTHESIS METHODS

Frequency Modulation (FM)

FM is a form of synthesizer developed in the early 1970s by Dr John Chowning of Stanford University, then later developed further by Yamaha, leading to the release of the now-legendary DX7 synthesizer: a popular source of bass sounds for numerous dance musicians.

Unlike analogue, FM synthesizer produces sound by using operators, which are very similar to oscillators in an analogue synthesizer but can only produce simple sine waves. Sounds are generated by using the output of the first operator to modulate the pitch of the second, thereby introducing harmonics. Like an analogue synthesizer, each FM voice requires a minimum of two oscillators in order to create a basic sound, but because FM only produces sine waves the timbre produced from just one carrier and modulator isn't very rich in harmonics.

In order to remedy this, FM synthesizers provide many operators that can be configured and connected in any number of ways. Many will not produce musical results, so to simplify matters various algorithms are used. These algorithms are preset as combinations of modulator and carrier routings. For example, one algorithm may consist of a modulator modulating a carrier, which in turn modulates another carrier, before modulating a modulator that modulates a carrier to produce the overall timbre. The resulting sound can then be shaped and modulated further using LFOs, filters and envelopes using the same subtractive methods as in any analogue synthesizer.

This means that it should also be possible to emulate FM synthesizer in an analogue synthesizer with two oscillators, where the first oscillator acts as a modulator and the second acts as a carrier. When the keyboard is played, both oscillators produce their respective waveforms with the frequency dictated by the particular notes that were pressed. If the first oscillator's output is routed into the modulation input of the second oscillator and further notes are played on the keyboard, both oscillators play their respective notes but the pitch of the second oscillator will change over time with the frequency of the first, essentially creating a basic FM synthesizer. Although this is, in effect, FM, it is usually called 'cross modulation' in analogue synthesizers.

Due to the nature of FM, many of the timbres created are quite metallic and digital in character, particularly when compared to the warmth generated by the drifting of analogue oscillators. Also due to the digital nature of FM synthesizer, the facia generally contains few real-time controllers. Instead, numerous buttons adorn the front panel forcing you to navigate and adjust any parameters through a small LCD display.

Notably, although both FM and analogue synthesizers were originally used to reproduce realistic instruments, neither can fabricate truly realistic timbres. If the goal of the synthesizer system is to recreate the sound of an existing instrument, this can generally be accomplished more accurately using digital sample-based techniques.

SAMPLES AND SYNTHESIS

Unlike analogue or FM, sample synthesizer utilizes samples in place of the oscillators. These samples, rather than consisting of whole instrument sounds, also contain samples of the various stages of a real instrument along with the sounds produced by normal oscillators. For instance, a typical sample-based synthesizer may contain five different samples of the attack stage of a piano, along with a sample of the decay, sustain and release portions of the sound. This means that it is possible to mix the attack of one sound with the release of another to produce a complex timbre.

Commonly, up to four of these individual 'tones' can be mixed together to produce a timbre and each of these individual tones can have access to numerous modifiers including LFOs, filters and envelopes. This obviously opens up a whole host of possibilities not only for emulating real instruments, but also for creating complex sounds. This method of synthesis has become the de facto standard for any synthesizer producing realistic instruments. By combining both samples of real-world sounds with all the editing features and functionality of analogue synthesizers, they can offer a huge scope for creating both realistic and synthesized sounds.

GRANULAR SYNTHESIS

One final form of synthesizer that has started to make an appearance with the evolution of technology is granular synthesizer. It is rare to see a granular synthesizer employed in hardware synthesizers due to its complexity, but software synthesizers are being developed for the public market that utilize it. Essentially, it works by building up sounds from a series of short segments of sounds called 'grains'. This is best compared to the way that a film projector operates, where a series of still images, each slightly different from the last, are played sequentially at a rate of around 25 pictures per second, fooling the eyes and brain into believing there is a smooth continual movement.

A granular synthesizer operates in the same manner with tiny fragments of sound rather than still images. By joining a number of these grains together, an overall tone is produced that develops over a period of time. To do this, each grain must be less than 30 ms in length as, generally speaking, the human ear is unable to determine a single sound if they are less than 30–50 ms apart. This also means that a certain amount of control has to be offered over each grain. In any one sound there can be anything from 200 to 1000 grains, which is the main reason why this form of synthesizer appears mostly in the form of software. Typically, a granular synthesizer will offer most, but not necessarily all, of the following five parameters:

- Grain length: This can be used to alter the length of each individual grain. As previously mentioned, the human ear can differentiate between two grains if they are more than 30–50 ms apart, but many granular synthesizers usually go above this range, covering 20–100 ms. By setting this length to a higher value, it's possible to create a pulsing effect.
- Density: This is the percentage of grains that are created by the synthesizer. Generally, it can be said that the more the grains created, the more complex a sound will be, a factor that is also dependent on the grain shape.
- Grain shape: Commonly, this offers a number between 0 and 200 and represents the curve of the envelopes. Grains are normally enveloped so that they start and finish at zero amplitude, helping the individual grains mix together coherently to produce the overall sound. By setting a longer envelope (a higher number) two individual grains will mix together, which can create too many harmonics and often result in the sound exhibiting lots of clicks as it fades from one grain to the other.
- Grain pan: This is used to specify the location within the stereo image where each grain is created. This is particularly useful for creating timbres that inhabit both speakers.
- Spacing: This is used to alter the period of time between each grain. If the time is set to a negative value, the preceding grain will continue through the next created grain. This means that setting a positive value inserts space between each grain; however, if this space is less than 30 ms, the gap will be inaudible.

The sound produced with granular synthesizer depends on the synthesizer in question. Usually, the grains consist of single frequencies with specific waveforms or occasionally they are formed from segments of samples or noise that have been filtered with a band-pass filter. Thus, the constant change of grains can produce sounds that are both bright and incredibly complex, resulting in a timbre that's best described as glistening. After creating this sound by combing the grains, the whole sound can be shaped by using envelopes, filters and LFOs.

CHAPTER 2

Compression, Processing and Effects

'Compression plays a major part of my sound. I have them patched across every output of the desk...'

Armand Van Helden

Armed with the basic understanding of synthesis, we can examine the various processors and effects that are available. This is because the deliberate abuse of these processors and effects play a vital role in not only the sound design process but also the requisite feel of the music, so it pays to understand what they are, how they affect the audio and how a mixing desk can determine the outcome of the effect. Consequently, this chapter concentrates on the behaviours of the different processors and effects that are widely used in the design and production of dance music, including reverb, chorus, phasers, flangers, delay, EQ, distortion, gates, limiters and, perhaps the most important of all, compressors.

Of all the effects and processors available, a compressor is possibly the most vital tool to achieve that atypical sound heard on so many dance records, so a thorough understanding of it is essential. Without compression, drums appear wimpy in comparison to the chest thudding results heard in professionally produced music, mixes can appear lacking in depth and basses, and vocals and leads can lack any real presence. Despite its importance, however, the compressor is the least understood processor of them all.

COMPRESSION THEORY

The whole reason a compressor was originally introduced was to reduce the dynamic range of a performance, which is particularly vital when working with any form of music. Whenever you record any sound into a computer, sampler or recording device, you should aim to capture the loudest signal possible so that you can avoid artificially increasing the volume afterwards. This is because if you record a source that's too low in volume and then attempt to artificially increase it later, not only will it increase the volume of the recorded source, it'll also increase any background noise.

To prevent this, you need to record a signal as loud as possible but the problem is that vocals and 'real' instruments have a huge dynamic range. In other words, the vocals, for example, can be quiet in one part and suddenly become loud in the next (especially when moving from verse to chorus). Consequently, it's impossible to set a good average recording level with so much dynamic movement since if you set the recording level to capture the quiet sections, when it becomes louder the recording will go into the red clip. Conversely, setting the recorder so that the loud sections do not clip, any quieter sections will be exposed to more background noise.

Of course, you could sit by the recording fader and increase or decrease the recording levels depending on the section being recorded but this would mean that you need lightening reflexes. Instead, it's much easier to employ a compressor to control the levels automatically. By routing the source sound through a compressor and then into the recorder, you can set a threshold on the compressor so that any sounds that exceed this are automatically pulled down in volume, thus allowing you to record at a more substantial volume overall.

A compressor can also be used to control the dynamics of a sound while mixing. For example, a dance track that uses a real bass guitar will have a fairly wide dynamic range, even if it was compressed during the recording stage. This will cause problems within a mix because if the volume is adjusted so that the loudest parts fit well within the mix, the quieter parts may disappear behind other instrumentation. Conversely, if the fader is set so that quieter sections can be heard over other instruments, the loud parts could be too prominent. Using compression more heavily on this sound during the mixing stage, the dynamic range can be restricted, allowing the sound to sit better overall within the final mix.

Although these are the key reasons why compressors were first introduced, it has further, far-reaching applications for the dance musician and a compressor's action has been abused to produce the typical dance sound.

Since the signals that exceed the threshold are reduced in gain, the parts that do not exceed the threshold aren't touched, so they remain at the same volume as they were before compression. In other words, the difference in volume between the loudest and the quietest parts of the recording are reduced, which means that any uncompressed signals will become louder relative to the compressed parts. This effectively boosts the average signal level, which in turn not only allows you to push the volume up further but also makes it sound louder (Figures 2.1 and 2.2).

Note that after reducing the dynamic range of audio it may be perceived to be louder without actually increasing the gain. This is because we determine the overall volume of music from the average volume (measured in root mean square, RMS), not from the transient peaks created by kick drums.

Nevertheless, applying heavy compression to certain elements of a dance mix can change the overall character of the timbre, often resulting in a warmer, smoother and rounder tone, a sound typical of most dance tracks around today.

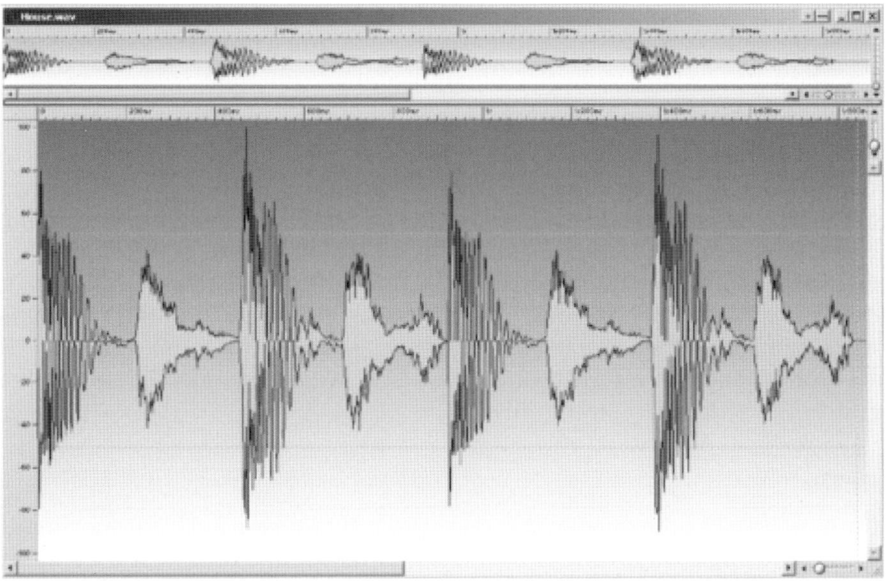

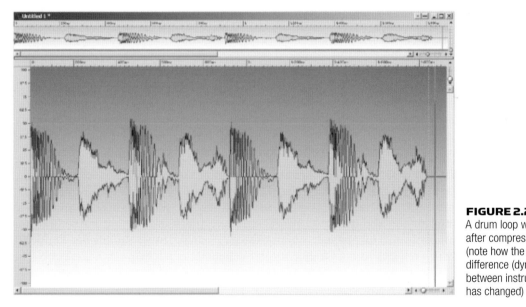

FIGURE 2.1
A drum loop waveform
before compression

FIGURE 2.2
A drum loop waveform
after compression
(note how the volume
difference (dynamics)
between instruments
has changed)

While there are numerous publications stating the 'typical' compression settings to use, the truth is that there are no generic settings for any particular genre
of music and its use depends entirely on the timbres that have been used. For
instance, a kick drum sample from a record will require a different approach than
a kick drum constructed in a synthesizer, while if sampled from a CD, synthesizer
or film different approaches are required again. Therefore, rather than attempt

to dictate a set list of useless compression settings, you can achieve much better results by knowing exactly what effect each control will have on a sound and how these are used to acquire the sounds typical of each genre.

Threshold

The first control on a compressor is the threshold which when touched upon sets the signal level where the compressor will begin squashing the incoming signal. These are commonly calibrated in dB and will work in direct relationship with a gain reduction meter to inform you of how much the compressor is affecting the incoming signal. In a typical recording situation this control is set so that the average signal level always lies just below the threshold, and if any exuberant parts exceed it, the compressor will jump into action and the gain will be reduced to prevent any clipping.

Ratio

The amount of gain reduction that takes place after a sound exceeds the threshold is set using a ratio control. Expressed in ratios, this control is used to set the dynamic range the compressor affects, indicating the difference between the signals entering the compressor that exceed the threshold to the levels that come out of the other end.

For example, if the ratio is 4:1, every time the incoming signal exceeds the threshold by 4 dB, the compressor will squash the signal so that there is only a 1 dB increase at the output of the compressor. Similarly, if the ratio set is 6:1, an increase at the compressor's output of 1 dB will occur when the threshold is exceeded by 6 dB and likewise for ratios of 8:1, 10:1 and so on. Subsequently, the gain reduction ratio always remains constant no matter how much compression takes place. In most compressors these range from 1:1 up to 10:1 and may, in some cases, also offer infinity:1.

From this we can determine that if a sound exceeds a predefined threshold, the compressor will squash the signal by the amount set with the ratio control. The problem with this approach, however, is that we gain a significant amount of information about sounds from their initial attack stage, and if the compressor jumps in instantaneously on an exceeded signal, it will squash the transients which reduces its high frequency (HF) content.

For instance, if you are to set up a compressor to squash a snare drum, the compressor will clamp down on the attack stage which in effect diminishes the initial transients reducing it to a 'thunk'. What's more, this instantaneous action will also appear when the sound drops below the threshold again as the compressor stops processing the audio. This can be especially evident when compressing low-frequency (LF) waveforms such as basses since compressors can apply gain changes during a waveform's period.

In other words, if a low-frequency waveform, such as a sustained bass note, is squashed the compressor may treat the positive and negative states of the waveform as different signals and continually activate and deactivate. The result of this is an unpleasant distortion of the waveform. To prevent this from occurring, compressors will feature attack and release parameters.

Attack/Release

Both these parameters behave in a manner similar to those on a synthesizer but control how quickly the volume is pulled down and how long it takes to rise back to its nominal level after the signal has fallen below the threshold. In other words, the attack parameter defines how long the compressor takes to reach maximum gain reduction while the release parameter determines how long the compressor will wait *after* the signal has dropped below the threshold before processing stops.

This raises the obvious question that if the attack is set so that it doesn't clamp down on the initial attack of the source sound, it could introduce distortion/clipping before the compressor activates. While this is true, in practice very short, sharp signals do not always overload an analogue recorder since these usually have enough headroom to let small transients through without introducing any unwanted artefacts. This isn't the case with digital recorders, though, and any signals that are beyond the limit of digital can result in clipping, so it's quite usual to follow a compressor with a limiter or, if the compressor features a knee mode, set it to use a soft knee.

Soft/Hard Knee

All compressors will utilize either soft or hard knee compression but some will offer the option to switch between the two modes. These are not controllable parameters but dictate the shape of the envelope's curve, and hence the characteristic of how the compressor behaves when a signal approaches the threshold. So far we've considered that when a signal exceeds the threshold the compressor will begin to squash the signal. This immediate action is referred to as hard knee compression. Soft knee, on the other hand, continually measures the incoming signal, and when it approaches 3–14 dB (dependant on the compressor) towards the current threshold, the compressor starts to apply the gain reduction gradually.

Generally this will initially start with a ratio of 1:1, and as the signal grows ever closer to the threshold, it's gradually increased until the threshold is exceeded, whereby full gain reduction is applied. This allows the compressor's action to be less evident and is particularly suitable for use on acoustic guitars and wind instruments where you don't necessarily want the action to be evident.

It should be noted that the action of the knee is entirely dependent on the compressor being used and some can be particularly long starting 12 dB before the threshold while others may start 3 dB before. As a matter of interest, 6–9 dB soft knees are considered to offer the most natural compression for instruments.

Peak/RMS

Not all compressors feature knees, so short transient peaks can sometimes catch the compressor unaware and 'sneak' past unaffected. This is obviously going to cause problems when recording digitally, so many compressors will implement a switch for Peak or RMS modes. Compressors that do not feature these two modes will operate in RMS, which means that the compressor will detect and control signals that stay at an average level rather than the short sharp transient peaks. As a result, no matter how fast the attack may be set, there's a chance that the transients will overshoot the threshold and not be controlled. This is because by the time the compressor has figured out that the sound has exceeded the threshold it's too late – the peak's been and gone again. Therefore to control short transient sound such as drum loops, it's often prudent to engage the peak mode. With this the compressor becomes sensitive to short sharp peaks and clamps down on them as soon as they come close to the threshold, rather than after they exceed it. By doing so, the peak can be controlled before it overshoots the threshold and creates a problem.

While this can be particularly useful when working with drum and percussion sounds it can create havoc with most other timbres. Keep in mind that many instruments can exhibit a particularly short, sharp initial attack stage, and if the compressor perceives these as possible problems, it'll jump down on them before they overshoot. In doing so, the high-frequency elements of the attack will be dulled which can make the instrument appear less defined, muddled or lost within the mix. Therefore, for all instruments bar drums and percussion, it's advisable to stick with the RMS mode.

Make-Up Gain

The final control on a compressor is the make-up gain. If you've set the threshold, ratio, attack and release correctly, the compressor should compress effectively and reduce the dynamics in a sound but this compression will also reduce the overall gain by the amount set by the ratio control. Therefore, whenever compression takes place you can use the make-up gain to bring the signal back up to its pre-compressed volume level.

Side Chaining

Alongside the physical input and output connections on hardware compressors, many also feature an additional pair of inputs known as side chains. By inputting an audio signal into these, a sound's envelope can be used to control the action that the compressor has on the signal entering the normal inputs. A good example of this is when a radio DJ begins talking over a record and the volume of the record lowers so that their voice becomes audible, then when they stop speaking the record returns to its original volume. This is accomplished by feeding the music through the compressor as normal but with the microphone connected into the side chain. This supersedes the compressor's normal operation and uses the signal from the side chain rather than the threshold as the trigger. Thus, the compressor is triggered when the microphone is spoken into, compressing (in effect lowering the volume of the music) by the amount set with the ratio control. This technique should only be viewed as an example to explain the process, though, and more commonly side chaining is usually used to make space in a mix for the vocals.

In a typical mix, the lead sound will occupy the same frequencies as the human voice, resulting in a cluttered mid-range if the two are to play together. This can be avoided if the lead mix is fed into the main inputs of the compressor while the vocal track is routed into the side chain. With the ratio set at an appropriate level (dependent on the tonal characteristics of the lead and vocals) the lead track will dip when the vocals are present, allowing them to pull through the mix.

Hold Control

Most compressors that feature a side chain are likely to also have an associated 'hold' control on the facia or employ an automated hold function. This is employed because a side chain measures the envelope of the incoming signal, and if both the release and attack are too fast, the compressor may respond to the cycles of a low-frequency waveform rather than the actual envelope. As touched upon previously, this can result in the peaks and dips of the waveform, activating and deactivating the compressor resulting in distortion. By using a hold the compressor is forced to wait a finite amount of time (usually 40–60 ms on automated hold) before beginning the release phase, which is longer than the period of a low-frequency waveform.

STANDARD COMPRESSION

Despite the amount of control offered by the average compressor, they are relatively simple to set up for recording audio. As a generalized starting point, it's advisable to set the ratio at 4:1 and lower the threshold so that the gain reduction meter reads between –8 and –10 dB on the loudest parts of the signal. After this, the attack parameter should be set to the fastest speed possible and the release set to approximately 500 ms. Using these as preliminary settings, they can then adjusted further to suit any particular sound.

It's advisable that compression is applied sparingly during the recording stage because once applied it cannot be removed. Any exuberant parts of the performance should be prevented from forcing the recorder's meters into the red while also ensuring that the compressor is as transparent as possible. Solid-state compressors are more transparent than their valve counterparts and so are better suited for this purpose.

As a general rule of thumb, the higher the dynamic range of the instrument being recorded the higher the ratio and the lower the threshold settings need to be. These settings help to keep the varying dynamics under tighter control and prevent too much fluctuation throughout the performance. Additionally, if the choice between hard or soft knee is available, the structure of the timbre should be taken into account. To retain a sharp, bright attack stage, hard knee compression with an attack setting that allows the initial transient to sneak through unmolested should be used, provided of course that the transient is unlikely to bypass the compression. In these instances, and to capture a more natural sound, soft knee compression should be used.

Finally, the release period should be set as short as possible but not so short that the effect is noticeable when the compressor stops processing. After setting the release at 500 ms, the time should be continually reduced until the processing is noticeable and then increased slowly until it isn't.

Some compressors feature an automode for the release that uses a fast release on transient hits and a slower time for smaller peaks, making this task easier.

The settings shown in Table 2.1 are naturally only starting points and too much compression should be avoided during the recording stage, something that can only be accomplished by setting both the ratio and the threshold control carefully. This involves setting the compressor to squash audio but ensuring that it stops processing and that the gain reduction meter drops to 0 dB (i.e. no signal is being compressed) during any silent passages.

As a more practical example, with a simple four to the floor kick running through the compressor and the ratio and threshold controls set so that the gain reduction reads –8 dB on each kick, it's necessary to ensure that the gain reduction meter returns to 0 dB during any silent periods. If it doesn't, then the loop is being overcompressed. If the gain reduction only drops to –2 dB during

Table 2.1   **Compression Settings**

Compression Settings	Ratio	Attack Parameter (ms)	Release Parameter (ms)	Gain Reduction (dB)	Knee
Starting settings	5–10:1	1–10	40–100	–5 to –15	Hard
Drum loop	5–10:1	1–10	40–100	–5 to –15	Hard
Bass	4–12:1	1–10	20 or auto	–6 to –13	Hard
Leads	2–8:1	3–10	40 or auto	–8 to –10	Hard
Vocals	2–7:1	1–7	50 or auto	–3 to –10	Soft
Brass instruments	4–10:1	1–7	30 or auto	–8 to –13	Hard
Electric guitars	8–10:1	2–7	50 or auto	–5 to –12	Hard
Acoustic guitars	5–9:1	5–20	40 or auto	–5 to –12	Hard

the silence between kicks, then it makes sense that only 6 dB of gain reduction is actually being applied. This means that every time the compressor activates it has to jump from 0 to 8 dB, when in reality it only needs to jump in by 6 dB. This additional 2 dB of gain will distort the transient that follows the silence, making it necessary for the gain reduction to be adjusted accordingly.

PRACTICAL COMPRESSION

While it is generally worth avoiding any evident compression during the recording stage, deliberately making the compressor's action evident forms a fundamental part of creating the typical sound of dance music. To better describe this, we'll use a drum loop to experiment upon. This is a typical dance drum loop consisting of a kick, snare, closed and open hi-hats (Figure 2.3).

The data CD contains the drum loop.

It's clear from looking/listening to the drum loop that the greatest energy – that is the loudest part of the loop – is derived from the kick drum. With this in mind, if a compressor is inserted across this particular drum loop and the threshold is set just below the peak level of the loudest part, each consecutive kick will activate the compressor.

If you have access to a wave editor, open the file in the wave editor and open a compressor plug-in. If you work in hardware, set up the compressor across

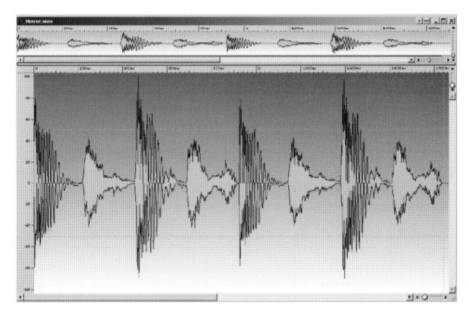

FIGURE 2.3
The waveform of the drum pattern

the drum loop as an insert (if you don't understand insert effects yet, it is discussed in the next chapter; go there and return here when you understand the principles).

Now, set the threshold just below the peak level of the kick drum. You can do this by watching the gain reduction meter and ensuring it moves every time a kick occurs. Set the ratio to 4:1, the attack time fast and then while playing back the loop, experiment with the release time. Note that as the release is shortened, the loop begins to pump more dramatically. This is a result of the compressor activating on the kick, then quickly releasing when the kick drops below the threshold. The result is a rapid change in volume, producing a pumping effect as the compressor activates and deactivates on each kick.

> The data CD contains the drum loop compressed.

This is known as 'gain pumping' and, although frowned upon in some areas of music, is deliberately used in dance and popular music to give a track a more dynamic feel. The exact timing of the release will depend entirely on the tempo of the drum loop and must be short enough for the compressor to recover before the next kick. Similarly, the release must be long enough for this effect to sound natural, so it's best to keep the loop repeated over four bars and experiment with the attack and release parameters until the required sound is achieved.

If we expand on this principle further and add a bass line, pad and chords to the previously compressed loop and compress it again in the same way (i.e. with a short release), the entire mix will pump energetically. As the kick is still controlling the compressor (and provided that the release isn't too short or long), every time a kick occurs the rest of the instruments will drop in volume, which accentuates the overall rhythm of the piece.

Gain pumping is also useful when applied across the whole mix, even though each element in the mix may already have been compressed in isolation. Gain pumping across the whole mix is used to balance areas in the track where instruments are dropped in and out. When fewer instruments play, the gain of the mix will be perceived as lower than when all the instruments play simultaneously. The overall level can be controlled by strapping a compressor across the mixing desk's main stereo bus (more on this in later chapters), to make the mix pump with energy.

Gain pumping across the entire mix ('mix pumping') should be applied with caution because if the mix is pumped too heavily it will sound strange. Setting a 20–30 ms attack with a 250 ms release and a low threshold and ratio to reduce the range by 2 dB or so should be sufficient to produce a mix that has the 'right' feel.

That said, there is a trend emerging where gain pumping is becoming an actual part of the music, such as Eric Prydz's Valerie, but this is applied in a different manner, using the compressor's side chain. To accomplish this effect, you require a mix and an individual kick loop that is in tempo with the mix.

> The data CD contains parts required for this example.

If you don't have any available, track 3 of the CD contains these parts. Place the mix into one channel of a sequencer and drop the kick drum onto a second channel. Set up a compressor on the kick drum channel, use the kick as a side chain and feed the mix into the main compressor's inputs. Set the ratio to 4:1, with a fast attack and release, and if the compressor features it, set it to Peak (or turn RMS off). Begin playback of both channels and slowly reduce the threshold; the entire mix will pump with every kick. This can be used more creatively to create a gated pad effect.

> The data CD contains an example of gain pumping.

This technique can also be used in hip-hop, rap, house and big beat to help create room in the mix. Since these often have particularly loud bass elements that play consecutively with the kick, the different sounds can conflict, muddying the bottom end of the mix. This can be prevented by feeding the kick drum separately into the side-chain inputs of the compressor, with the ratio set to 3:1, with a fast attack and medium release (depending on the sound). If the bass is fed into the compressor's main inputs, every time the kick occurs the bass will drop in volume, making space for the kick thereby preventing any conflicts.

Compression can also be used on individual sounds to change the tonal content of a sound. For example, by using heavy compression (a low threshold and high ratio) on an isolated snare drum, the snare's attack avoids compression but the decay is squashed which brings it up to the attack's gain level. This technique is often employed to create the snare 'thwack' typical of trance, techno and house music styles. Similarly, if a deeper, duller, speaker-mashing, kick drum 'thud' is required the compressor's attack should be set as short as possible so that it clamps down hard on the initial attack of the kick. This eradicates much of the initial high frequency content, and as the volume is increased with the compressor's make-up control, a deeper and much more substantial 'thud' is produced.

It is, however, important to note that the overall quality of compression depends entirely on the type of compressor being used. Compressors are one of two types: solid-state or valve. Solid-state compressors use digital circuitry throughout and will not tend to pump as heavily or sound as good as those that utilize valve technology. Some valve compressors will be solid state in the most part, using a valve only at the compressor's make-up gain. Solid-state compressors are usually more transparent than their valve-based counterparts and are used during the recording stages. Valve compressors are typically used after recording to add warmth to drums, vocals and basses, an effect caused by small amounts of second-order harmonic distortion that are introduced into the final gain circuitry[1] of the compressor. This distortion is a result of the random movement of electrons which, in the case of valves, occurs at exactly twice the frequency of the amplified signal. Despite the fact that this distortion only contributes 0.2% to the amplified signal, the human ear (subjectively!) finds it appealing.

More importantly, the characteristic warmth of a valve compressor differs according to the model of valve compressor that is used, as each will exhibit different characteristics. These differences and the variations from compressor to compressor are the reasons why many dance producers will spend a fortune on the right compressor and why it isn't uncommon for producers to own a number of both valve and solid-state types.

Most dance producers agree that solid-state circuitry tends to react faster, producing a more defined, less forgiving sound, while valve compressors add warmth that improves the overall timbre.

While it isn't essential to know why these differences exist from model to model, it's worth knowing which compressor is most suited to a particular style of work. Failing that, it also makes for excellent conversation (if you're that way inclined), so what follows is a quick rundown of the five most popular methods of compression:

- Variable MU
- Field effect transistor (FET)
- Optical
- VCA
- Computer-based digital

Variable MU

The first compressors ever to appear on the market were called variable MU units. This type of compressor uses valves for the gain control circuitry and does not have an adjustable ratio control. Instead of an adjustable control, the ratio is increased in proportion to the amount of the incoming signal that exceeds the threshold. In other words, the more the level overshoots the threshold the more the ratio increases. While these compressors do offer attack and release stages, they're not particularly suited towards material with fast transients, even

---

[1]John Ambrose Fleming originally developed the valve in 1904 but it was 2 years later that Lee De Forest constructed the first Triode configuration. Edwin Howard Armstrong then used this to create the first ever valve amplifier in 1912.

with their fastest attack settings. Due to the valve design, the valves run out of dynamic range relatively quickly, so it's unusual to acquire more than 15–20 dB of gain reduction before the compressor runs out of energy. Nevertheless, variable MU compressors are renowned for their distinctive, phat, lush character, and can work magic on basses and pumping dance mixes. The most notorious variable MU compressors are made by Manley and can cost in excess of £3500.

FET

FET compressors use a field effect transistor to vary the gain. These were the first transistors to emulate the action of valves. They provide incredibly fast attack and release stages making them an excellent choice for beefing up kick and snare drums, electric guitars, vocals and synthesizer leads. While they suffer from a limited dynamic range, if they're pushed hard they can pump very musically and are perfectly suited for gain pumping a mix. The only major problem is getting your hands on one. Original FETs are as rare as rocking horse manure, and consequently second-hand models are incredibly expensive. Reproduction versions of the early FETs, such as the UREI 1176LN Peak Limiter (approximately £1800) and the LA Audio Classic II (approximately £2000), are a worthwhile alternative.

Optical

Optical (or 'opto') compressors use a light bulb that reacts to the incoming audio by glowing brighter or dimmer depending on the incoming sound (seriously!). A phototransistor tracks the level of illumination from the bulb and changes the gain. Because the phototransistor must monitor the light bulb before it takes any action, some latency is created in the compressor's response, so the more heavily the compression is applied the longer the envelope times tend to be. Consequently, most optical compressors utilize soft knee compression. This creates a more natural attack and release but also means that the compressor is not quick enough to catch many transients. Despite this, optical compressors are great for compressing vocals, basses, electric guitars and drum loops, providing that a limiter follows the compression. (Limiters will be explained later.)

There are plenty of opto compressors to choose from, including the ADL 1500 (approximately £2500), the UREI LA3 and UREI Teletronix LA-2A (approximately £2900 each), the Joe Meek C2 (approximately £250) and the Joe Meek SC2.2 (approximately £500). Both Joe Meek units sound particularly smooth and warm considering their relatively low prices, and for the typical gain pump synonymous with dance then you could do worse than to pick up the SC2.2.

Notably, all Joe Meek's compressors are green because after designing his first unit he decided to spruce it up by colouring it with car aerosol paint and green was the only colour he could find in the garage at the time.

VCA

VCA compressors offer the fastest envelope times and highest gain reduction levels of any of the compressors covered so far. These are the compressors most likely to be found in a typical home studio. As with most things, the quality of a VCA compressor varies wildly in relation to its price tag. Many of the models aimed at the more budget conscientious musician reduce the high frequencies when a high gain reduction is used, regardless of whether you're clamping down on the transients or not. When used on a full mix these also rarely produce the pumping energy that is typical of models that are more expensive. Nevertheless, these types of compressor are suitable for use on any sound. The most widely celebrated VCA compressor is the Empirical Labs Stereo Distressor (approximately £2500), which is a digitally controlled analogue compressor with VCA, solid-state and op amps. This allows switching between the different methods of compression to suit the sound. Two versions of the Distressor are available to date: the standard version and the British version. Of the two, the British version produces a much more natural, warm tone (I'm not just being patriotic) and is the preferred choice of many dance musicians.

Computer-Based Digital

Computer-based digital compressors are possibly the most precise compressors to use on a sound. Because these compressors are based in the software domain, they can analyse the incoming audio before it actually reaches the compressor, allowing them to predict and apply compression without the risk of any transients sneaking past the compressor. This means that they do not need to utilize a peak/RMS operation. These digital compressors can emulate both solid-state, transparent compression and the more obvious, warm, valve compression at the fraction of the price of a hardware unit. In fact, the Waves RComp can be switched to emulate an optical compressor. Similarly, the PSP Vintage Warmer and Sonalksis TBK3 can add an incredible amount of valve warmth.

The look-ahead functions employed in computer-based compressors can be emulated in hardware with some creative thought, which can be especially useful if the compressor has no peak function. Using a kick drum as an example, make a copy of the kick drum track and then delay it in relation to the original by 50 ms. By then feeding the delayed drum track into the compressor's main inputs and the original drum track into the compressor's side chain, the original drum track activates the compressor just before the delayed version goes through the main inputs, in effect creating a look-ahead compressor!

Ultimately, it is advisable not to get too carried away when compressing audio as it can be easy to destroy the sound while still believing that it sounds better.

This is because louder sounds are invariably perceived as sounding better than those that are quieter. If the make-up gain on the compressor is set at a higher level than the inputted signal, even if the compressor was set up by your pet cat, it will still sound better than the non-compressed version. The incoming signal must be set at exactly the same level as the output of the compressor so that when bypassing the compressor to check the results, the difference in volume doesn't persuade you that it sounds better.

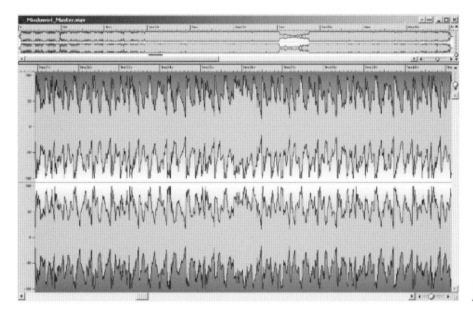

FIGURE 2.4
A mix with excursion

Furthermore, while any sounds that are above the threshold will be reduced in gain, those below it will be increased when the make-up gain is turned up. While this has the advantage of boosting the average signal level, a compressor does not differentiate between music and unwanted noise. So 15 dB of gain reduction will reduce the peak level to 15 dB while the sounds below this remain the same. Using the make-up gain to bring this back up to its nominal level (i.e. 15 dB) any signals that were below the threshold will also be increased by 15 dB, and if there is noise present in the recording, it may become more noticeable.

Most important of all, dance music relies heavily on the energy of the overall 'punch' produced by the kick drum, which comes from the kick drum physically moving the loudspeaker's cone in and out. The more the cone is physically moved, the greater the punch of the kick. This degree of movement is directly related to the size of the kick's peak in relation to the rest of the music's waveform. If the difference between the peak of the kick and the main body of the music is reduced too much through heavy compression, it may increase the average signal level but the kick will not have as much energy since the dynamic range is restricted, meaning that all the music will move the cone by the same amount. So, you should be cautious as to how much you compress otherwise you may lose the excursion which results in a loud yet flat and unexciting track with no energetic punch from the kick (Figures 2.4 and 2.5).

LIMITERS

After compression is applied, it's common practice to pass the audio through a limiter, just in case any transient is not captured by the compressor. Limiters

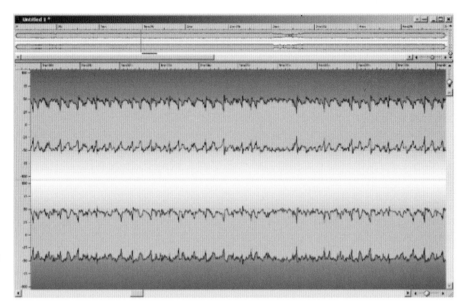

FIGURE 2.5
A mix with no excursion (all the contents of the mix are almost at equal volume)

work along similar principles to compressors but rather than compress a signal by a ratio, they stop signals from ever exceeding the threshold in the first place. This means that no matter how loud the inputted signal becomes, it will be squashed down so that it never violates the current threshold setting. This is referred to as 'brick wall' because no sounds can ever exceed the threshold. Some limiters, however, allow a slight increase in level above the threshold in an effort to maintain a more natural sound.

A widespread misconception is that if the compressor offers a ratio above 10:1 and is set to this it will act as a limiter but this isn't necessarily always the case. As we've seen, a compressor is designed to detect an average signal level (RMS) rather than a peak signal, so even if the attack is set to its fastest response, there's a good chance that signal peaks will catch the compressor unaware. The circuitry within limiters, however, does not employ an attack control, and as soon as the signal reaches the threshold, it is brought under control instantaneously. Therefore, if recording a signal that contains plenty of peaks, a limiter placed directly after the compressor will clamp down on any signals that creep past the compressor and prevent clipping.

Most limiters are quite simple to use and only feature three controls: an input level, a threshold and an output gain, but some may also feature a release parameter. The input is used to set the overall signal level entering the limiter while the threshold and output gain, like a compressor, are used to set the level where the limiter begins attenuating the signal and controlling the output level. The release control is not standard on all limiters, but if included, it's straightforward and allows the time it takes the limiter to return to its nominal state after limiting to be set. As with compression, however, this must be set cautiously, giving the limiter time to recover before the next signal is received to avoid distorting the subsequent transients.

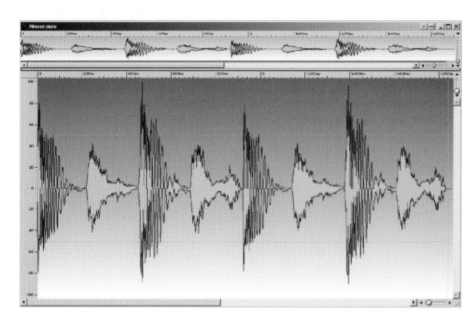

FIGURE 2.6
Drum loop before
limiting

The main purpose of a limiter is to prevent transient signals from overshooting the threshold. Although there is no need for an additional attack control, some software plug-ins will make use of one. This is because they employ look-ahead algorithms that constantly analyse the incoming signal. This allows the limiter to begin the attack stage just before the peak signal occurs. In most cases, this attack isn't user definable and a soft or hard setting will be provided instead. Similar to the knee setting on a compressor, a hard attack activates the limiter as soon as a peak is close to overshooting. On the other hand, a soft attack has a smoother curve with a 10 or 20 ms timing. This reduces the likelihood that any artefacts are introduced into the processed audio by jumping in on the audio too quickly. These software look-ahead limiters are sometimes referred to as ultramaximizers.

As discussed, the types of signals that require limiting are commonly those with an initial sharp transient peak. As a result, limiters are generally used for removing the 'crack' from snare drums, keeping the kick drum under control, and are often used on a full track to produce a louder mix during the mastering process. Like compressors, though, limiters must be used cautiously because they work on the principle of reducing the dynamic range. That is, the harder a sound is limited, the more dynamically restricted it becomes. Too much limiting can result in a loud but monotonous sounding signal or mix. On average, approximately 3–6 dB is a reasonable amount of limiting, but the exact figure depends entirely on the sound or mix. If the sound has already been quite heavily compressed, it's best to avoid boosting any more than 3 dB at the limiting stage, otherwise any dynamics deliberately left in during the compression stage may be destroyed (Figures 2.6 and 2.7).

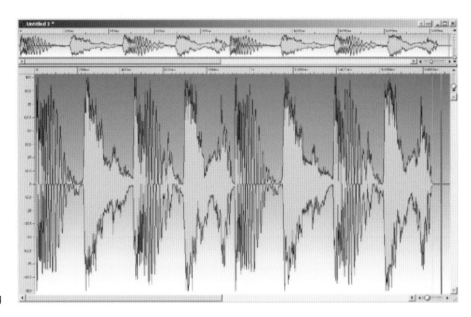

FIGURE 2.7
Drum loop after limiting

NOISE GATES

Noise gates can be described as the opposite of compressors. This is because while a compressor attenuates the level of any signal that exceeds the threshold, a gate can attenuate or remove any signals that are below the threshold. The main purpose of this is to remove any low-level noise that may be present during a silent passage. For instance, a typical effect of many commercial dance tracks is to introduce absolute silence or perhaps a drum kick just before the reprise so that when the track returns fully, the sudden change from almost nothing into everything playing at once creates a massive impact. The problem with this approach, though, is that if there is some low-level noise in the recording it will be evident when the track falls silent (i.e. noise between the kicks), which not only sounds cheap but reduces the impact when the rest of the instruments jump back in. In these instances, by employing a gate it can be set so that whenever sounds fall below its threshold the gate activates and creates absolute silence. While in theory this sounds simple enough, in practice it's all a little more difficult.

Firstly, we need to consider that not all sounds stay at a constant volume throughout their period. Indeed, some sounds can fluctuate wildly in volume, which means that they may constantly jump above and below the threshold of the gate. What's more, if the sound was close to the gates threshold throughout, with even a slight fluctuation in volume it'll constantly leap above and below the threshold resulting in an effect known as chattering. To prevent this, gates will often feature an automated or user-definable hold time. Using this, the gate can be forced to wait for a predetermined amount of time after the signal has fallen below the threshold before it begins its release stage, thus avoiding the problem.

The action of this hold function is sometimes confused with a similar gate process called hysteresis but the two processes, while accomplishing the same goal, are very different. Whereas the hold function forces the gate to wait for a predefined amount of time before closing, hysteresis adjusts the threshold's tolerance independently for opening and closing the gate. For example, if the threshold was set at, say, −12 dB, the audio signal must breach this before the gate opens but the signal must fall a few extra dB below −12 dB before the gate closes again. Consequently, while both hold and hysteresis accomplish the same goal in preventing any chatter, it is generally accepted that hysteresis sounds much more natural than simply using a hold control.

A second problem develops when we consider that not all sounds will start and stop abruptly. For instance, if you were gating a pad that gradually rose in volume, it would only be allowed through the gate after it exceeds the predefined threshold. If this threshold happened to be set quite high, the pad would suddenly jump in rather than fade in gradually as it was supposed to. Similarly, rather than fade away, it would be abruptly cut off as it fell below the threshold again. Of course, you could always lower the threshold, but that may allow noise to creep in, so gates will also feature attack and release parameters. These are similar in most respects to a compressor's envelope in that they allow you to determine the attack and release times of the gate's action. Using these on our example of a pad, by setting the release quite long as soon as the pad falls below the threshold the gate will enter the release stage and gradually fade out rather than cut them off abruptly. Likewise, by lengthening the attack on the gate, the strings will fade in rather than jump in unexpectedly.

The third, and final, problem is that we may not always want to silence any sounds that fall below the threshold. Suppose that you've recorded a rapper (or any vocalist for that matter) to drop into the music. He or she will obviously need to breathe between the verses, and if they're about to scream something out, they'll need a large intake of breath before starting. This sharp intake of breath will make its way onto the vocal recording, and while you don't want it to be too loud, at the same time you don't want it totally removed either otherwise it'll sound totally unnatural – the audience instinctively know that vocalists have to breathe!

Consequently, we need a way of lowering the volume of any sounds that fall below the threshold rather than totally attenuating them, so many gates (but not all!) will feature a range control. Fundamentally, this is a volume control that's calibrated in decibels allowing you to define how much the signal is attenuated when it falls below the threshold. The more this is increased, the more the signal will be reduced in gain until – set at its maximum setting – the gate will silence the signal altogether. Using this range control on the imaginary rapper, you could set it quite low so that the volume of the breaths is not too loud but not too quiet either. By setting the threshold so that only the vocals breach it and those below are reduced in volume by a small amount, it will sound much more natural. Furthermore, by setting the gate's attack to around 100 ms or so as he/she breathes, it will begin at the volume set by the range and then slowly swell in to the vocal, which produces a much more natural effect.

For this application to work properly, the release time of the gate must be set cautiously. If it's set too long, the gate may remain open during the silence between the vocals, which doesn't allow a new attack stage to be triggered when they begin to sing again. On the other hand, if it's set too short it can result in the 'chattering' effect described earlier. Consequently, it's prudent to use the shortest possible decay time possible, yet long enough to provide a smooth sound. Generally, this is usually somewhere between 50 and 200 ms.

Employed creatively, the range control can also be used to modify the attack transients of percussive instruments such as pianos, organs or lead sounds (not on drums, though, these are far too short).

One final aspect of gates is that many will also feature a side-chain connection. In this context they're often referred to as 'key' inputs but nevertheless this connection behaves in a manner similar to a compressor's side chain. Fundamentally, they allow you to insert an audio signal in the key input which can then be used to control the action of the gate, which in turn affects the audio travelling through the noise gates normal inputs. This obviously has numerous creative uses but the most common use is to programme a kick drum rhythm and feed the audio into the key input. Any signals that are then fed through the gate's normal inputs will be gated every time a kick occurs. This action supersedes the gate's threshold setting but the attack, release, range, hold or hysteresis controls are often still available allowing you to contour the reaction of the gate on the audio signal.

Another use of this key input is known as 'ducking' and many gates will feature a push button allowing you to engage it. When this is activated, the gate's process is reversed so that any signals that enter the key input will 'duck' the volume of the signal running through the gate. A typical use for this is to connect a microphone into the key input so that every time you speak, the volume of the original signal travelling through the gate is ducked in volume. Again, this supersedes the threshold control, but all of the other parameters are still available allowing you to contour the action of the signal being ducked, although it should be noted that the attack time turns into a release control and vice versa. Also as a side note, some of the more expensive gates will feature a MIDI in port that prevents you from inserting an audio signal into the key input; instead, MIDI note on messages can be used to control the action of the gate.

TRANSIENT DESIGNERS

Transient designers are quite simple processors that generally feature only two controls: an attack and a sustain parameter, both of which allow you to shape the dynamic envelope of a sound. Fundamentally, this means that you can alter the attack and sustain characteristics of a pre-recorded audio file the same as you would when using a synthesizer. While this may initially not seem to be too impressive, it has a multitude of practical uses.

Since we determine a significant amount of information about a sound through its attack stage, modifying this can change the appearance of any sound. For example, if you have a sampled loop and the drum is too loud, by reducing its attack (and lengthening the sustain so that the sound doesn't vanish) it will be moved further back into the mix. In a more creative application, if a groove has been sampled from a record it allows you to modify the drum sounds into something else.

Similarly if you've sampled or recorded vocals, pianos, strings or any instrument for that matter, the transient designer can be used to add or remove some of the attack stage to make the sound more or less prominent while strings and basses could have a longer sustain applied. Similarly, by reducing the sustain parameter on the transient designer you could reduce the length of the notes. Notably, noise gates can also be used to create the effect of a transient designer and, like the previously discussed processors, these can be an invaluable tool to a dance producer; we'll be looking more closely at the uses of both in the genre chapters.

REVERB

Reverberation (often shortened to reverb or just verb) is used to describe the natural reflections we've come to expect from listening to sounds in different environments. We already know that when something produces a sound, the resulting changes in air pressure emanate out in all directions but only a proportion of this reaches our ears directly. The rest rebounds off nearby objects and walls before reaching our ears; thus, it makes common sense that these reflected waves would take longer to reach your ears than the direct sound itself.

This creates a series of discrete echoes that are all closely compacted together and from this our brains can decipher a staggering amount of information about the surroundings. This is because each time the sound is reflected from a surface, that surface will absorb some of the sound's energy, thereby reducing the amplitude. However, each surface also has a distinct frequency response, which means that different materials will absorb the sound's energy at different frequencies. For instance, stone walls will rebound high-frequency energy more readily than soft furnishings which absorb it. If you were in a large hall it would take longer for the reverberations to decay away than it would if you were in a smaller room. In fact, the further away from a sound source you are, the more reverberation there would be in comparison to the direct sound in reflective spaces, until eventually, if the sound was far enough away and the conditions were right you would hear a series of distinct echoes rather than reverb.

There should be little need to describe all the differing effects of reverb because you'll have experienced them all yourself. If you were blindfolded, you would still be able to determine what type of room you're in from the sonic reflections. In fact, reverb is such a natural occurrence that if it's totally removed (such as is an anechoic chamber) it can be unsettling almost to the point of nausea. Our eyes are informing the brain of the room's dimensions, but the ears are informing it of something completely different.

Ultimately, while compression is the most important processor, reverb is the most important effect because samplers and synthesizers do not generate natural reverberations until the resulting signals are exposed to air. So, in order to create some depth in a mix you often need to add it artificially. For example, the kick may need to be at the front of a mix but any pads could sit in the background. Simply reducing the volume of the pads may make them disappear into the mix, but by applying a light smear of reverb you could fool the brain into believing that the sound is further away from the drums because of the reverberation that's surrounding it.

However, there's much more to applying reverb than simply increasing the amount that is applied to the sound. As we've seen, reverb behaves very differently depending on the furnishings and wall coverings, so all reverb units will offer many more parameters and using it successfully depends on knowing the effects all these will have on a sound. What follows is a list of the available controls on a reverb unit, but it should be noted that in many cases all of these will not be available – it depends on the quality of the unit itself.

Ratio (Sometimes Labelled as Mix)

The ratio controls the ratio of direct sound to the amount of reverberation applied. If you increase the ratio to near maximum, there will be more reverb than direct sound, while if you decrease it significantly, there will be more direct sound than reverb. Using this, you can make sounds appear further away or closer to you.

Pre-Delay Time

After a sound occurs, the time separation between the direct sound and the first reflection to reach your ears is referred to as the pre-delay. This parameter on a reverb unit allows you to specify the amount of time between the start of the unaffected sound and the beginning of the first sonic reflection. In a practical sense, by using a long pre-delay setting the attack of the instrument can pull through before the subsequent reflections appear. This can be vital in preventing the reflections from washing over the transient of instruments, forcing them towards the back of a mix or muddying the sound.

Early Reflections

Early reflections are used to control the sonic properties of the first few reflections we receive. Since sounds reflect off a multitude of surfaces, subtle differences are created between subsequent reflections reaching our ears. Due to the complex nature of these first reflections, only the high-end processors feature this type of control, which allows you to determine the type of surface the sound has reflected from.

Diffusion

This parameter is associated with the early reflections and is a measure of how far the early reflections are spread across the stereo image. The amount of stereo width associated with the reflections depends on how far the sound source is. If a sound is far away then much of the stereo width of the reverb will dissipate but there will be more reverberation than if it was upfront. If the sound source is quite close, however, then the reverberations will tend to be less spread and more monophonic. This is worth keeping in mind since many artists wash a sound in stereo reverb to push it into the background and then wonder why the stereo image disappears and doesn't sound quite 'right' in context with the rest of the mix.

Density

Directly after the early reflections come the rest of the reflections. On a reverb unit this is referred to as the density. Using this control it's possible to vary the number of reflections and how fast they should repeat. By increasing it, the reflections will become denser giving the impression that the surface they have reflected from is more complex.

Reverb Decay Time

This parameter is used to control the amount of time the reverb takes to decay away. In large buildings the reflections will *generally* take longer to decay into silence than in a smaller room. Thus, by increasing the decay time you can effectively increase the size of the 'room'. This parameter must be used cautiously, however, as if you use a large decay time on a motif the subsequent reflections from previous notes may still be decaying when the next note starts. If the motif is continually repeated, it will be subjected to more and more reflections until it eventually turns into an incoherent mush of frequencies.

The amount of time it takes for a reverb to fade away (after the original sound has stopped) is measured by how long it takes for the sound pressure level to decay to one-millionth of its original value. Since one-millionth equates to a 60 dB reduction, reverb decay time is often referred to as RT60 time.

HF and LF Damping

The further reflections have to travel the less high frequency content they will have since the surrounding air will absorb them. Additionally, soft furnishings will also absorb higher frequencies, so by reducing the high frequency content (and reducing the decay time) you can give the impression that the sound is in a small enclosed area or has soft furnishings. Alternatively, by increasing the decay time and removing smaller amounts of the high frequency content you can make the sound source appear further away. Further, by increasing the lower frequency damping you can emulate a large open space. For instance, while singing in a large cavernous area there will be a low end rumble with the reflections but not as much high-frequency energy.

Despite the amount of controls a reverb unit may offer, it is also important to note that units from different manufacturers will sound very different to one another as each manufacturer will use different algorithms to simulate the effect. Although it is quintessential to use a good reverb unit (such as a Lexicon hardware unit or the TC Native Plug-ins included on the CD), it's not uncommon to use two or three different models of reverb in one mix.

CHORUS

Chorus effects attempt to emulate the sound of two or more of the same instruments playing the same parts simultaneously. Since no two instrumentalists could play exactly in time with one another, the result is a series of phase cancellations. This is analogous to two synthesizer waveforms slightly detuned and playing simultaneously together; there will be a series of phase cancellations as the two frequencies move in and out of phase with one another. A chorus unit achieves this same effect by delaying the incoming audio signal slightly in time while also dynamically changing the time delay and amplitude as the sound continues.

To provide control over this modulation a typical chorus effect will offer three parameters all of which can be directly related to the LFO parameters on a typical synthesizer. The first allows you to select a modulation waveform that will be used to modulate the pitch of the delayed signal, while the second and third parameters allow you to set the modulation rate (referred to as the frequency) and the depth of the chorus effect (often referred to as delay). However, it should be noted that because the modulation rate stays at a constant depth, rate and waveform it doesn't produce the 'authentic' results you would experience with real instrumentalists. Nevertheless, it has become a useful effect in its own right and can often be employed to make oscillators and timbre appear thicker, wider and much more substantial.

PHASERS AND FLANGERS

Phasers and flangers are very similar effects with subtle differences in how they are created, but work on a principle comparable to the chorus effect. Originally phasing was produced by using two tape machines that played slightly out of sync with one another. As you can probably imagine, this created an irregularity between the two machines, which resulted in the phase relationship of the audio being slightly different, in effect producing a hollow, phase-shifted sound.

This idea was developed further in the 1950s by Les Paul as he experimented by applying pressure onto the 'flange' (i.e. the metal circle that the tape is wound upon) of the second tape machine. This effectively slowed down the speed of the second machine and produced a more delayed swirling effect due not only to the phase differences but also to the speed. With digital effect units, both these work by mixing the original incoming signal with a delayed version but also by feeding some of the output back into the input. The only difference between the two is that flangers use a time delay circuit to produce the effect while a phaser uses a phase shift circuit.

Nevertheless, both use an LFO to modulate either the phase shifting of the phaser or the time delay of the flanger. This creates a series of phase cancellations since the original and delayed signals are out of phase with one another. The resulting effect is that phasers produce a series of notches in the audio file that are harmonically related (since they are related to the phase of the original audio signal) while flangers have a constantly different frequency because they use a time delay circuit. Consequently both flangers and phasers share the same parameters. They both feature a rate parameter to control the speed of the LFO effect along with a feedback control to set how deeply the LFO affects the audio. Notably, some phasers will only use a sine wave as a modulation source but most flangers will allow you to not only change the shape but also control the number of delays used to process the original signal. Today, both these effects have become a staple in the production of dance, especially House, with the likes of Daft Punk using them on just about every record they've ever produced.

DIGITAL DELAY

To the dance music producer, digital delay (often referred to as digital delay line – DDL) is one of the most important effects to own as if used creatively it can be one of the most versatile. The simplest units will allow you to delay the incoming audio signal by a predetermined time which is commonly referred to in milliseconds or sometimes in note values. The number of delays produced by the unit is often referred to as the feedback, so by increasing the feedback setting you can produce more than one repeat from a single sound. This works by sending some of the delayed output back into the effects input so that it's delayed again, and again, and again and so forth. Obviously this means that if the feedback is set to a very high value the level of the repeats end up collecting together rather than gradually dying away until eventually you'll end up with a horrible howling sound.

While all delay units will work on this basic premise, the more advanced units may permit you to delay the left and right channels individually and pan them to the left and right of the stereo image. They may also allow you to pitch shift the subsequent delays, employ filters to adjust the harmonic content of the delays, distort or add reverb to the results and apply LFO modulation to the filter. Of all these additional controls (most of which should require no explanation) the modulation is perhaps the most creative application to have on a delay unit as it allows you to modulate the filters cut-off or pitch, or both, with an LFO. The number and type of waveforms on offer vary from unit to unit but fundamentally most will feature at least a sine, square and triangle wave. Similar to a synthesizer's LFO parameters, they will feature rate and depth controls allowing you to adjust how fast and by how much it should modulate the filter cut-off or pitch parameters.

One of the most common uses for a delay in dance music is not necessarily to add a series of delays to an audio signal but to create an effect known as granular delay. As we touched upon in Chapter 3, we cannot perceive individual sounds if they are less than 30 ms apart, which is the principle behind granular synthesis. However, if a sound is sent to a delay unit and the delay time is set to less than 30 ms and combined with a low feedback setting, the subsequent delays collect together in a short period of time which we cannot perceive as a delay. The resulting effect is that the delayed timbre appears much bigger, wider and upfront. This technique is often employed on leads or, in some cases, basses if the track is based around a powerful driving bass line.

EQ

At its most basic, EQ is a frequency-specific volume control tone control that allows you to intensify or attenuate specific frequencies. For this, three controls are required

- A frequency control allowing you to home in on the frequency you want to adjust.
- A 'Q' control allowing you to determine how many frequencies either side of the centre frequency you want to adjust.
- A gain control to allow you to attenuate or intensify the selected frequencies.

Notably not all EQ units will offer this amount of control and some units will have a fixed frequency or a fixed Q, meaning that you can only adjust the volume of the frequencies that are preset by the manufacturer. EQ plays a much larger role in mixing than it does in sound design, so this has only been a quick introduction and we'll look much more deeply into its effects when we cover mixing and mastering in later chapters.

DISTORTION

The final effect for this chapter, distortion, is pretty much self-explanatory; it introduces an overdrive effect to any sounds that are fed through it. However, while the basic premise is quite simple, it has many more uses than to simply grunge up a clean audio signal. As touched upon in Chapter 3, a sine wave does not contain any harmonics except for the fundamental frequency, and therefore applying effects such as flangers, phasers or filters will have very little influence. However, if distortion were applied to the sine wave it would introduce a series of harmonics into the signal giving the aforementioned effects something more substantial to work with.

CHAPTER 3

Cables, Mixing Desks and Effects Busses

'I don't just use a (mixing) desk to mix sounds together
I use it as a creative tool...'

Juan Atkins

There is much more to understanding the various processors and effects used within the creation of dance music; you also need to know how to access them through a typical mixing desk to gain the correct results. While most producers today rely on a computer to handle the recording, effecting, processing and mixing, there will nevertheless come a time when you have to employ external units into your rig, whether synthesizers, processors, effects or even a sampler, and therefore you'll start by looking at the cables used to connect these to your mixing desk, laptop or computer.

Any competent studio is only as capable as its weakest link, so if low-quality cables are used to connect devices together, the cables will be susceptible to introducing interference, which results in noise. This problem arises because any cables that carry a current, no matter how small, produce their own voltage as the current travels along them. The level of voltage that is produced by the cable will depend on its resistance and the current passing through it, but this nevertheless results in a difference in voltage from one end of the cable to the other.

Because all studio equipment (unless it's all contained inside a Digital Audio Workstation) requires cables to carry the audio signal to and from the mixing desk, the additional voltage introduced by the cables is then transmitted around the instruments from the mixing desk and through to earth. This produces a continual loop resulting in an electromagnetic field that surrounds all the cables. This field introduces an electrical hum into the signal, an effect known as 'ground hum'. The best way to reduce this is by using professional-quality 'balanced' cables, although not all equipment, particularly equipment intended for the home studio, uses this form of cable. Home studio equipment tends to use 'unbalanced' cables and connectors.

The distinction between balanced and unbalanced cable is determined by the termination connectors at each end. Cables terminated with mono jack or

Mono jack

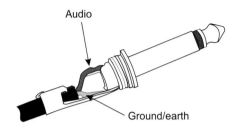

Stereo jack

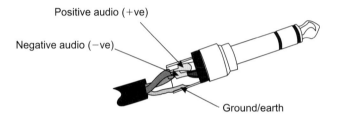

Male XLR

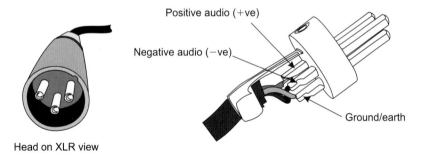

Head on XLR view

FIGURE 3.1
Mono, stereo jack and XLR connectors

phono connectors are unbalanced, while stereo Tip–Ring–Sleeve (TRS) jack connections or extra long run (XLR) connections will be found on balanced cables. Examples of these are shown in Figure 3.1.

All unbalanced cables are made up of two internal wires contained within the outer plastic or rubber core of the wire (known as the earth screen). One of these internal wires carries the audio signal and is connected to the tip of the connector while the other carries the ground signal and is connected directly to the connector sleeve. The signal is therefore grounded at each end of the cable helping to prevent any interference from the device itself, but is still susceptible to electromagnetic interference as it is transmitted from one device to another. Because of this, most professional studios use balanced cables with XLR or TRS terminating connections, if the equipment they connect to supports it.

> TRS connections do not necessarily mean that the cable is balanced as they can be used to carry a stereo signal (left channel, right channel and earth) but in a studio environment they are more commonly used to transfer a mono signal.

Balanced cables contain three wires within the outer screen. In this configuration a single wire is still used as a ground but the other two wires carry the audio signal, one of which is a phase-inverted version of the original. When this is received by a device, the phase-inverted signal is put back into phase with the original and the two are added together. As a result, any interference introduced is cancelled out when the two signals are summed together. This is similar to the way two oscillators that are in phase cancel each other out, as described in here in this chapter. That's the theory. In practice, although this reduces the problem, phase cancellation rarely removes all of the interference.

Another favourable advantage of using balanced cables is that they also utilize a more powerful signal level. Commonly referred to as a professional standard, a balanced signal uses a signal level of +4 dBu rather than the semi-professional signal level of −10 dBV. The reasons for this are more the subject matter of electrical engineering than music, and although it's not necessarily important to understand this, a short explanation is given in the text box. If you're not interested you can skip this box because all you really need to know is that +4 dBu signals have a hotter, louder signal than −10 dBV signals and are generally preferred as the signal is over 11 dB hotter, so the chance of capturing a poor signal is reduced.

Before light-emitting diode (LED) and liquid crystal display (LCD) displays appeared on musical gear, audio engineers used volume unit (VU) meters to measure audio signals. These were featured on any hardware that could receive an input. This meant that every VU meter had to give the same reading for the same signal level no matter who manufactured the device. If this were not the case, different equipment within the same studio would have different signal levels. Consequently, engineers decided that if 1 milliwatt (mW) was travelling through the circuitry then the VU meter should read 0 dB. Hence, 0 dB VU was referred to as 0 dBm (with m standing for milliwatt).

Today's audio engineering societies are no longer concerned with using a reference level of milliwatt because the power levels today are much higher, so the level of 0 dBm is now obsolete and we use voltage levels instead. To convert this into an equivalent voltage level, the impedance has to be specified, which in this case is 600 Ohms. For those with some prior electrical knowledge it can be calculated as follows:

$$P = V^2/R$$

$$0.001\,W = V^2/600\,W$$

$$V^2 = 0.001\,W \times 600\,W$$

$$V = \sqrt{0.001\,W \times 600\,W}$$

For the layman, the sum of this equation equals 0.775 Volts and that's all you need to know.

The value of 0.775 is now used as the reference voltage and is referred to in dBu rather than dBm. Although it was originally referred to as dBv it was often confused with the reference level of dBV (notice the upper case V), so the suffix u is used in its place. This is only the reference level, though, and all professional equipment will output a level that is +4 dB, which is where we derive the +4 dBu standard. Consequently, on professional equipment, the zero level on the meters actually signifies that it is receiving a +4 dBu signal.

However, some hardware engineers agreed that it would be simpler to use 1 Volt as the reference instead, which is where the dBV standard originates. Unlike professional equipment, which uses the +4 dBu output level, unbalanced equipment outputs at 0.316 Volts, equivalent to −10 dBV. Therefore, on semi-professional equipment the zero level on the meters signifies that they are receiving a −10 dBV signal. If the professional and semi-professional signals are compared, the professional voltage of 0.775 V is considerably higher than the 0.316 Volts generated by consumer equipment. When converted to decibel this results in an 11.8 dB difference between the two.

Despite the difference in the levels of the two signals, in many cases it is still possible to connect a balanced signal to an unbalanced sampler/soundcard/mixer with the immediate benefit that the signal that is captured is 11.8 dB hotter. Although this usually results in the unbalanced recorder's levels jumping off the scale, whether the signal is actually distorting should be based on whether this distortion is audible. Most recorders employ a safety buffer by setting the clipping meters below the maximum signal level. This is, of course, a one-way connection from a balanced signal to an unbalanced piece of hardware, and it isn't a good idea to work with an unbalanced signal connecting to a balanced recorder because you'll end up with a poor input signal level and any attempts to boost the signal further could introduce noise.

Within the typical home studio environment, it is most likely that the equipment will be of the unbalanced type; therefore, the use of unbalanced connections is unavoidable. If this is the case, it's prudent to take some precautions that will prevent the introduction of unwanted electromagnetic interference but this can be incredibly difficult.

While the simplest solution would be to disconnect the earth from the power supply, in effect breaking the ground loop, this should be avoided at all costs. Keep in mind that the whole point of an earth system is to pass the current directly to ground rather than you if there's a fault. Human beings make remarkably good earth terminals and as electricity will always take the quickest route it can find to reach earth, it can be an incredibly painful (and sometimes deadly) experience to present yourself as a short cut.

A less suicidal technique is to remove the earth connection from one end of audio cable. This breaks the loop, but it has the disadvantage that it can make the cable more susceptible to radio frequency (RF) interference. In other words, the cable would be capable of receiving signals from passing police cars, taxis, mobile phones and any nearby citizens' band radios. While this could be useful if you want to base your music around *Scanners* previous work, it's something you'll want to avoid.

Although in a majority of cases hum will be caused by the electromagnetic field, it can also be the result of a number of other factors combined. To begin with it's worthwhile ensuring that the mains and audio cables are wrapped separately from one another and kept as far away from each other as possible. Mains cables create a higher electromagnetic field due to their large current, and if they are bound together with cables carrying an audio signal, serious hum can be introduced.

Transformers also generate high electromagnetic fields that cause interference, and although you may not think that you have any in the studio, both amplifiers and mixing desks use them. Consequently, amplifiers should be kept at a good distance from other equipment, especially sensitive equipment such as microphone pre-amps. If the amplifiers are rack-mounted, there should be a minimum space of 4-Rack Units between the amplifier and any other devices. This same principle also applies to rack-mounted mixing desks, which should ideally be placed in a rack of their own or kept on a desk. If the rack that is used is constructed from metal and metal screws hold the mixing desk in place; the result is the same as if the instruments were grounding from another source and yet more hum is introduced. Preferably, plastic washers and screw housings should be used, as these isolate the unit from the rack.

If, after all these possible sources have been eliminated, hum is still present, the only viable way of removing or further reducing it is to connect devices together digitally, or invest in a professional mains suppressor. This should be sourced from a professional studio supplier rather than from the local electrical hardware or car superstore, as suppressors sold for use within a studio are specifically manufactured for this purpose, whereas a typical mains suppressor is designed to suppress only the small amounts of hum that are typically associated with normal household equipment.

Digital connections can be used as an alternative to analogue cables if the sampler/soundcard/mixer allows it. This has the immediate benefit that no noise will be introduced into the signal with the additional benefit that this connection can also be used by the beat-slicing software to transmit loops to and from an external hardware sampler. Sampler-to-software connectivity is usually accomplished via a direct SCSI interface, so there is little need to be concerned about the digital standards, but on occasion it may be preferable to transmit the results through true digital interfaces such as Alesis digital audio tape (ADAT), Tascam digital interface (T-DIF), Sony/Philips digital interface (S/PDIF) or Audio Engineering Society/European Broadcasting Union (AES–EBU).

The problem is that digital interfacing is more complex than analogue interfacing because the transmitted audio data must be decoded properly. This means that the bit rate, sample rate, sample start and end points, and the left and right channels must be coded in such a way that the receiving device can make sense of it all. The problem with this is that digital interfaces appear in various forms, including (among many others) Yamaha's Y formats, Sony's SDIF, Sony and Phillips S/PDIF and the AES–EBU standard, none of which are cross-compatible.

In an effort to avoid these interconnection problems, the American Audio Engineering Society (AAES) combined with the EBU and devised a standard connection format imaginatively labelled the AES–EBU standard. This requires a three-pin XLR connection, similar to the balanced analogue equivalent, although the connection is specific to the digital.

Like 'balanced' analogue connections AES–EBU is expensive to implement, so Sony and Philips developed a less expensive 'unbalanced' standard known as S/PDIF. This uses either a pair of phono connectors or an optical TOS-link interface (Toshiba Optical Source). Most recent samplers and soundcards use a TOS-link or phono connection to transmit digital information to and from other devices.

The great thing about standards is that there are plenty of them…

With compatible interfaces between two devices, both the receiving and the transmitting device must be clocked together so that they are fully synchronized. This ensures that they can communicate with one another. If the devices are not synchronized, the receiving device will not know when to expect an incoming signal, producing 'jitter'. The resulting effect this has on audio is difficult to describe but rather than fill up the book's accompanying CD with the ensuing ear-piercing noises, it's probably best explained as an annoying high-frequency racket mixed with an unstable stereo image. To avoid this, most professional studios use an external clock generator to synchronize multiple digital units together correctly. This is similar in most respects to a typical multi-MIDI interface, bar the fact that it generates and sends out word clock (often abbreviated to WCLK but still pronounced word clock) messages simultaneously to all devices to keep them clocked together.

The WCLK message works by sending a one-bit signal down the digital cable, resulting in a square wave that is received in all the attached devices. When the signal is decompiled by the receiving device, the peaks and troughs of the square wave denote the left and right channels while the width between each pulse of the wave determines the clock rate.

These stand-alone WCLK generators can be expensive, so within a home studio set-up the digital mixing desk or soundcard usually generates the WCLK message, with the devices daisy-chained together to receive the signal. For instance, if the soundcard is used to generate the clock rate, the WCLK message could be sent to an effects device, through this into a microphone pre-amplifier, and so forth before the signal returns back into the soundcard, creating a loop. The principle is similar to the way that numerous MIDI devices are daisy-chained together, as discussed in Chapter 1. As with daisy-chained MIDI devices, though, the signal weakens as it passes through each device; therefore, if the signal is to pass through more than four devices, it's prudent to use a WCLK amplifier to keep the signal powerful enough to prevent any jitter.

Provided that the clock rate is correctly transmitted and received, another important factor to consider is the sample rate. When recording, both the transmitting and the receiving devices must be locked to the same sample rate, otherwise the recorder may refuse to go into record mode. Also, unless you are recording the final master, any Serial Copyright Management System (SCMS) features should be disabled.

SCMS was implemented onto all consumer digital connections to reduce the possibility of music piracy. This allows only one digital copy to be made from the original. It does this by inserting a 'copyright flag' into the first couple of bytes that are transmitted to the recording device. If the recorder recognizes this flag it will disable the device's record functions. This is obviously going to cause serious problems if you need to edit a second-generation copy because unless the recorder allows you to disable SCMS, you will not be allowed to record the results digitally. Thus, if you plan to transfer and edit data digitally, it is vital to ensure that you can disable the SCMS system.

If you own a digital audio tape (DAT) machine that does not allow you to disable the SCMS protection system then it's possible to get hold of SCMS strippers which remove the first few flags of the signal, in effect disabling the SCMS system.

MIXING DESK STRUCTURE

While some mixing desks appear relatively straightforward some professional desks, such as the Neve Capricorn or the SSL, look more ominous. However, whatever the level of complexity all types of hardware- and software-based mixing desks operate according to the same principles. With a basic understanding of the various features and how the channels, busses, subgroups, EQ and aux send and returns are configured, you can get the best out of your equipment, no matter how large or small the desk is.

The fundamental application of any mixing desk is to take a series of inputs from external instruments and provide an interface that allows you to adjust the tonal characteristics of each perspective instrument. These signals are then culminated together in the desk and fed into the loudspeaker monitors and/ or recorder to produce the finished mix. As simple as this premise may be, though, simply looking at a desk reveals that there's actually a lot more going on and to better understand this we need to break the desk down to each individual input channel.

Typically, a mixing desk can offer anywhere from two input channels to over a hundred, depending on its price. Each of these channels (referred to by engineers as a 'strip') is designed to accept either a mono or stereo signal from one musical source and provide some tonal editing features for that one strip. Clearly, this means that if you have five external hardware instruments, each with a mono output, you would need a desk with an absolute minimum total of five 'strips' so that you could input each into a separate channel of the desk. For reasons we'll touch upon, you should always aim to have as many channels strips in a mixer as you can afford, no matter how few instruments you may own.

Generally speaking, the physical inputs for each channel are located at the rear of the desk and will consist of a 1/4″ jack or XLR connection, or both. This latter configuration doesn't mean that two inputs can be directed into the same channel strip simultaneously; rather it allows you to choose whether that particular channel accepts a signal from a jack *or* an XLR connector. Most desks will have the capacity to accept a signal of any level, whether it's mic-level ($-60\,dBu$) or line level ($-10\,dBV$ or $+4\,dBu$) and whether the cables are balanced or unbalanced.

Some of the older mixing desks may describe these inputs as being Low-Z or Hi-Z but all this is describing is the input impedance of that particular channel. If Hi-Z is used then the impedance is higher to accept unbalanced line-level signals, while if Low-Z is used the impedance is lower to accept balanced ones.

As all mixing desks operate at line level to keep the signal to noise ratio at a minimum, once a signal enters the desk it is first directed to the pre-amplifier stage to bring the incoming signal up to operating level of the desk. Although this pre-amp stage isn't particularly necessary with line-level signals as most desks use this as their nominal operating level, it is required to bring the relatively low levels generated by most microphones to a more respectable level for mixing. This pre-amp will have an associated rotary gain control on the facia of the desk (often called pots – an acronym for potentiometer) labelled 'trim' or 'gain'. As its name would suggest, this allows you to adjust the volume of the signal entering the channels input by increasing the amount of amplification at the pre-amp.

To reduce the possibility of the mixer introducing noise into the channel you should ensure that the signal entering the mixer is as high as possible rather than input a low-level signal and boost it at pre-amp stage of the mixer. Keep in mind that a good signal entering the mixer before the pre-amp is more likely to remain noise-free as it passes through the rest of the mixer.

Some mixers may also offer a 'pad' switch at the input stage, which is used to attenuate the incoming signal before it reaches the pre-amplifier. The amount of attenuation applied is commonly fixed at 12 or 24 dB and is used to prevent any hot signals entering the inputs from overdriving the desk. A typical example of this may be where the mixer operates at the 'semi-professional' −10 dBV and one of the instruments connected to it outputs at the 'professional' +4dBu. As we've previously touched upon, this would mean that the input level at the desk would be 11.8 dB higher than the desk's intended operational level. Obviously this will result in distortion, but by activating the pad switch, the incoming signal could be attenuated by 24 dB and the volume could then be made back up to unity gain with the trim pot on that particular channel.

> The term Unity Gain means that one or all of the mixer channels volume faders are set at 0 dB, the loudest signal possible before you begin to move into the mixers headroom and possible distortion.

On top of this, some high-end mixers may also offer a phantom power switch, a ground lift switch and/or a phase reverse switch, the functions of which are described below:

- Phantom power is used to supply capacitor microphones with the voltage they require to operate. This also means that the mixer has to use its own amplifier to increase its signal level, and generally speaking, the amp circuits in mixers will not be as good as those found in stand-alone microphone pre-amplifiers.
- A ground lift switch will only be featured on mixers that accept balanced XLR connections, and when this is activated, it disables pin number one on the connector from the mixer's ground circuitry. Again, as touched upon this will eliminate any ground hum or extraneous noises.
- Phase reverse – sometimes marked by a small circle with a diagonal line through it – switches the polarity of the incoming signal, which can have a multitude of uses. Typically, they're included on mixing desks to prevent any recordings taken simultaneously with a number of microphones from interfering with one another, since the sound of one microphone can weaken the sound from a second microphone that's positioned further away. Subsequently, phase-reverse switches are only usually found next to XLR microphone inputs since it's easy to implement on these – the switch simply swaps over the two signal pins.

Following this input section, the signal is commonly routed through a mute switch (allowing you to instantly mute the channel) and into the insert buss.[1]

---

[1]Busses refer to the various signal paths within a mixer that the inputted audio can travel through.

These allow you to insert processors such as compressors or noise gates to clean up or bring a signal under control before it's routed into the EQ. On a mixing desk these are the most important aspect as they allow you to modify the tonal content of a sound for creative applications or, more commonly, so that the timbre fits into a mix better. As a result, when looking for a mixing desk it's worthwhile seeing how well equipped the EQ section is as this will give a good impression of how useful the desk will be to you. Most cheap consumer mixers will only offer a high and low EQ section, which can be used to increase or reduce the gain at a predetermined frequency. More expensive desks will offer sweepable EQs that allow you to select and boost any frequency you choose and offer filters so that you can remove all the frequencies above or below a certain frequency. This section may also offer a pre- or post-EQ button that allows you to re-route the signal path to *bypass* the EQ section (pre-EQ) and move directly to the pre-fader buss, or move *through* the EQ section (post-EQ) and then into the pre-fader buss. Obviously, this allows you to bypass the EQ section once in a while to determine the effect that any EQ adjustments have had on a sound.

The pre-fader buss allows you to bypass the channel's volume fader and route the signal at its nominal volume to the auxiliary buss or go through the faders and then to the auxiliary buss. Fundamentally, an aux buss is a way of routing the channel's signal to physical outputs usually labelled as 'aux outs' located at the rear of the desk. The purpose behind this is to send the channel's signal out to an external effect and then return the results back into the desk (we'll look more closely at aux and insert effects in a moment). The number of aux busses featured on a desk varies widely from 1 to over 20, and not all desks will give the option of pre- or post-fader aux sends and may be hardwired to pre-fader. This means that you would have no control over the level that's sent to the aux buss via the channel fader; instead the aux send control would be used to control the level.

After this aux buss section, the signal is passed through onto the volume control for the channel. These can sometimes appear in the form of rotary controllers on cheaper desks but generally they use faders with a 60 or 100 mm throw.[2] Although these shouldn't particularly require any real explanation, it's astounding how many users believe that they are used to increase the amplification of a particular channel. This isn't the case at all, since if you look at any desk the faders are marked 0 dB close to the top of their throw instead of at the bottom. This means that you're not *amplifying* the incoming signal by increasing the fader's position; rather you're allowing more of the original signal travel through the fader's circuitry that increases the gain on that channel.

After the faders, the resulting signal then travels through the panning buss (allowing you to pan the signal left and right) and into a subgroup buss. The number of subgroup busses depends entirely on the price and model of mixer but essentially these allow you to group a number of fader positions together

---

[2] 'Throw' refers to the two extremes of a fader's movement from minimum to maximum.

and control them all with just one fader control. A typical application of this is if the kick, snare, claps, hi-hats and cymbals each have a channel of their own in the mixer. By then setting each individual element to its respective volume and required EQ (in effect mixing down the drum loop so that it sounds right) all

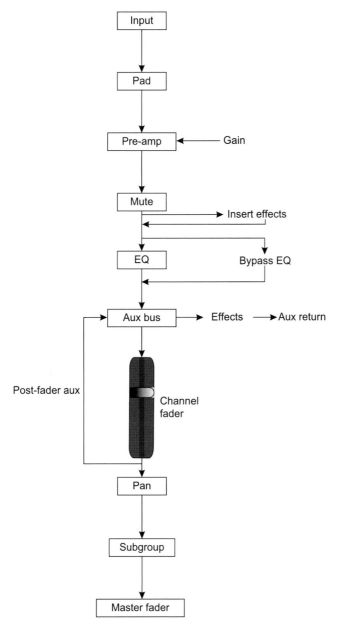

FIGURE 3.2
The typical structure of a mixing desk

of these levels can be routed to a single group track whereby moving just one subgroup fader, the volume of the entire drum sub-mix can be adjusted to suit the rest of the mix.

Finally, these subgroup channels, if used, along with the rest of the 'free' faders, are combined together into the main stereo mix bus, which passes to the main mix fader, allowing you to adjust the overall level of the mix. It's at this stage that things can become more complicated since the direction and options available to this stereo buss are entirely dependent on the mixer in question.

In the majority of smaller, less expensive mixers the buss will simply pass through a pan control and then out to the mixer's physical outputs. Semi-professional desks may pass through the pan control *and* an EQ before sending the mix out to the main physical outputs. Professional desks may go through panning, EQ and then split the buss into any number of other stereo busses, allowing you to send the mix to not only the speakers but also the headphones, another pair of monitors (situated in the recording room) and a recording device.

Ultimately, the more EQ and routing options a desk has to offer, the more creative you can become. But, at the same time, the more features on offer, the more expensive it will be. Naturally, most of these concerns are circumvented with audio sequencers as they generally offer everything a professional desk does with the only limitation being the number of physical connections dictated by the soundcard you have fitted. Nevertheless, this doesn't deter many from relying entirely on sequencers as any external instruments can always be recorded as audio and placed on their own track and have software effects applied.

ROUTING EFFECTS AND PROCESSORS

Understanding the internal buss structure of a typical mixing desk (hardware or software) is only part of the puzzle because when it comes to actually processing signals, the insert/aux buss system they are transferred through will often dictate the overall results. Any 'external' signal processing can be divided into two groups: processors and effects. The difference between these two is relatively simple but important to recognize.

All effects will utilize a wet/dry control (wet is the affected signal and dry is the unaffected signal) that allows you to configure how much of the original signal remains unaffected and how much is affected. A typical example of this would be for reverb whereby you don't necessarily want to run the entire audio through the effect otherwise it could appear swamped in decays; instead you would want to affect the signal by a small amount only. For instance, you may keep 75% of the original signal and apply just 25% of the reverb effect. Conversely, all processors, such as compressors, noise gates and limiters, are designed to work with 100% of the signal and thus have no wet/dry parameter. This is simply because in many instances there would be little point trying to control the dynamics of just some of the signal because the rest of it would still retain its dynamic range. Nevertheless, due to the different nature of these two processes, a mixing desk uses two different buss systems to access them: an insert buss for processors and an auxiliary buss for effects. The difference and reasons behind using these two busses will become clear as we look at both.

AUXILIARY BUSS

Nearly all mixers will feature the aux buss after the EQ rather than before since it's considered that effects are applied at the final stages of a mix to add the final 'polish'. At this point, a percentage of the signal can be routed through the aux bus and to a series of aux outs located at the back of the mixer. Each aux out is connected to an effects unit and the signal is then returned into the desk. With the aux potentiometer on the channel strip set at zero, the audio signal ignores the bus and moves directly onto the fader. However, by gradually increasing the aux pot you can control how much of the signal is split and routed to the aux buss and onto the effect. The more this is increased the more audio signal will be directed to the aux buss. The aux return (the effected signal) is then returned to a separate channel on the mixing desk, which can then be mixed with the original channel.

As you can see from the above diagram, two separate audio leads are required. One has to carry the signal from the mixer and into the effects while the other has to return the effect back into the desk. Additionally, the effected signal cannot be returned into the original channel since there will still be some dry audio at the fader; instead they are usually returned into specific aux returns. These signals are then bussed through the mixer into additional mixer 'subgroup' channels. These are most commonly a group of volume faders or pots (one for each return), which permit you to balance the volume of the effected signal with the original dry channel.

Note that when accessing effects from a mixing desk, the mix control on the effects unit should generally be set to 100% wet since you can control the amount of wet/dry signal with the mixing desk itself.

While retuning the effected signal to its predefined aux channel return may seem sensible, very few artists will actually bother using the aux returns at all,

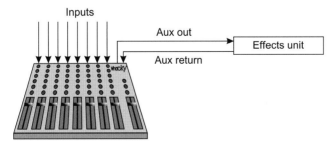

FIGURE 3.3
An aux send configuration

instead preferring to return the signal to a normal free mixing channel. This opens up a whole new realm of possibilities since the returning effect has access to the channels pre-amp, an insert, EQ, pan and volume. For example, a returned reverb effect could be pushed into distortion by increasing the mixers' pre-amp, or the EQ could be used as a low- and high-frequency filter if the effects unit doesn't feature them.

Another benefit of using aux sends is that each channel on the mixer will have access to the same auxiliary busses, meaning that signals from other channels could also be sent down the same auxiliary buss to the effects unit. This approach can be especially useful when using computer-based audio sequencers since opening multiple instances of the same effect can use up a proportionate amount of the CPU. Instead, you can simply open up one effect and send each channel to the same effect, which also saves time spent in setting a number of effects units up, all of which share the same parameters. What's more, in a hardware situation, as only a portion of the signal is being sent to the effect, there is less noise and signal degradation, and as the effects is returned to a separate channel, you have total control over the wet and dry levels through the mixer.

One final aspect of aux busses is that they can sometimes be configured to operate the aux bus either pre- or post-fader. Of the two, pre-fader is possibly the least used since the signal is split into two before it reaches the fader. This means that if you were to reduce the fader you would not reduce the gain of the signal being bussed to the effect, so with every fader adjustment you would also need to readjust the aux pot to suit. This can have its uses as it allows you to reduce the volume of the dry signal while leaving the wet at a constant volume, but generally post-fader is much more useful while mixing. Using post-fader, reducing the channel's fader will also systematically reduce the auxiliary send too, saving the need to continually adjust the aux buss send level whenever you change the volume.

INSERT BUSS

Unlike aux busses, insert busses are positioned just before the EQ and are designed for use with processors rather than with effects. This is because processors require the entire signal and there is no point in applying compression or gating to only part of a signal! Typically, a processor takes the output of a microphone pre-amp and feeds it directly into a compressor to prevent any clipping. The resulting compressed signal would then be connected into the mixing desk's channel so that the signal flows out of the microphone pre-amp into the compressor and then finally into the desk for mixing. This, however, is a rather convoluted way of working because to compress the output of a synthesizer, for example, requires that you scrabble around at the back of the rack and rewire all the necessary connections. In addition, if the output from the synthesizer was particularly low there is no way of increasing its gain into the compressor.

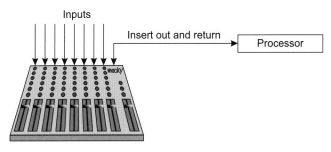

FIGURE 3.4
An insert configuration

This can be avoided if the mixing desk features insert points, which commonly appear directly after the pre-amp stage. Consequently, the mixer's pre-amp stage can be used to increase the signal level from the synthesizer before it's passed on to the insert bus. This is then passed into the external compressor where the signal is squashed before being returned to the same mixing channel.

On occasion compressors may be accessed through the aux send bus rather than the insert bus. With this configuration, you can increase the overall level of the signal by mixing the compressed with the uncompressed signal. This can help to preserve the transients of the signal.

This is accomplished using only one physical connection – an insert cable – on the mixing desk. Fundamentally, this is an ordinary audio cable with a T + S (Tip–Sleeve) jack at one end and a R + S (Ring–Sleeve) jack at the other. The tip is used to transmit the signal from the desk's channel to the compressor and is returned back into the same mixing channel through the ring connection of the same cable. Most mixers return this processed signal before the EQ and volume bus, allowing shaping of the overall signal after processing.

Tonal shaping after processing won't necessarily create any problems if the processor happens to be a noise gate, but if it's a compressor, it raises the issue that if the EQ were raised *after* compression, the signal would be increased past the level of the previously controlled dynamics. To understand the issue, we first need to consider what would happen if the compressor was returned *after* the EQ section.

To begin with, the EQ section would be used to correct the tone of the incoming signal on that particular channel before it was sent to the compressor. This compressor would then be adjusted so that it controls the dynamics of the timbre *and* the subsequent EQ before the signal is returned to the desk. Due to the nature of compression, however, the tonal content of the sound will be modified, so it would need to be EQ'd again. This subsequent EQ will reintroduce the peaks that were previously controlled by the compressor, so the compressor must be adjusted again. This would alter the tonal content, so it would need EQ'ing yet again, and so on. Thus, an ever-increasing circle of EQ, compression, EQ, compression must continue until the signal is overcompressed or pushing beyond unity gain, distorting the desk. By inserting the compressor *before* the EQ section, this continual circle can be avoided.

With the compressor positioned before the EQ, the issue of destroying the previously controlled dynamics when boosting the EQ is still raised, but this shouldn't be the case provided the EQ is correctly used. As touched upon in earlier chapters, it's in our nature to perceive louder sounds to be infinitely better than quieter, even if the louder sound is tonally worse than the quieter one. Thus, when working with EQ it's necessary to reduce the channel's fader while boosting any frequencies so that they remain at the same volume as the un-EQ'd version. Used in this way, when bypassing the modified EQ to compare it with the unmodified version, a difference in volume cannot cloud your judgement. As a result, the signal level of the modified EQ is the same as the unmodified version, so the dynamics from the compressor remain the same.

This approach maintains a more natural sound, so it is worth experimenting by placing the EQ before the compressor. For example, using this configuration the EQ could be used to create a frequency-selective compressor. In this set-up, the loudest frequencies control the compressor's action. By boosting quieter frequencies so that they instead breach the compressor's threshold it's possible to change the dynamic action of a drum loop or motif. In fact, it's important to note that the order of any processing or effects, if they're inserted, can have a dramatic influence on the overall sound.

From a theoretical mixing point of view, effects should not be chained together in series. This is because although it is perfectly feasible to route reverb, delay, chorus, flangers and phasers into a mix as inserts, chaining effects in this way can introduce a number of problems. If an effect is used as an insert, 100% of the signal will be directed into the external effects unit, and because many units are introduced with low levels of noise while also degrading the signal's overall quality, both of the effected signal and the noise will be returned into the desk. What's more, the control over the relationship between dry audio and wet effects would be from the effects interface and, in many instances, greater control will be required.

Many effects use a single control to adjust the balance between dry and wet levels, so as the wet level is increased, the dry level decreases proportionally. Equally, increasing the dry level proportionally decreases the wet level. While this may not initially appear problematic, if you decided to thicken out a trance lead with delay or reverb but wanted to keep the same amount of dry level in the mix that you already have, it isn't going to be possible. As soon as the wetness factor is increased the dry level will decrease proportionally which may result in a more wet than dry sound. This can cause the transients of each hit to be washed over by the subsequent delays or reverb tail from the preceding notes, reducing the impact of the sound.

Nevertheless, by using effects as inserts and chaining them together in series it's possible to create new effects because both dry and wet results from each preceding effect would be transformed by the ones that follow them. This opens up a whole world of experimental and creative opportunity that could easily digest the rest of this book, so rather than list all of the possible combinations (as if I could) we'll examine the reasoning behind why, theoretically at least, some processors and effects should precede others.

Gate > Compressor > EQ > Effects

Normally, to maintain a 'natural' sound a gate should always appear first in the line of any processor or effects since they're used to remove unwanted noise from the signal before it's compressed, EQ'd or effected. While it is perfectly feasible to place the compressor before the gate as it would make little difference to the actual sound it's unwise to do so. This is because the compressor reduces the dynamic range of signal and, as a gate works by monitoring the dynamic range and removing artefacts below a certain volume, placing compression first, the gate would be more difficult to set up and may remove some of the signal you wish to keep. For reasons we've already touched upon, the EQ should then appear *after* compression and the effects should follow the EQ section as they're usually the last aspect in the mixing chain.

Gate > Compressor > Effects > EQ

Again the beginning aspects of this arrangement will keep the signal natural but by placing the EQ *after* the effects it can be used to sculpt the tonal qualities produced by the effect. For example, if the effect following the compression is distortion, the compressor will even out the signal level making the distortion effect more noticeable on the decays of notes. Additionally, since distortion will introduce more harmonics into the signal, some of which can be unpleasant, it can be carefully sculpted with the EQ unit to produce a more controlled pleasing result.

Gate > Compressor > EQ > Effects > Effects

The beginning of this signal chain will produce the most natural results but the order of the effects afterwards will determine the outcome. For instance, if you were to place reverb before distortion, the reverb tails will be treated to distortion, but if it were placed afterwards, the effect would not be as strong since the reverbs tail would not be treated. Similarly, if delay were placed after distortion, the subsequent delays would be of the distorted signal, while if the delay came first, the distortion would be applied to the delays producing a different sound altogether. If flanger were added to this set-up, things become even more complicated since this effect is essentially a modulated comb filter. By placing it after distortion the flanger would comb filter the distorted signal producing a rather spectacular phased effect, yet if it were placed before, the effect would vary the intensity of the distortion.

To go further, if the flanger were placed after distortion but before reverb, the flange effect would contain some distorted frequencies but the reverbs tail would wash over the flanger, diluting the effect but producing a reverb that modulates as if it were controlled with an LFO. The possibilities here are, as they say, endless, and it's worth experimenting by placing the effects in a different order to create new effects.

Gate > Effects > EQ > Effects > Compressor

While the aforementioned method of placing one effect after the other can be useful, the subsequent results can be quite heavy-handed, but by placing an EQ after the first series of effects, the tonal content can be modified so that there isn't an uncontrolled mess of frequencies entering the second series of effects, muddying the effect further. Additionally, by placing a compressor at the end of the arrangement, any peaking frequencies introduced by a flanger, following distortion or similar arrangement, can be brought back under control with the compressor sat at the end of the line.

Compressor > EQ > Effects > Gate

We've already discussed the effects of placing a compressor before the gate and EQ, but using this configuration, the compressor could be used to control the dynamics of sounds before they were EQ'd and subsequently affected. However, by placing the gate *after* the effects would mean that the effected signals could be treated to a gate effect. Although there is a long list of possible uses for this if you use a little imagination, possibly the most common technique is to apply reverb to a drum kick or trance/techno/house lead and then use the following gate to remove the reverb's tail. This has the effect of thickening out the sound without turning the result into a washed-over mush.

Gate > Effects > Compressor > EQ

Although it is generally accepted that the compressor should come before effects, placing it directly after can have its uses. For instance, if a filter effect has been used to boost some frequencies and this has been followed by chorus or flanger, there may be clipping so that the compressor can be used to bring these under control before they're shaped tonally with the EQ. Notably, though, placing compression after distortion will have little effect since distortion effects tend to reduce the dynamic range anyway.

Above all, though, keep in mind that setting effects in an order that would not theoretically produce great results most probably will in practice. Indeed, it's the artists who are willing to experiment and often produce the most memorable effects on records, and with many of today's audio sequencers offering a multitude of free effects, experimentation is cheap but the results may be priceless.

AUTOMATION

A final yet vital aspect of mixing desks appears in the form of mix automation. We've already touched upon the importance of movement within programmed sounds, but it's also important to manipulate a timbre constantly throughout the music. Techno would be nowhere if it were not possible to program a mixer or effects unit to gradually change parameters during the course of the track. With a mixer that features automation, these parameter changes can be recorded as data (usually MIDI) into a sequencer which when played back to the mixer forces it to either jump or gradually move to the new settings.

Originally mix automation was carried out by three or four engineers sat by the faders/effects and adjusting the relevant parameters when they received a nod from the producer. This included riding the desk's volume faders throughout the mix to counter any volume inconsistencies. As a result any parameter changes had to be performed perfectly in one pass since the outputs of the desk are connected directly into a recording device. If you made a mistake, then you had to do it all again, and again, and again until it was right. In fact, this approach was the only option available when dance music first began to develop and all tracks of that time will have the filters or parameters tweaked live while recording direct to tape. This type of approach is almost impossible today, however, since the development of dance music has embraced the latest forms of mixer automation so much so that it isn't unusual to have five or six parameters changing at once, or for the mixer and effects units to jump to a whole new range of settings for different parts of the track.

Mix automation is only featured on high-end mixing desks due to the additional circuitry involved, and the more parameters that can be automated the more expensive the desk will generally be. Nevertheless, mix automation appears in two forms: VCA and motorized. Each of these perform the same functions but in a slightly different way. Whereas motorized automation utilizes motors on each fader so that they physically move to the new positions, VCA faders remain in the same position while the relative parameters change. This is accomplished by the faders generating MIDI information rather than audio. This is then transferred to the desk's computer which adjusts the volume. While this does mean that you have to look at a computer screen for the 'real' position of the faders, it's often preferred by many professional studios since motorized faders can be quite noisy when they move.

Alongside automating the volume, most desks will also permit you to automate the muting of each channel. This can be particularly useful when tracks are not currently playing in the arrangement, since muting the channel will remove the possibility of hiss on the mixer's channel. On top of this, some of the considerably expensive desks also allow you to automate the send and return system and store snapshots (or 'scenes') of mixes. Essentially, these are a capture of the current fader, EQ and mute settings which can be recalled at any time by sending the respective data to the desk. Despite the fact that these features are fairly essential to mixing, they are only available on expensive desks and it is not possible to automate any effects parameters. For this type of automation, software sequencers offer much better potential.

Most audio capable MIDI sequencers will offer mix automation, but unlike hardware, this is not limited to just the volume, muting and send/return system. It's usually possible to also automate panning, EQ along with all the parameters on virtual studio technology (VST) instruments and plug-in effects or processors. These can usually be identified by an R (Read automation) and W (Write automation) appearing somewhere on the plug-in's interface. By activating the Write button and commencing playback any movements of the plug-in or mixing desk will be recorded, which can then be played back by activating the Read button. What's more, many sequencers also offer the opportunity to finely edit any recorded automation data with an editor. This can be invaluable to the dance musician since you can finely control volume or filter sweeps easily.

CHAPTER 4
Programming Theory

'It's ridiculous to think that you're music will sound better if you go out and buy the latest keyboard. People tend to crave more equipment when they don't know what they want from their music...'

A Guy Called Gerald

Armed with an understanding of basic synthesis, effects and routing of sounds through a mixing desk we can look more closely at sound design using all the elements previously discussed.

Sound design is one of the most vital elements of creating dance music since the sounds will more often than not determine the overall genre of music. However, although it would be fair to say that quite a few timbres are gleaned from other records or sample CDs, there are numerous advantages to programming your own sounds.

Not only do you have much more control over the parameters when compared to samples, there are no key-range/pitch-shifting restrictions and no copyright issues. What's more, you'll get a lot more synthesizer for your money than if you simply stick with the presets, and you'll also open up a whole new host of avenues that you would otherwise have missed.

To many, the general approach to writing music is to get the general arrangement/idea of the piece down in MIDI form and then begin programming the sounds to suit the arrangement. It doesn't necessarily have to be a complete song before you begin programming but it is helpful if most of the final elements of the mix are present. This is because when it comes to creating the timbres it's preferable to have an idea of the various MIDI files that will be playing together so that you can determine each file's overall frequency content.

Keep in mind that the sounds of any instrument/programmed timbres will occupy a specific part of the frequency range. If you programme each individually without any thought to the other instruments that will be playing

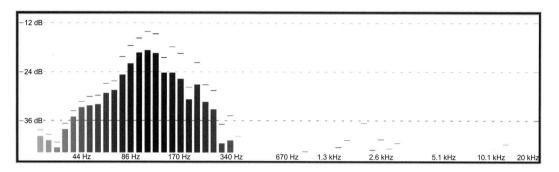

FIGURE 4.1
As the above spectral analysis reveals, the groove and motif take up specific frequencies while leaving others free for new instruments

alongside it, you can wind up programming a host of complex timbre's that sound great on their own but when placed in the mix they all conflict with one another creating a cluttered mix. With the MIDI down from the start you can prioritize the instruments depending on the genre of music.

Knowing which frequencies are free after creating the most important parts is down to experience but to help you on the way it's often worth employing a spectral analyser. These are available in software or hardware form and display the relative volume of each frequency band of every sound that is played through them. Thus, playing the groove and lead motif into the analyser will give you an idea of the frequencies that are free (Figure 4.1).

In some cases, once these fundamental elements are created you may find that there is too much going on in the frequency range so that you can remove the superfluous parts or place them elsewhere in the arrangement. Alternatively, you could EQ the less important parts to thin them out so that they fit in with the main programmed elements. Although they may then sound peculiar on their own, it doesn't particularly matter so long as they sound right in the context of the mix and they're not played in isolation during the arrangement. If they are, then it's prudent to either use two different versions of the timbre, one for playing in isolation and one for playing with the rest of the mix, or more simply leave the timbre out when the other instruments are playing. Indeed, the main key to producing a good track/mix is to play back sections of the arrangement and programme the sounds so that they all fit together agreeably, and it isn't unusual at this stage to modify the arrangement to accomplish this. By doing so, when it comes to mixing you're not continually fighting with the EQ to make everything fit together appropriately. If it sounds right at the end of the programming stage, not only the mixing will be much easier but the entire mix will sound more professional due to less cluttering. As a result, it shouldn't need explaining that after you've programmed each timbre you should leave it playing back while you work on the next in line, progressively programming the less important instruments until they are all complete.

TIMBRE EFFECTS

One of the main instigators of creating a cluttered mix or sounds that are too power-ful is a synthesizers/samplers effects algorithm. All synthesizers and many sample patches are designed to sound great in isolation to coax you into parting with money, but when these are all combined into the final mix, the effects tails, delays, chorus etc. will all combine together to produce a muddy result. Much of the impact of dance music comes from noticeable spaces within the mix and a great house motif, for instance, is often created sparingly by keeping the lengths of the samples short so that there are gaps between the hits. This adds a dynamic edge because there are sudden shifts from silence to sound. If, however, there are effects such as delay or reverb applied to the sound from the source, the gaps in between the notes are smeared over which considerably lessens the overall impact. Consequently, before even beginning to programme you should create a user bank of sounds by either copying the presets you like into the user bank or creating your own and turning any and all effects off.

Naturally, some timbres will benefit heavily from effects, but if this is heav-ily effected to make it wider and more in your face, it's prudent to refrain from using effects on any other instruments. A lead soaked in reverb/delay/chorus etc. will have a huge impact if the rest of the instruments are dry, whereas if all the instruments are soaked in effects, the impact will be significantly lessened. A good mix/arrangement works in contrast – you can have too much of a good thing and it's better to leave the audience gasping for more than gasping for breath. Also, if many of the instruments are left dry, and if when it comes to mixing you feel that they need some effecting, you can always introduce the effects at the mixing desk.

Of course, this approach may not always be suitable since the effects signature may actually contribute heavily toward the timbre. If this is the case, then you'll need to consider your approach carefully. For example, if reverb is contribut-ing to the sound's colour, ask yourself whether the subsequent tails are really necessary as these will often reduce the impact. If they're not required then it's prudent to run the timbre through a noise gate which is set to remove the reverb tail. This way the colour of the sound is not affected but the successive tails are removed which also prevents it from being moved to the back of the mix. Similarly, if delay is making the timbre sound great with the programmed motif, try emulating the delay by ghosting the notes in MIDI and using velocity. This can often reduce the additional harmonics that are introduced through the delay running over the next note, and as many delay algorithms are in stereo, it allows you to keep the effect in mono (more on this in a moment). Naturally, if this overrun is contributing to the sound's overall colour then you will have no option to leave it in but you may need to reconsider other sounds that are accompanying the part to prevent the overall mix from becoming cluttered.

Another facet of synthesizers and samplers that may result in problems is through the use of stereo. Most tone modules, keyboards, VST instruments and samplers will greatly exaggerate the stereo spread to make the individual sounds appear more impressive. These are created in one of the two ways, either through the use of effects or layering two different timbres together that are spread to the left and right speakers. Again, this makes them sound great in isolation but when placed into a mix they can overpower it quickly, resulting in a wall of incomprehensible sound. To avoid this, you should always look towards programming sounds in mono unless it is a particularly important part of the music and requires the width (such as a lead instrument). Even then, the bass kick drum and snare, while forming an essential part of the music, should be in mono as these will most commonly sit dead centre of the mix so that the energy is shared by both speakers. Naturally, if you're using a sampled drum loop then they will most probably be in stereo but this shouldn't be the cause of any great concern. Provided that the source of the sample (whether it's a record or sample CD) has been programmed and mixed competently, the instruments will have been positioned efficiently with the kick located in the centre and perhaps the hi-hats and other percussive instruments spread thoughtfully across the image. This will, however, often dictate the positioning and frequencies of other elements in the mix.

PROGRAMMING THEORY

Fundamentally, there are two ways to programme synthesizers. You can use careful analysis to reconstruct the sound you have in mind or alternatively you can simply tweak the parameters of a preset patch to see what you come up with. It's important to keep in mind that *both* these options are viable ways of working and despite the upturned noses from some musicians at the thought of using presets, a proportion of professional dance musicians do use them. The Chemical Brothers have used unaltered presets from the Roland JP8080 and JV1080; David Morales has used presets from the JV2080, the Korg Z-1, E-mu Orbit and Planet Phat; Paul Oakenfold has used presets from the Novation SuperNova, E-mu Orbit and the Xtreme lead; and Sasha, Andy Gray, Matt Darey and well, just about every trance musician on the planet has used presets from the Access Virus. At the end of the day, if the preset fits into your music then you should feel free to use it.

When it comes to synth programming, it's generally recommended that those with no or little prior experience begin by using simple editing procedures on the presets so that you can become accustomed to not only the effects each has on a sound, but also the character of the synth you're using. Despite the manufacturer's bumf that their synth is capable of producing any sound, they each exhibit a different character and you have to accept that there will be some sounds that cannot be constructed unless you have the correct synth. Alternatively, those with a little more experience can begin stripping all the modulation away and building on the foundation of the oscillators.

Remember that a proportionate amount of instruments within a dance mix share many similarities. For example, a bass sound is just that – a bass sound. As we'll cover a little later, many specific instruments are programmed roughly in the same manner, using the same oscillators, and it's the actual synthesizer used, along with just the modulation from the envelope generators (EGs),

LFOs and filters that produces the different tones. Thus, it can often be easier to simply strip all the modulation away from the preset patch so you're left with just the oscillators and then build on this foundation.

CUSTOMIZING SOUNDS

Unless you've decided to build the entire track around a preset or have a fortunate coincidence, presets will benefit from some editing. From earlier chapters, the effects that each controller will impart on a sound should be quite clear but for those new to editing presets it can be difficult knowing the best place to start.

First and foremost, the motif/pad/drums etc. should be playing into the synthesizer or sampler in its entirety. A common mistake is to just keep banging away at middle C on a controller keyboard to audition the sounds. While this may give you an idea of the timbre, remember that velocity and pitch can often change the timbre in weird and wonderful ways and you could miss a great motif and sound combination if you were constantly hitting just one key. What's more, the controls for the EGs, LFOs, filters and effects (if employed) will adjust not only the sound but the general mood of the motif. For example, a motif constructed of 1/16th notes will sound very different when the amplifier's release is altered. If it's shortened, the motif will become more 'stabby', while if it's lengthened it will flow together more. Alternatively, if you lengthen the amplifier's attack, the timing of the notes may appear to shift which could add more drive to the track.

Also consider that the LFOs will restart their cycle as you audition each patch, but if they're not set to restart at key press, they will sound entirely different on a motif than a single-key strike. Plus, if filter key follow is being used, the filter's action will also change depending on the pitch of the note. Perhaps most important of all, though, both your hands are free to experiment with the controls, allowing you to adjust two parameters simultaneously such as filter and resonance or amp attack and release.

Once you've come across the timbre that shares some similarities as the tonality as you require for the mix, most users tend to go for the filter/resonance combination. However, it's generally best to start by tweaking the amplifier envelope to shape the overall volume of the sound before you begin editing the colour of the timbre. This is because the amp EG will have a significant effect on the MIDI motif and the shape of the sound over time so it's beneficial to get this 'right' before you begin adjusting the timbre. We've already covered the implications the amp EG has on a sound but here we can look at it in a more practical context.

The initial transient of a note is by far the most important aspect of any sound. The first few moments of the attack stage provide the listener with a huge amount of information and can make the difference between a timbre sounding clear-cut, defined and up-front or more atmospheric and sat in the background. This is because we perceive timbres with an immediate attack to be louder than those with a slower attack and short stabby sounds to be louder than those with a quick attack but long release. Thus, if a timbre seems too sloppy when played back from MIDI reducing the attack and/or release stage will make it appear much more defined. That said, a common mistake is to shorten the attack and release stage of every instrument to make all the instruments appear distinct but this should be avoided.

Dance music relies heavily on contrast and not all timbres should start and/or stop immediately. You need to think in terms of not only tone but also time. By using a fast attack and release on some instruments, employing a fast attack and slow release on others and using a slow attack and fast release on other will create a mix that gels together more than if all the sounds used the same envelope settings.

Naturally, what amp EG settings to use on each instrument will depend entirely on the mix in question but very roughly speaking, trance, big beat and techno often benefit from the basses, kicks, snares and percussion having a fast attack and release stage with the rest of the instruments sharing a mix of slow attack/ fast release and a fast attack/slow release. This latter setting is particularly important in attaining the 'hand in the air' trance leads.

Conversely, house, drum 'n' bass, lo-fi and chill out/ambient benefit from longer release settings on the basses almost to the point that the notes are all connected together. Sounds that sit on top of this then often employ a short attack and release to add contrast. Examples of this behaviour can be heard in the latest stem of house releases that utilize the funky guitar riffs. The bass (and often a phased pad sat in the background using a quick attack and slow release) almost flow together while the funk guitars and vocals that sit on top are kept short and stabby.

This leaves the decay and sustain of the amp EG. Sustain can be viewed as the body of the sound after the initial pluck of the timbre has ended, while the decay controls the amount of 'pluck'. By decreasing the decay the pluck will become stabby and prominent and increasing it will create a more drawn-out pluck. What's more, by adjusting the shape of the decay stage from the usual linear fashion to a convex or concave structure the sound can take on a more 'thwacking' sucking feel or a more rounded transient, respectively. Indeed, a popular sound design technique is to use a lengthy decay (and attack) and reduce the release and sustain parameters to zero. This creates the initial transient of the sound, which is then mixed with a different timbre with only the release and sustain.

Once the overall shape of the riff has been modified to add contrast to the music, the filter envelopes and filter cut-off/resonance can be used to modify the tonal content of the timbre. As the filter's envelope will react to the current filter settings, it's beneficial to adjust these first. Most timbres in synthesizers will utilize a low-pass filter with a 12 dB transition as these produce the most musically useful results; however, it is worthwhile experimenting with the other filter types on offer such as band-pass and high-pass. For instance, by keeping the timbre on a low-pass filter but double tracking the MIDI to another synthesizer and using a high-pass or band-pass to remove the low-frequency elements you're left with just the fizzy overtones. This can then be mixed with the original timbre to produce a sound that's much richer in harmonic content. If this is then sampled, the sampler's filters can be used to craft the sound further.

Also, keep in mind that the filter's transition will have a large effect on the timbre. As mentioned most synthesizers will utilize a 12 dB transition but a 24 dB transition used on some timbres of the mix will help to introduce some contrast as the slope is much sharper. A typical use for this is if two filter sweeps are occurring simultaneously, by setting one to 12 dB and the other to 24 dB the difference in the harmonic movement can produce tonally interesting result and add some contrast to the music. Alternatively, by double tracking an MIDI file and using a 12 and 24 dB on the same timbre the two transitions will interact creating a more complex tone that warps and shifts in harmonic content.

With the tone correct, you can move onto the filter's EG. These work on the same principle as the amp's EG but rather than control volume, they control the filter's action over time. This is why it's important to modify the amp's envelope before any other parameters since if the amp's attack is set at its fastest and the filter's attack is set quite long the timbre may reach the amp's release portion before the filter has fully been introduced.

While mentioning the attack and decay of the amplifier envelope we touched upon the pluck being determined by the decay rate but the filter's attack and decay also play a part in this. If the attack is set to zero, the filter will act upon the sound on key press, but if it's set slightly longer than the attack on the amp EG, the filter will sweep into the note creating a *quack* at the beginning of the note. Similarly, as the decay setting determines how quickly the filter falls to its sustaining rate, by shortening this you can introduce a harder/faster plucking character to the note. Alternatively, with longer progressive pad sounds, if the filter's attack is set so that it's the same length as the amp EGs attack and decay rates, the sound will sweep in to the sustain stage whereby the decay and sustain of the filter begin creating harmonic movement in the sustain.

However, keep in mind that while any form of sonic movement in a sound makes it much more interesting, it's the initial transient that's the most important as it provides the listener with a huge amount of information about the timbre. Thus, before tweaking the release and sustain parameters, concentrate on getting the transient of the sound right first.

Of course, the filter's envelope doesn't always provide the best results and the synthesizer's modulation matrix can often produce better results. For instance by using a sawtooth LFO set to modulate the filter's cut-off the harmonic content will rise sharply but fall slowly and the speed at which all this takes place will be governed by the LFO's rate. In fact, the LFO is one of the most underrated yet important aspect of any synthesizer as it introduces movement within a timbre which is the real key behind producing great results.

Static sounds will bore the ear very quickly and make even the most complex motifs appear tedious, so it's practical to experiment by changing the LFOs waveforms and its destinations within the patch. Notably, if an LFO is used, remember that its rate is a vial aspect. As the tempo of dance is of paramount importance it's sensible to sync the timing of any modulation to the tempo. Without this not only can an arrangement become messy but in many tracks the LFO speeds up with the tempo and this can only be accomplished by syncing the LFO to the sequencer's clock.

PROGRAMMING BASICS

While customizing presets to fit into a mix can be rewarding it's only useful if the synth happens to have a timbre similar to that you're looking for, and if not, you're have to programme from the ground up. Before going any further, though, there are a few things that should be made clear.

Firstly, while you may listen to a record and wonder how they programmed that sound or groove it's important to note that the artist may not have actually programmed them and they may be straight off a sample CD, or in some instances another record. If you want to know how Stardust created the groove on 'Music Sounds Better with You', how David Morales programmed 'Needin' You' or Phats and Small managed to produce 'Turn Around', you should first listen to the original music they sampled to produce them. In many cases it will become extremely clear how they managed to inject such a good groove or sound – they sampled from previous hits or funk records.

What follows is a list of the most popular sounds and grooves that have actually been derived from other records. It's by no means exhaustive as to catalogue them all would require approximately 30 pages but instead it covers the most popular and well-established grooves.

Dance Artist	Title of Track	Original Artist	Original Title of Track
Stardust	*Music Sounds Better With you*	Chaka Khan	*Fate*
Armand Van Helden	*You Don't Know Me*	Carrie Lucas	*Dance With You*
David Morales	*Needin' You*	Rare Pleasure The Chi-lites	*Let Me Down Easy* *My First Mistake*
Daft Punk	*Digital Love*	George Duke	*I Love You More*
Phats and Small	*Turn Around*	Tony Lee Change	*Reach Up* *The Glow Of Love*
Cassius	1999	Donna Summer	*If It Hurts Just A Little*
The Bucketheads	*The Bomb*	Chicago	*Street Player*

Dance Artist	Title of Track	Original Artist	Original Title of Track
DJ Angel	*Funk Music*	Salsoul Orchestra	*Take Some Time Out For Love*
Blueboy	*Remember Me*	Marlena Shaw	*Woman of the Ghetto*
Full Intention	*Everybody Loves the Sunshine*	Roy Ayers	*Everybody Loves the Sunshine*
Todd Terry	*Keep on Jumpin*	Lisa Marie Ex Musique	*Keep on Jumpin* *Keep on Jumpin*
Byron Stingly	*Come On Get Up Everybody*	Sylvester	*Dance (Disco Heat)*
Soulsearcher	*I Can't Get Enough*	Gary's Gang	*Let's Lovedance Tonight*
GU	*I Need GU*	Sylvester	*I Need You*
Peppermint Jam Allstars	*Check it Out*	MFSB	*TSOP (The Sound of Philadelphia)*
Nuyorican Soul	*Runaway*	Salsoul Orchestra	*Runaway*
The Bucketheads	*I Wanna Know*	Atmosfear	*Motivation*
Michael Lange	*Brothers and Sisters*	Bob James	*Westchester Lady*
Basement Jaxx	*Red Alert*	Locksmith	*Far Beyond*
Deaf 'n' Dumb Crew	*Tonite*	Michael Jackson	*Off the Wall*
Moodyman	*I Can't Kick this Feeling When it Hits*	Chic	*I Want Your Love*
Cassius	*Feeling for You*	Gwen McCrae	*All This Love I'm Givin*
Spiller	*Batucada*	Sergio Mendes	*Batucada*
Eddie Amadour	*House Music*	Exodus	*Together Forever*
DJ Modjo	*Lady, Hear Me Tonight*	Chic	*Soup for One*
Spiller	*Groovejet*	Carol Williams	*Love Is You*
Prodigy	*Out Of Space*	Max Romeo And the Upsetters	*Chase The Devil*
Moby	*Natural Blues*	Vera Hall	*Troubled So Hard*

Dance Artist	Title of Track	Original Artist	Original Title of Track
FatBoy Slim	*Praise You*	Camille Yarbrough	*Take Yo Praise*
Dee-Lite	*Groove Is In The Heart*	Herbie Hancock	*Bring Down The Birds*
Massive Attack	*Be Thankful*	William De-Vaughn	*Be Thankful For What You've Got*
Massive Attack	*Safe From Harm*	Billy Cobham	*Stratus*
Eminem	*My Name is*	Labbi Siffre	*I Got The*
De La Soul	*3 Is the Magic Number*	Bob Dorough	*The Magic Number*
The Notorious BIG	*Mo Money Mo Problems*	Diana Ross	*I'm Coming Out*
A Tribe Called Quest	*Bonita Applebum*	Carly Simon	*Why*
De La Soul	*Say No Go*	Hall & Oats Neneh Cherry	*I Can't Go For That Buddy X*
Groove Armada	*At The River*	Patti Page	*Old Cape Cod*
Dream Warriors	*My Definition Of A Bombastic Jazz Style*	Quincy Jones	*Soul Bossa Nova*
Gang-Star	*Love Slick*	Young Hot Unlimited	*Ain't There Something Money Can't Buy*

Secondly, there is no 'quick fix' to creating great timbres. It comes from practical experience and plenty of experimentation with not only the synthesizer parameters but also effects and processors. The world's leading sound designers and dance artists didn't simply read a book and begin designing fantastic sounds a day later; they learnt the basics and then spent months and, in many cases, years learning from practical experience and taking the time to study what each synthesizer, effect, processor and sampler can and cannot accomplish. Thus, if you want to programme great timbres, patience and experimentation is the real key. You need to set aside time from writing music and begin to learn exactly what your chosen synthesizer is capable of and how to manipulate it further using effects or processors.

Finally, and most important of all there is no such thing as a 'hit' sound. While some genres of music are created from using a particular type of sound and it can be fun emulating the timbres from other popular tracks, using them will not instantly make your own music an instant hit. Instead it will often make it look like an imitation of a great record and will be subsequently judged alongside it rather than on its own merits. Despite the promise of many books offering the secret advice to writing hit sounds or music, there is no secret formula; there are no special synthesizers, no hit making effects and no magic timbres. What's more, copying a timbre exactly from a previous hit dance track isn't going to make your music any better as dance floor tastes change very quickly.

As a case in point, the 'pizz' timbre was bypassed by every dance musician on the planet as an unworthy sound until Faithless soaked the Roland JD-990's 'pizz' in reverb and used it for their massive club hit 'Insomnia'. Following this, a host of 'pizz' saturated tracks appeared on the scene and it became so popular that it now appears on just about every dance-based module around. But as tastes have changed, the timbre has now almost vanished into obscurity and is skipped past by most musicians.

Keep in mind that Joe Public, your potential listening audience, is fickle, insensitive and short of attention span, and while some timbres may be doing the rounds today, next week/month they could be completely different. As a result, the following is not devoted to how to programme a precise timbre from a specific track as it would most likely date the book before I've even completed this chapter. Instead, it will concentrate on building the basic sounds that are synonymous with dance music and as such will create a number of related 'presets' that you can then manipulate further and experiment with.

Creativity may be tiring and difficult at times but the rewards are certainly worth it…

PROGRAMMING TIMBRES

As previously touched upon, despite the claims from the synthesizer's manufacturer just one synthesizer is not capable of producing every type of sound suited toward every particular genre. In fact, in many instances you will have to accept that it is only possible to create some timbres on certain instruments no matter how good at programming you may be. Just as you wouldn't expect different models of speakers/monitors to sound exactly alike the same is true of synthesis.

Although the oscillators, filters and modulation are all based on the same principles, they will all sound very different. This is why some synthesizers are said to have certain character and many older analogue keyboards demand such a high price on the second-hand market. Musicians are willing to pay for the particular sound characteristics of a synthesizer. As a result, although you may follow the sound design guidelines in this chapter to the letter there is no guarantee that they will produce exactly the same sound. Consequently, before you even begin programming your own timbres, the first step to understanding programming is to learn your chosen synthesizers inside out.

Indeed, it's absolutely crucial that you set time aside to experiment to learn the character of the synthesizer's oscillators by mixing a sine with a square, a square with a saw, a sine and a saw, a triangle and a saw, a sine and a triangle and so forth, and noting the results it has on the overall timbre. This is the *only* way you'll be able to progress toward creating your own sounds short of manipulating presets.

You *have* to know the characteristics of the synthesizers you use and how to exploit its idiosyncrasies to create sounds. All professional artists have purchased one synthesizer and learnt it inside out before purchasing another. This is obviously more difficult today with the multitude of virtual instruments appearing for ridiculously low prices but despite how tempting it may be to own every synthesizer on the market you need to limit yourself to a few choice favourites. You'll never have the chance to programme timbres if you have to learn the characteristics of 50 different virtual instruments.

Gerald's advice at the beginning of this chapter is a very wise counsel indeed and you have to accept that there are no shortcuts to creating great dance music, short of ripping from sample CDs.

PROGRAMMING PADS

Although pads are not usually the most important timbre in dance music, we'll look at these first. This is simply because an understanding of how they're created will help increase your knowledge of how the various processes of LFO and envelopes can all work together to produce evolving, interesting sounds. Of course, there are no predetermined pads to use within dance music production and therefore there are no definitive methods to create them. But, while it's impossible to suggest ways of creating a pad to suit a particular style of music, there are, as ever, some guidelines that you can follow.

Firstly, pads in dance music are employed to provide one of the following three things:

- To supply or enhance the atmosphere in the music – especially the case with chill out/ambient music.
- To fill a 'hole' in the mix between the groove of the music and lead or vocals.
- To be used as a lead itself.

Depending on which of these functions the pad is to provide will also determine how it should be programmed. Although many of the sounds in dance will utilize an immediate attack stage on the amp and filter's EG so that the sound starts immediately, it is only really necessary to use them on pads if the sound is providing the lead of the track. As discussed, we determine timbres that start abruptly to be perceivably louder than those that do not but more interestingly we also tend to perceive sounds with a slow attack stage to be 'less important' to the mix, even though in reality this may not be the case at all. As a consequence, when pad sounds are used as 'backing' instruments, they should not start abruptly but filter in or start slowly, while if they're used as leads, the attack stage should be quite abrupt as this not only helps it cut through the mix but also gives the impression that it's an important aspect of the music.

A popular, if somewhat cliché and overused, technique to demonstrate this is the gated pad. Possibly the best example of this in use (which incidentally doesn't sound cliché) is from Sasha's club hit Xpander which to many clubbers is still viewed as one of the greatest trance tracks of all time. A single evolving, shifting pad is played as one long progressive note throughout the track and a noise gate is employed to rhythmically cut the pad. When the noise gate releases and lets the sound back through, a plucked lead is dropped in to accentuate the return of the pad. The gated pad effect can be constructed in one of the following three ways:

- A series of MIDI notes are sent to a (MIDI compatible) gate to determine where the pad should cut off. The length of the 'gate' is determined by the length of the MIDI note.
- A short percussive sound is inserted into the key input of the gate, so at every kick the gate is activated. The length of the gate is determined by the hold and release parameters on the gate.
- A series of CC11 (expression) commands are sent to the synthesizer creating the pad which successively closes and opens the expression to produce a gated sound. The first CC11 command is set at zero to turn the pad off, while this is followed a few ticks later by another CC11 command set to 127 to switch it back on again. The length of the gate is controlled by the distance between the CC11 off and on commands in the sequence.

Naturally, for this to work in a musical context the pad must evolve throughout because if this technique was used on a sustaining sound with no movement you may as well just retrigger the timbre for its attack stage whenever required rather than gate it. In fact, it's this movement that's the real secret behind creating a good pad. If the timbre remains static throughout without any timbral variations then the ear soon becomes bored and switches off. This is the main reason why analogue synths are often recommended for the creation of pads since the random fluctuations of the oscillators pitch and timbre provide an extra sense of movement that when augmented with LFOs and/or envelopes produces a sound that has constant movement that you can't help but be attracted too. There are various ways you can employ this movement ranging from LFOs to using envelopes to gradually increase or decrease the harmonic content while it plays. To better explain this, we'll look at the methodology behind how the envelopes are used to create the beginnings of a good pad.

By setting a fast attack and the decay quite long on an amplifier envelope we can determine that the sound will take a finite amount of time to reach the sustain portion. Provided that this sustain portion is set just below the amp's decay stage, it will decay slowly to sustain and then continually 'loop' until the MIDI note is released whereby it'll progress onto the release stage. This creates the basic premise of any pad or string timbre – it continues on and on until the key is released. Assuming that the pad has a rich harmonic structure, movement can be added by gradually increasing a low-pass filter's cut-off while the pad is playing – there will be a gradual rise in the amount of harmonics contained within the pad.

If a positive filter envelope was employed to control the filter's action and the envelope amount was set to fully modulate the filter, by using a long attack, short decay, low sustain and fast release, the filter's action would be introduced slowly before going through a fast decay stage and moving onto the sustain. This would create an effect whereby the filter would slowly open through the course of the amplifier's attack, decay and sustain stage before the filter entered a short decay stage during the 'middle' of the amp's sustain stage. Conversely, by using the same filter envelope settings but applied negative the envelope is inverted creating an effect of sweeping downwards rather than upwards. Nevertheless, the amp and filter envelopes working in different time scales create a pad that evolves in harmonic content over a period of time. Notably, this is a one-way configuration because if the function of these envelopes were reversed (in that the filter begins immediately but the amplifier's attack was set long) the filter would have little effect since there would be nothing to filter until the pad is introduced.

This leads onto the subject of producing timbres with a high harmonic content and this is accomplished by mixing a series of oscillators together along with detuning and modulating. The main principle here is to create a sound that features plenty of harmonics for the filters to sweep, so this means that saw, triangle, noise and square waves produce the best results although on some occasions a sine wave can be used to add some bottom end presence if required. A good starting point for any pad is to use two saw, triangle or pulse wave oscillators with one detuned from the other by −3 or −5 cents. This introduces a slight phasing effect between the oscillators helping to widen the timbre and make it more interesting to the ear. To further emphasize this detuning, a saw, triangle, sine or noise waveform LFO set to gently and slowly modulate the pitch or volume of one of the oscillators will produce a sound with a more analogue feel while also preventing the basic timbre from appearing too static.

If the pad is being used as a lead or has to fill in a large 'hole' in the mix then it's worthwhile adding a third oscillator and detuning this by +3 or +5 to make the timbre more substantial. The choice of waveform for the third oscillator depends on the waveform of the original two but in general it should be a different waveform. For instance, if two saws are used to create the basic patch, adding a third detuned triangle wave will introduce a sparkling effect to the sound while replacing this triangle with a square wave would in effect make the timbre exhibit a hollow character. These oscillators then, combined with the previously discussed envelopes for both the filter and the amp, form the foundation of every pad sound, and from here, it's up to the designer to change the envelope settings, modulation routings and the waveform used for the LFOs to create a pad sound to suit the track. What follows is a guide to how most of the pads used in dance music are created but it is by no means the definitive list and it's always experimenting to produce different variations.

Rise and Fall Pad

To construct this pad, use two sawtooth oscillators with one waveform detuned by 3 cents. Apply a fast attack, short decay, medium sustain and a long release for the amp envelope and employ a low-pass filter with the cut-off and resonance set quite low. This should result in a static buzzing timbre. From this, set the filter's envelope to a long attack and decay but use a short release and no sustain and set the filter envelope to maximum positive modulation. Finally use the filter's key follow so that it tracks the pitch of the notes being played. This results in the filter sweeping up through the pad before slowly settling down. If the pad is to continue playing during the sustain portion for a long period of time then it's also worth modulating the pitch of one of the oscillators with a triangle LFO and modulating the filters cut-off or resonance with a square wave LFO. Both these should be set to a medium depth and a slow rate.

RESONANT PADS

Resonant pads can be created by mixing a triangle and square wave together and detuning one of the oscillators from the other by 5 cents. Similar to the previous pad, the amp's attack should be set to zero with a short decay, medium sustain and long release, but set the filter's envelope to a long attack, sustain and release with a short decay. Using a low-pass filter set the cut-off quite low but set the resonance to around 3/4 so that the timbre appears quite resonant. Finally, modulate the pitch of the triangle oscillator with a sine wave LFO set to a slow rate and a medium depth and use the filters key follow. The LFO modulation creates a pad that exhibits the natural analogue 'character' while the filter tracks the pitch and sweeps in through the attack and decay of the pad and then sustains itself through the amp's sustain. Again, if the pad's sustain is going to continue for a length of time it's worthwhile employing a sine, pulse or triangle wave LFO to modulate the filter's cut-off to help maintain interest.

SWIRLING PADS

The swirling pad is typical of some of Daft Punk's work and consists of two sawtooth oscillators detuned from one another by 5 cents. A square LFO is often applied to one of the saws to gently modulate the pitch while a third oscillator set to a triangle wave is pitched approximately 6 semitones above the two saws to add a 'glistening' tone to the sound. The amp envelope uses a medium attack, sustain and release with a short decay while the filter envelope uses a fast attack and release with a long decay and medium sustain. A low-pass filter is used to modify the tonal content and this is set to a low-cut with the resonance set to midway. Finally, chorus *and then* phaser or flanger effects are applied to the timbre to produce a swirling effect. It's important that the flanger or phaser is inserted *after* the chorus rather than before so that it modulates the chorus effect as well.

THIN PADS

All of the pads we've covered so far are quite heavy, and in some instances, you may need a lighter pad to sit in the background. For this, pulse oscillators are possibly the best to use since they do not contain as many harmonics as saws or triangles and you can use an LFO to modulate the pulse width to add some interest to the timbre. These types of pads simply consist of one pulse oscillator that uses a medium attack, sustain and release with a fast decay on the amp envelope. A low-pass or high-pass filter is used, depending on how deep or bright you want the sound to appear and there is usually little need for a filter envelope since gently modulating the pulse width with a sine, sawtooth or noise waveform produces all the movement required. If you decide that it does need more movement, however, a very slow triangle LFO set to modulate the filter cut-off or resonance set to a medium depth will usually produce enough movement to maintain some interest but try not to get too carried away. The purpose of this pad is to sit in the background and too much movement may push it to the front of the mix resulting in all the instruments fighting for their own place in the mix.

The data CD contains audio examples of these timbres being programmed.

PROGRAMMING DRUMS

The majority of drum timbres used in the creation of dance chiefly originated from four machines that are now out of production – the Roland TR909, the Roland TR808, the Simmons SDS-5 and the E-mu Drumulator. Consequently, while these machines (or to use their proper term drum synthesizers) are often seen as requisites for producing most genres of dance music they're very highly sought after and demand an absurd sum of money on the second-hand market, if you can find them. Because of this, most musicians will use software or hardware emulations, and although, to the author's knowledge, there are no alternatives to the Simmons SDS-5 and the E-mu Drumulator, both the TR machines are available from numerous software and hardware manufacturers. Indeed, due to the importance of using these kits when producing dance, most keyboards and tone modules today will feature the requisite TR808 and 909 kits and there are plenty of software plug-ins and stand-alone sequencers that offer them. The most prominent of these is Propellerhead's ReBirth which imitates both TR machines and a couple of TB303s (more on these later), but Propellerhead's Reason, Cakewalk's Fruity-loops and D-lusion's Drum station all offer samples/synthesis of the original machines too.

KICK DRUMS

In the majority of cases, the Roland TR909 kick drum is the most frequently used kick in dance music but the way it is created and edited with the synth parameters and effects can sometimes determine the genre of music it suits the most. In the original machines this kick was created using a sine wave with an EG used to control its pitch. To add more of a transient to this sine wave, a pulse and a noise waveform were combined together and filtered to produce an initial click. This was then combined with the sine wave to produce the typical 909 kick sound. Notably, due to the age of these machines, there was not a huge amount of physical control over the timbre they created but by building your own kick you can play with a larger number of parameters once the basic elements are down. In fact, this applies to all drum sounds, not just the kick, and is much better than sampling them from other records or sample CDs since you can't alter the really important parameters.

When constructing a kick in a synthesizer, the frequency of the sine wave will determine how deep the kick becomes, and while anywhere between 30 and 100 Hz will produce a good kick, it does depend on how deep you want it to be. A sine wave at 30–60 Hz can be used to create an incredibly deep bowel moving thud typical of hip-hop, a frequency of 50–80 Hz can provide a starting block for lo-fi and a frequency of 70–100 Hz can form the beginning of a typical club kick. An attack/decay EG can then be used to modulate the pitch of the sine wave. These envelopes are most common on drum machines but are also available on some synthesizers such as the Roland JP8080 and a number of soft synthesizers. Similarly, this action can be recreated on any synthesizer by dropping both sustain and release parameters to zero.

Naturally, the depth of the pitch modulation needs to be set to maximum so that the effect of the pitch envelope can be heard and it should be set to a positive depth so that the pitch moves downwards and not upwards (as it would if the pitch envelope were negative). Additionally, the attack parameter should be set as fast as possible so that the pitch modulation begins the instant the key is struck. If your synthesizer doesn't have access to a pitch envelope, then the same effect can be produced by pushing the resonance up to maximum so that the filter begins to self-oscillate. The filter's envelope will then act in a manner similar to a pitch envelope, so the attack, sustain and release will need to be set at zero and the decay can be used to control the kick's decay.

Once this initial timbre is created it's prudent to synthesize a clicking timbre to place over the transient of the sine wave. This can help to give the kick more presence and help it to pull through a mix. Possibly the best way to accomplish this is to use a square wave pitched down and use a very fast amplifier attack and decay setting to produce a short sharp click. The amount that this wave is pitched down will depend entirely on the sound you want to produce, so it's sensible to layer it over the top of the sine wave and then pitch it up or down until the transient of the kick sounds right for the mix. This, however, is only to acquire a basic kick sound, and it's now open to tweaking with all the parameters you have at your disposal, which incidentally are many more than the humble 909 or 808 have to offer.

Firstly, as we've already touched upon the transient where we attain most of the information about a sound, adjusting the filter cut-off and resonance of the square wave will dramatically change the entire character of the kick. A high-resonance setting will produce a kick with a more analogue character, while increasing the amplifier's decay will produce a kick that sounds quite 'boxy'. Increasing the filter's cut-off will result in a more natural sounding kick, while increasing the pulse width will create a more open, hollow timbre. Additionally, the oscillator producing the sine wave, can also be affected with the synthesis parameters on offer. Using the pitch and pitch envelope parameters you can adjust how the pitch reacts on the sine wave, but more importantly, you can determine how 'boomy' the kick is through the pitch decay. In this context, this will set the time that it takes to drop from the maximum pitch change to the sine wave's normal pitch. Thus, by increasing this, the pitch of the sine wave doesn't fall as quickly, permitting the timbre to continue for longer creating a 'boomy' feel. Similarly, decreasing it will shorten its length making it appear snappier. More interestingly, though, if you can adjust the properties of the envelope's decay slope you can use it to produce kicks that are fatter or have a smacking/slapping texture.

If the decay's slope remains linear, the sound will die in a linear fashion producing the characteristic 909 kick sound. However, if a convex slope is used in its place, as the pitch decays it will 'bow' the pitch at a number of frequencies, which results in a kick that's more 'rounded' and much fatter. On the other hand, if the slope is concave, the pitch will curve 'inwards' during the decay period producing the sucking/smacking timbre similar to the E-mu Drumulator. By increasing the length of the pitch decay further these effects can be drawn out producing kicks that are suited towards all genres of dance. It should be noted here, however, that not all synthesizers allow you to edit the envelope's slope in such a manner but some software samplers (such as Steinberg's HALion) will allow you to modify the slope of a sample. Thus, if you were to sample a kick with a lengthy decay, you could then modulate the various stages of the envelope.

Alternatively, if the synthesizer is quite substantial it may allow you to modulate not only the sine wave but certain aspects of the envelope with itself. Fundamentally this means that the pitch envelope modulates not only the sine wave but also its own parameters. Using this you can set the synthesizer's modulation destination to affect the oscillator's pitch and its own decay parameter by a negative or positive amount, which results in the decay becoming convex or concave, respectively. This effect is often referred to as 'recursive' modulation, but as mentioned, this is only possible on the more adept synthesizers. Nevertheless, whether recursive modulation is available or not, the key, at this stage, is to experiment with different variations of pitch decay, the filter section on the square wave and both oscillators. For instance, replacing the sine wave with a square produces a 'clunky' sound, while replacing it with a triangle produces a sound similar to the Simmons SDS-5.

While these methods will produce a kick that can be modified to suit all genres, for hip-hop you can sometimes glean better results using a self-oscillating filter and a noise gate. If you push the resonance up until it breaks into self-oscillation it will produce a pure sine wave. This is often purer than the oscillators themselves and you can use this to produce a deep tone that's suitable for use as a kick in the track (usually 40 Hz). If you then programme a 4/4 loop in a MIDI sequencer and feed the results into a noise gate it can be used as an EG. While playing the loop, lower the threshold so that only the peaks of the wave are breaching, set the attack to zero and use the release to control the 'kicks' delay. This kick can be modified further by adjusting the hold time, as this will allow more of the peak through before entering the release stage.

Once the basic kick element is down it will most likely benefit from some compression but the settings to use will depend on the genre of music. Typically, house, techno and trance will benefit from the compressor using a fast attack so that the transient is crushed by the compression. This produces the emblematic club kick while setting the compressor so that the attack misses the transient but grips the decay stage allowing it to be raised in gain to produce the characteristic hip-hop, big beat or drum 'n' bass timbre.

SNARE DRUMS

The snare drum in most dance music is derived (somewhat unsurprisingly) from the TR909, or, in the case of house, the E-mu Drumulator or the Roland SDS-5. All these, however, were synthesized in much the same way by using a triangle oscillator mixed in with pink or white noise that was treated to positive pitch movements. This can, of course, be emulated in any synthesizer by selecting a triangle wave for the first oscillator and using either pink or white noise for the second. The choice between whether to use pink or white noise depends on the overall effect you wish to achieve but, by and large, pink noise is used for house, lo-fi and ambient snares while white is used for drum 'n' bass, techno, garage, trance and big beat. This is simply because pink noise contains more low-frequency content and energy than white and hence produces a thicker, wider sounding timbre.

To produce the initial snare sound much of the low-frequency content will need to be removed, so it's sensible to employ a high-pass, band-pass filter or notch filter depending on the type of sound you require. Notching out the middle frequencies will create a clean snare sound that's commonly used in breakbeat while a band pass will add crispness to the timbre making it suitable for techno. Alternatively, using a high-pass filter with a medium resonance setting will create the house 'thunk' timbre. As with the kick drum, snares need to start immediately on key press and remain fairly short, so the amp's EG will need setting to a zero attack, sustain and release while the decay can be used to control the length of the snare itself. In some cases, if it's possible in the synthesizer, it's prudent to employ a different amp EG for both the noise and the triangle wave. This way the triangle wave can be kept quite short and swift by using a fast decay while the noise can be made to ring a little further by increasing its decay parameter. The more this is increased, the more atmospheric snare will appear allowing you to move from the typical techno snare, through big beat and trance before finally arriving at ambient. What's more, this approach may also allow you to use a convex envelope on the noise to produce a smacking timbre similar to the Nine Inch Nails (NIN) 'Closer' snare.

If two amp EGs are not available, small amounts of reverb can help to lengthen the timbre, and if this is followed with a noise gate with a fast attack and short hold time, the decay can be used to control the amount of ambience in the loop. Even if artificial ambience isn't required, employing a gate can be particularly important when programming house, drum 'n' bass and hip-hop loops as the snare is often cut short in these genres to produce a more dynamic loop. Additionally, for drum 'n' bass the snare can then be pitched further up the keyboard to produce the characteristic bright 'snap'.

This initial snare can be further modified using a pitch envelope to modulate both oscillators and can be applied either positive or negative depending on the genre of music. Most usually, small amounts of positive pitch modulation are applied to force the pitch downwards as it plays but some house tracks will employ a negative envelope to create a snare that exhibits a 'sucking' nature resulting in a thwacking sound. If you decide to use this technique, however, it's often worth removing the transient of the snare in a wave editor and replacing it with one that uses positive pitch modulation.

The two combined then produce a sound that has a good solid strike but decays upwards in pitch at the end of the hit. If this isn't possible in the synthesizer/wave editor then a viable alternative is to sweep the pitch from low to high with a sawtooth or sine LFO (provided that the saw starts low and moves high) set to a fast rate or programme a series of control change (CC) messages to sweep it from the sequencer. Once this is accomplished, small amounts of compression set so that only the decay is squashed (i.e. slow attack) will help to bring it up in volume so that it doesn't disappear into the rest of the mix.

HI-HATS

Hi-hats can be synthesized in a number of ways depending on the type of sound you require but in the 'original' drum machines they were created with nothing more than filtered white noise. This can be accomplished in most synthesizers by selecting *white* noise as an oscillator and setting the filter envelope to a fast attack, sustain and release with a medium-to-short decay. Finally, set the filter to a high pass and use it roll off any frequencies that are too low to create a high hat. The length of the decay parameter will determine whether the hi-hat is open or closed (open hats have a longer decay period).

While this is the best way to produce a typical analogue hi-hat sound it can sound rather cheap and nasty on some synthesizers, and even if it produces the timbre, it can still appear quite dreary. As a result, a much better approach is to use either ring or FM as this produces a hi-hat that sounds sparkling and animated, helping to add some energy to the music. Ring modulation is possibly the easiest solution of the two simply consisting of modulating a high-pitched triangle wave with a lower pitched triangle. FM consists of modulating a square or sine wave with a high-pitched triangle oscillator. The result is a high-frequency noise waveform that can then be modified with a volume envelope set to a zero attack, sustain and release with a short-to-medium decay. If FM is used and the modulator source is modified with a pitch envelope and the amount of FM is increased or reduced, the resulting waveform can take on more interesting properties, so it's worthwhile experimenting with both these parameters (if available). Once this basic timbre is constructed, shortening the decay creates a closed hi-hat while lengthening it will produce an open hat. Similarly, it's also worth experimenting by changing the decay slope to convex or concave to produce fatter or thinner sounding hats.

Notably, unlike most percussive instruments, compression should not be used on hi-hats as all but the best compressors will reduce the higher frequencies even if the attack is set so that it skips by the attack stage. This obviously results in a dull sounding top end of a mix, so any form of dynamic restriction should be avoided on high-frequency sounds, which include shakers, cymbals, cowbells, claves and claps.

SHAKERS

Shakers are constructed in a fashion similar to the high hats. That is, they are created from white noise with a short attack, sustain and release with a medium decay on the amplifier and filter envelope. A high-pass filter is then used to remove any low-end artefacts to produce a timbre consisting entirely of higher frequencies. Once you have the basic 'hi-hat' timbre, the decay can be lengthened to produce a longer sound, which is then treated to an LFO modulating the high-pass filter to produce some movement. The waveform, rate and depth of the LFO depend entirely on the overall sound you want to produce, but as a general starting point, a sine wave LFO with a fast rate and medium depth produces the typical 'shaker' timbre. Again once this initial timbre has been constructed, like all drum sounds, it's worth experimenting by changing the envelope's attack and decay slope from linear to convex or concave.

CYMBALS

Again, cymbals are created in a fashion similar to hi-hats and shakers as they're constructed from noise, but rather than use the noise from an oscillator, it's commonly generated from ring or FM. Using two square waves played high on the keyboard, detune them so that they're approximately two octaves apart and set the amp's EG to a fast attack with no release or sustain and a medium decay (in fact similar to hi-hats and shakers). Once the tone is shaped, both need to be fed into a ring or cross modulator to produce the typical analogue cymbal noise timbre.

The attack of the cymbal is particularly important, so it may be worth synthesizing a transient to drop over the top or using a filter envelope to produce more of an initial crash. The filter envelope is set to a fast attack, sustain and release with a medium-to-short decay as this produces an initial 'hit' but you can synthesize an additional transient using an oscillator that produces *pink* noise with the same filter settings as previously mentioned. If ring modulation is not available on the synthesizer, a similar sound can be created using FM. This consists of modulating a high-pitched square wave with another, lower pitched, square or a triangle wave. As with the FM used to create high hats, by modifying the source with pitch modulation or increasing/decreasing the amount of FM you can create a number of different crash cymbals.

CLAPS

Claps are perhaps the most difficult 'percussive' element to synthesize because they consist of a large number of 'snaps' all played in a rapid, sometimes pitch shifting, sequence. Although in the interests of theory we'll look at how they're created, generally speaking you're much better off recording yourself clapping (remember to run the mic through a compressor first, though, as the transient will often clip a recorder) and treating them to a chorus or harmonizing effect or, simpler still, sampling some from a CD or record.

Generally, claps are created from white noise passed through a high-pass filter. The filter and amp EGs (as should be obvious by now) are set to a fast attack with no release or sustain and a decay set to suit the timbre you require – midway is a good starting point. The filter cut-off, however, often benefits from being augmented by a sawtooth LFO set to a very fast rate and maximum depth to produce a 'snapping' type timbre. This produces the basic tone but you'll also need to use an arpeggiator to create the successive snaps to follow. For this, programme a series of staccato notes into an MIDI sequencer and use these to trigger the arpeggiator set to one octave or less so that it constantly repeats the same notes in a fast succession. These can be pitched downwards, if required, using a pitch envelope set to a positive depth on the oscillator but it must be set so that it doesn't retrigger on each individual note otherwise the timbre turns into mushy pitch-shifted clutter. Although claps are difficult to synthesize it is often worth the effort required, however, as adjusting the filter section and/or the decay slopes of the amp and filter's EG opens up a whole new realm of clap timbres.

COWBELLS

Fundamentally, cowbells are quite easy to synthesize and can be constructed in a number of ways. You can use two square oscillators or a triangle and a square depending on the sound you require. If you require a sound with more body then it's best to use two square waves, but if you prefer a brighter sound then a square mixed with a triangle will produce better results.

For a cowbell with more body, set two oscillators to a square wave and detune them so that the first square plays at C#5 (554 Hz) while the other plays at G#5 (830 Hz). Follow this by setting the amp envelope to a fast attack with no release or sustain and a very short decay. This resulting tone is then fed into a band-pass filter which can be used to shape the overall colour of the sound. Alternatively, if you want to create a cowbell that exhibits a brighter colour to sit near the top of a mix it's preferable to use just one square wave mixed with a triangle wave. The frequency of the square should be set around 550 Hz and the triangle should be detuned so that it sits anywhere from half an octave to an octave from the square, depending on the timbre you require. Both these are then ring modulated and the result is fed through a high-pass filter allowing you to remove the lower frequencies introduced by the ring modulation. Once these basic timbres are created, the amp's EG can be lengthened or shortened to suit the current rhythm.

CONGAS

Congas are constructed from two oscillators with a dash of FM to produce a 'clunky' timbre. These can be easily constructed in any synth (that features FM) by setting the first oscillator to a sine wave and the second to any noise waveform. The sine wave amp's EG is set to a very fast attack and decay with no release or sustain to produce a click which is then used as FM for the noise waveform. The noise waveforms amp EG needs to be set to a fast attack with no release or sustain and the decay set to taste.

There is little need to use a filter on the resulting sound, but if it seems to have too much bottom or top end then use a low-pass or high-pass filter to remove the upper or lower frequencies consecutively. In fact, by employing a high-pass filter and reducing it slowly you can create muted congas while adjusting the noise amp's decay slope to convex or concave can produce the typical slapped congas that are sometimes used in-house.

TAMBOURINES

Tambourines, like claps, are difficult to synthesize as they essentially consist of a number of hi-hats with a short decay each occurring one after the other in a rapid succession which is passed through a band-pass filter. This means that they can initially be constructed by using white noise, FM or ring modulation in the same manner as high hats. Once this basic timbre is down, the tone is augmented with a sawtooth LFO set to a fast rate and full depth. After this, you'll need to programme a series of staccato notes to trigger the synthesizer's arpeggiator to create a series of successive hits. The decay parameter of the amp's EG can then be used to manipulate the tambourines, character, while the results are fed into a band-pass filter which can be used to shape the type of tambourine being used. Typically, wide band-pass settings will recreate a tambourine with a large tympanic membrane and thinner settings will recreate a tambourine with a smaller membrane.

TOMS

Toms can be synthesized in one of the two ways depending on how deep and wide you want the tom drum to appear. Typically, they're synthesized by using the same methods as producing a kick drum but utilize a higher pitch with a longer decay on the amp's EG and some white noise mixed in to produce some ambience. The settings for the white noise oscillator generally remain the same as the amp's EG for the sine wave (zero attack, release and sustain with a medium decay). Alternatively, they can be produced by creating a snare timbre by mixing a triangle wave with a noise waveform but modulating the noise waveform with pitch so that it falls while the triangle wave continues unchanged.

PERCUSSION GUIDELINES

Although we've looked at the percussive instruments used throughout all dance genres, if you decide to become more creative, or simply want to experiment further with producing percussive hits there are some general guidelines that you can follow.

The main oscillator usually always consists of a sine or triangle wave with its pitch modulated by a positive pitch envelope. This creates the initial tone of the timbre while the second oscillator is used to create either the subsequent resonances of the skin after it's been hit or alternatively the initial transient. For the resonance, white or pink noise is commonly used, while to create the transient, a square wave is often used. The amp and filter envelope of the first oscillator is nearly always set to a zero attack, zero release and medium decay. This is so that the sound starts immediately on key press (the drummer's strike) while the decay controls how ambient the surrounding room is. If the decay is set quite long the sound will obviously take longer to decay away, producing an effect similar to reverb on most instruments. That said, if the decay is set too long on low-pitched percussive elements such as a kick it may result in a 'whooping' sound rather than a solid hit. If the second oscillator is being used to create the subsequent resonance to the first oscillator then the amp and filter settings are the same as the first oscillator whereas if it's being used to create the transient, the same attack, release and sustain settings are used but the decay is generally much shorter.

For more creative applications, it's worthwhile experimenting with the slope of the amp and filter's EG decay and occasionally attack. Also, by experimenting with frequency modulation and ring modulation it's possible to create a host of new drum timbres. For instance, if the second oscillator is producing a noise waveform, this can be used to modulate the main oscillator to reduce the overall tone of the sound. What's more, by using the filters the sound can be shaped to fit into the current loop. For example, using a high-pass filter you can remove the 'boom' from a kick drum which, as a consequence, produces a tighter and more punchy kick. The key is to experiment.

> The data CD contains audio examples of these timbres being programmed.

PROGRAMMING BASS

Synthesizer basses are a difficult instrument to encapsulate in terms of programming as we have no real expectations of how they should sound and they always sound different when placed into a mix anyway. As such, there are no definitive ways to construct a bass as pretty much anything goes provided that it fits with the music. Of course, as always, there are some guidelines that apply to all basses and many genres also tend to use a similar bass timbre, so here we'll concentrate on how to construct these along with some of the basic guidelines.

Generally speaking, most synthesizer bass sounds are quite simple in design as their main function is to supply some underpinning because it's the lead/vocals that provide the main focal point. As a result, they are not particularly complex to programme and you can make some astounding bass timbres using just one or two oscillators. Indeed, the big secret to producing great basses is not from the oscillators but from the filters and a thoughtful implementation of modulation to create some movement.

Whenever approaching a bass sound it's wise to have the bass riff programmed and playing back in your sequencer along with the kick drum and any precedence instruments. By doing so, it's much easier to hear if the bass is interfering with the kick and other priority instruments while you manipulate the parameters. For example, if the bass sound seems to disappear into the track or has no definite starting point, making it appear 'woolly', then you'll need to work with both the amp and the filter envelopes to provide a more prominent attack. Typically, the amplifier's attack should be set to its shortest time so that the note starts immediately on key press and the decay should be set so that it acts as a release setting (sustain and release are rarely used in bass timbres). This is also common with the filter's envelope. Setting the filter's attack stage too long will result in the filter slowly fading in over the length of the note, which can destroy the attack of the bass.

Notably, bass timbres also tend to have fairly complex attack, so if it needs a more prominent attack it's prudent to sometimes layer an initial pluck over the transient. This pluck can be created by synthesis in the same manner as creating a pluck for a drum's kick, but more commonly, percussion sounds such as cowbells and wood blocks pitched down are used to add to the transient. If this latter approach is used then it's advisable to reduce the length of the drum sample to less than 30 ms. This is because, as we touched upon when discussing Granular synthesis, we find it difficult to perceive individual sounds if they are less than 30 ms in length. This can usually be accomplished by reducing the amp's decay (remember that drum samples have no sustain or release) with the benefit that you can use amplitude envelopes to fade out the attack transient oscillator as you fade in the main body of the bass, helping to keep the two sounds from getting in each other's way. If this is not possible then simply reducing the volume of the drum timbre until it merges with the bass timbre may provide sufficient results.

Another important aspect of creating a good bass is sonic movement. No matter how energetic the bass riff may be in MIDI, if it's playing a simple tone with absolutely no movement our ears can get bored very quickly and we tend to 'turn off' from the music. In fact, this lack of movement is one of the main reasons why some grooves just don't seem to groove at all. If it's a boring timbre, the groove will appear just as monotonous. This movement can be implemented in a number of ways. Firstly, as touched upon in the genre chapters, programming CC messages or using velocity commands can breathe life into a bass provided, of course, that you programme the synthesizer to accept these controllers. Secondly, it's often worthwhile assigning the modulation wheel to control the filter cut-off, resonance, LFO rate or pitch. This way, after programming a timbre you can move the wheel to introduce further sonic movement and record these as CC data into the sequencer. And finally, you can employ LFO modulation to the timbre itself to introduce pitch or filter movement.

If, on the other hand, you decide to use a real bass guitar then unless you're already experienced it isn't recommended attempting to programme the timbre in a synthesizer or record one live. Unlike synthetic timbres we know how a real bass guitar should sound. If these are constructed in a synthesizer they often sound too synthetic while recording a bass guitar reasonably well requires plenty of experience and the right equipment. Thus, if you feel that the track would benefit from a real bass it is much easier to invest in a sample CD or alternatively Spectrasonics Trilogy – a VST instrument that contains multi-samples of a huge number of basses, real and synthetic.

DEEP HEAVY BASS

The heavy bass is typical of drum 'n' bass tracks (by artists such as Photek) and is the simplest bass timbre to produce as it's essentially a kick drum timbre with the amplifier's decay and release parameter lengthened. This means that you'll need to use a single oscillator set to a sine wave with its pitch positively modulated by an attack decay envelope. On top of this it may also be worth synthesizing a short stab to place over the transient of the sine wave. As with the kick, the best way to accomplish this is to use a square wave pitched down and use a very fast amplifier attack and decay with no release or sustain. Once this initial timbre is laid down, you can increase the sine wave's amp EG decay and sustain until you have the sound you require.

SUB-BASS

Following on from the large bass, another alternative for drum 'n' bass is the earth shaking, speaker melting sub-bass. Essentially, these are formed from a single sine wave with perhaps a small 'clunk' positioned at the transient to help it pull through a mix. The best results come from a self-oscillating filter using any oscillator, but if the filter will not resonate, a sine wave from any synth should provide good results. Obviously, the amplifier's attack stage should be set at zero so that the sound begins the moment the key is depressed, but the decay setting to use will depend entirely on the sound you require and the current bass motif (a good starting point is to use a fairly short decay, with no sustain or release). If the sine wave has been produced by an oscillator then the filter cut-off will have no effect as there are no harmonics to remove, but if it's been created with a self-oscillating filter, reducing the filter's cut-off will remove the high-end artefacts that may be present. Also it's prudent to set the filter's key follow, if available, to positive so that the further up the keyboard you play the more it will open. This will help to add some movement to the sound. If a click is required at the beginning of the note then, as with the drums, a square wave with a fast amplifier attack and decay and a high cut-off setting can be dropped onto the transient of the sine wave. Alternatively setting the filter's envelope to a zero attack, sustain and release along with a very quick decay and increasing the amount of the filter's EG until you have the transient can also provide great results.

This will produce the basic 'preset' tone typical of a deep sub-bass but is of course open to tweaking the initial click with filter cut-off and resonance along with modulating the sine wave to add some movement. For example, by modulating the sine wave's pitch by 2 cents with an EG set to a slow attack, medium decay and no sustain or release the note will bend slightly every time it's played. If there is no EG available, then a sine wave LFO with a slow rate and set to restart at key press will produce much the same results. That said you will have to adjust the rate by ear so that the tone pitches properly. As a variation of this sliding bass, rather than modulate the pitch, and provided that the sine wave has been created with a self-oscillating filter, it's worthwhile sliding the filter.

This can be accomplished by setting the filter envelope to a fast attack with no sustain or release and a halfway setting on the decay parameter. By then increasing the positive depth of the envelope to filter you can control how much the bass slides during playback.

On top of this, keep in mind that changing the attack, decay and release of the amp or/and filter EG from linear to convex or concave will also create new variations. For example, by setting the decay to a concave slope will create a 'plucking' timbre while setting it to convex will produce one that's more rounded. Similarly, small amounts of controlled distortion or very light flanging can also add movement. A more creative approach, though, is to use a vocal intonation programme with experimental settings. The more extreme these settings are, the more the pitch will be adjusted, but with some experimentation it can introduce interesting fluctuations.

MOOG BASS

The Minimoog was one of the proprietary instruments in creating basses for dance music and has been used in its various guises throughout big beat, trance, hip-hop and house. Again this synthesizer is out of production and although it's unlikely that you'll find one on the second-hand market as their highly prized possessions, if you do they'll demand an extraordinarily high price. Nevertheless, this type of timbre can be constructed on most analogue-style synthesizers (emulated or real) from either using a sine or triangle wave as the initial oscillator depending on whether you want a clean rounded sound (sine) or a more gritty timbre (triangle). On top of this, add a square wave and detune it from the main oscillator by either + or −3 and set the amplifier's envelope for both oscillators to its fastest attack, a medium decay, no sustain and no release. The square wave helps to 'thicken' the sound and give it a more woody character while the subsequent amplifier setting creates a timbre that starts on key press. The decay setting acts as the release parameter for the timbre. The filter envelope to use will depend entirely on how you want the sound to appear, but a good starting point is to set the low-pass filter cut-off to medium with a low resonance and use a fast attack with a medium decay. By then lengthening the attack or shortening the decay of the filter's envelope you can create a timbre that 'plucks' or 'growls'. Depending on the type of sound you require it may also benefit from filter key follow, so the higher up the keyboard you play the more the filter opens. If you decide not to employ this, however, it is worth modulating the pitch of one, or both, of the oscillators with an LFO set to a sine wave running at slow rate and make sure that this restarts with every key press. This will create the basic Moog bass patch but it should be noted that the original Moog synthesizer employed convex slopes on the envelopes, so if you want to emulate it exactly you will have to curve the attack, decay and release parameters.

TB303

1: Acid House Bass

The acid house bass was a popular choice during the late 1980s but has been making something of a comeback in-house, techno, drum 'n' bass and chill out/ambient. Fundamentally, it was first created using the Roland TB303 Bass Synthesizer which, like the accompanying TR909 and TR808, is now out of production and, as such, demands a huge price on the second-hand market. Similar to the 909 and 808, however, there are software emulations available, with the most notable being Propellerhead's ReBirth which includes two along with the TR808 and 909. As usual, though, this timbre can be recreated in any analogue synthesizer (emulated or real) but it should be noted that due to the parameters offered by the original synthesizer, there are thousands of permutations available. As a result, we'll just concentrate on creating the two most popular sounds and you can adjust the parameters of these basic patches to suit your own music.

The acid house ('donk') bass can be created using either a square or sawtooth oscillator depending on whether you want it to sound 'raspy' (saw) or more 'woody' and rounded (square). As with most bass timbres the sound should start immediately on key press it requires the amp's attack to be set to its fastest position but the decay can be set to suit the type of sound you require (as a starting point try a medium decay and no sustain or release). Using a low-pass filter set the cut-off quite low and then slowly increase the resonance so that it sits just below self-oscillation. This will create a quite bright ('donky') sound that can be further modelled using the filter's envelope.

As with the amp settings, the filter envelope should be set so that it fits your music, but as a general starting point, a zero attack, sustain and release with a decay that's slightly longer than the amp's EG decay will produce the typical house bass timbre. Filter key follow is often employed in these sounds to create movement but it's also prudent to modulate the filter's cut-off with velocity so that the harder the key is struck the more it opens. Generally speaking, by adopting this approach you'll have to tune the subsequent harmonics into the key of the song but this is not always necessary. In fact, many dance and drum 'n' bass artists have used this timbre but made it deliberately dissonant to the music to make it more interesting.

2: Resonant Bass

The resonant bass is similar in some respects to the tone produced by the acid bass but has a much more resonant character that almost squeals. This, however, is a little more difficult to accomplish on many synths as it relies heavily on the quality of the filters and ideally these should be modelled around analogue. Start with a sawtooth oscillator and set the amplifier's EG to zero attack and sustain with a medium release and a fast decay. This sets the timbre to start immediately on key press and then quickly jump from a short decay into the release portion, in effect, producing a bass with a quick pluck. Follow this by setting the filter's envelope to a zero attack and sustain with a short release and a short decay (both shorter than the amp's settings), and set the low-pass filter's cut-off and resonance to halfway. These settings on the filter envelope introduce resonance to the decay's 'pluck' at the beginning of the note. This creates the basic timbre but it's worth employing positive filter key follow so that the filter's action follows the pitch helping to maintain some interest in the timbre. On the subject of pitch, modulating the sawtooth using a positive or negative envelope set over a 2 semitone range can help to further enhance the sound. Typically, if you want the timbre to 'bow' and drop in pitch as it plays then it's best to use a positive envelope while if you want to create a timbre that exhibits a 'sucking' motion with the pitch raising it's best to use a negative envelope.

SWEEPING BASS (TYPICAL OF UK AND SPEED GARAGE)

The sweeping bass is typical of UK garage and speed garage tracks and consists of a tight yet deep bass that sweeps in pitch and/or frequencies. These are created with two oscillators, one set to a sine wave to add depth to the timbre while the other is set to a sawtooth to introduce harmonics that can be swept with a filter. These are commonly detuned from one another but the amount varies depending on the type of timbre required. Hence, it's worth experimenting by first setting them apart by 3 cents and increasing this gradually until the sound becomes as thick as you need for the track.

As with all bass sounds, they should start the moment the key is depressed so the amp EG's attack is set to zero along with sustain but the release and decay should initially be set midway. The decay setting provides the 'pluck' while the release can be modified to suit the motif being played from the sequencer. The filter cut-off is set to a low-pass as you want to remove the higher harmonics from the signal (opposed to removing the lower frequencies first) and this, along with the resonance, is adjusted so that they both sit approximately halfway between fully exposed and fully closed. Ideally, the filter should be controlled with a filter envelope using the same settings as the amp EG but to increase the 'pluck' of the sound it's beneficial to adjust the attack and decay so that they're slightly longer than the amplifier's settings. Finally, positive filter key follow should be employed so that the filter will track the pitch of the notes being played which helps to add more movement.

These settings will produce the basic timbre but it will benefit from pitch-shifting and/or filter movements. The pitch shifting is accomplished, somewhat unsurprisingly, by modulating both oscillators with a pitch envelope set to a fast attack and medium decay but, if possible, the pitch bend range should be limited to 2 semitones to prevent it from going too wild. If you decide to modulate the filter then it's best to use an LFO with a sawtooth that ramps upwards so that the filter opens, rather than decays, as the note plays. The depth of the LFO can be set to maximum so that it's applied fully to the waveform and the rate should be set so that it sweeps the note quickly. What's more, if the notes are being played in succession it's prudent to set the LFO to retrigger on key press, otherwise it will only sweep properly on the first note and any successive notes will be treated differently depending on where the LFO is in its current cycle.

TECHNO 'KICK' BASS

Although a bass is not always used in techno, if it is, it's usually kept short and sharp so as not to get in the way of the various rhythms. That said, as there are few leads employed the bass is programmed so that it's quite deep and powerful as (if they're used) they play a major role in the music.

To create this type of bass requires four oscillators stacked together, all using the same waveform. Typically sawtooth waveforms are used but square, triangle and sine waves can also work equally well so long as the oscillators used are all of the same waveform. One waveform is kept at its original pitch while the other three are detuned from this and each other as far as possible without sounding like individual timbres (i.e. less than 20 Hz). Obviously, the sound needs to begin the moment the key is struck, so the resulting timbre is sent to an amp EG with a zero attack along with a zero sustain and release with a medium decay setting. Typically, a techno bass also exhibits a *whump* at the decay stage, which can be introduced by modulating the pitch of all the oscillators with an attack/decay envelope. Obviously this uses a fast attack so that pitch begins at the start of the note but the decay setting should be set just short of the decay used on the amp EG. Usually, the pitch modulation is positive so that the 'whump' is created by moving the pitch downwards but it's worth experimenting by setting this to negative so that it sweeps upwards. Additionally, if the synthesizer offers the option to adjust the slope of the envelopes, a convex decay is used, but experimentation is the real key with this type of bass and, in some cases, a concave envelope may produce more acceptable results. Filter key follow is rarely used as the bass tends to remain at one key but if your motif moves up or down in the range, it's prudent to use a positive key follow to introduce some movement into the riff.

TRANCE 'DIGITAL' BASS

This bass is typical of those used in many trance tracks, and while it doesn't exhibit a particularly powerful bottom end, it does provide enough of a bass element without being too rich in harmonics so that it interferes with the characteristic trance lead. It requires two oscillators, both set to square waves and detuned from each other by 3 cents to produce the basic tone. A low-pass filter is used with the cut-off set so that it's almost closed and the resonance is pushed up so that it sits just below self-oscillation. The sound, as always, needs to start immediately on key press, so the amp's attack is set to zero along with both the release and the sustain. The decay should be set about midway between being fully exposed and fully closed. The filter envelope emulates these amp settings using a zero attack, sustain and release but the decay should be set so that it's slightly shorter than the amp's decay so that it produces a resonant pluck. Finally, the filter key follow is applied so that the filter follows the pitch across the bass motif. Once constructed, if the bass is too resonant it can be condensed by reducing the filter's decay to make the 'pluck' tighter, or alternatively, you can lower the resonance and increase the filter's cut-off.

'POP' BASS

The pop bass is commonly used in many popular music tracks, hence the name, but it is also useful for some house and trance mixes where you need some 'bottom-end' presence but at the same time don't want it to take up too much of the frequencies available in the mix. These are easily created in most synthesizers by using a sawtooth and a triangle oscillator, with the triangle transposed up by an octave from the sawtooth. The amp envelope is set to an on/off status whereby the attack, decay and release are all set to zero with the sustain set just below maximum. This means that the sound almost immediately jumps into the sustain portion, which produces a constant bass tone for as long as the key is depressed (remember that the sustain portion controls the volume level of sustain and not its length!). To add a 'pluck' to the sound, a low-pass filter is usually set very low to begin with while the resonance is increased as high as possible without forcing the filter into self-oscillation. The filter envelope is then set to a zero attack, release and sustain but the decay is set quite long so that it encompasses the sustain portion of the amp's EG, in effect producing a resonant pluck to the bass. By then increasing the depth of the filter envelope along with increasing the filter's cut-off and resonance you can control how resonant the bass becomes. Depending on the motif that this bass plays it may also be worth employing some filter key-tracking so that the filter follows the pitch of the bass.

GENERIC BASS GUIDELINES

Although here we've looked at the main properties that contribute towards the creation of bass timbres and covered how to construct the most commonly used basses throughout the dance genres, occasionally simply creating these sounds will not always produce the *right* results. Indeed, much of the time you find that the bass is too heavy without enough upper harmonics or that it's too light without the right amount of depth. In these cases it's useful to use different sonic elements from different synthesizers to construct a patch – a process known as layering. Essentially this means that you construct a bass patch in one synthesizer and then create one in another (and possibly another and so forth) and then layer them altogether to produce a single patch.

It's important to understand here, though, that these additional layers should not be sourced from the same synthesizer and the filter settings for each consecutive layer should be different. The reason behind this is due to the fact that all synthesizers sound tonally different from one another, even if they use the same parameter settings. For instance, if an analogue (or analogue emulated) synthesizer is used to create the initial patch, using an analogue emulation from another manufacturer to create another bass (or using a digital synthesizer) will create a timbre with an entirely different character. If these are layered on top of one another and the respective volumes from each are adjusted you can create a more substantial bass timbre. Ideally, to prevent the subsequent mixed timbres from becoming too overpowering in the mix it's quite usual to also employ different filter types on each synthesizer. For example, if the first synthesizer is producing the low-frequency energy but it lacks any top end, the second synthesizer should use a high-pass filter. This allows you to remove the low-frequency elements from the second synthesizer so that it's less likely to interfere with the harmonics from the first. This form of layering is often essential in producing a bass with the right amount of character and is one of the reasons why many professional artists and studios will have a number of synthesizers at their disposal.

It's also worth bearing in mind that it's inadvisable to apply any stereo-widening effects to bass timbres. This is because the bass should sit in the centre of the mix (for reasons we'll touch upon in the mixing chapter) but it's sometimes worthwhile applying controlled distortion to them to increase the harmonic content. As some basses are constructed from sine waves that contain no harmonics, they can become lost in the mix but by applying distortion, the upper harmonics introduced can help the bass cut through the mix. This distortion can then be accurately controlled using filters or EQ to mould the bass to the timbre you require. What's more, some effects such as flangers or phasers require additional harmonics to make any noticeable difference to the sound, something that can be accomplished by applying distortion before the flanger or phaser. Of course, if these effects are applied it's sensible to ensure that they're applied in mono, not stereo to prevent the bass becoming spread across the stereo image.

Finally, as with the drum timbres if you have access to a synthesizer that allows you to adjust the linearity of the envelope's attack and decay stage you should certainly experiment with this. In fact, a convex or concave decay on the filter and/or amp is commonly used to produce bass timbres with a harder pluck allowing it to pull through a mix better than a linear envelope.

> The data CD contains audio examples of these timbres being programmed.

PROGRAMMING LEADS

Saying that basses were difficult to encapsulate is only the tip of the sound design iceberg since trying to define what makes a good lead is impossible. Every track will use a different lead sound ranging from Daft Punk's distorted or phased leads to the various plucks used by artists such as Paul Van Dyk through to the hundreds of variations on the euphoric trance leads. Consequently, there are no definitive methods to creating a lead timbre but it is important to take plenty of time producing one that sounds *right*. The entire track rests on the quality of the lead and it's absolutely vital that this sounds precise. However, while it's impossible to suggest ways of creating new leads to suit any one particular track, there are some rough generalizations that can be applied.

Firstly, most lead instruments will utilize a fast attack on the amp and filter envelope so that it starts immediately on key press with the filter introducing the harmonics to help it to pull through a mix. The decay, sustain and release parameters, though, will depend entirely on the type of lead and sound you want to accomplish.

For example, if the sound has a 'pluck' associated with it then the sustain parameter, if used, will have to be set quite low on both amp and filter EGs so that the decay parameter can drop down to it to create the pluck. Additionally, the release parameter of the amp can be used to determine whether the notes of the lead motif flow together, or are more staccato (keep in mind that staccato notes appear louder than those that are drawn out). If the release is set so that the notes flow together it isn't unusual to employ portamento on the synthesizer so that notes rise or fall into the successive ones.

As the lead is the most prominent part of the track it's usually sits in the mid-range, and is often bursting with harmonics that occupy this area. The best approach to accomplish this is to build a harmonically rich sound by using sawtooth, square waves, triangle and noise waveforms stacked together and make use of the unison feature (if the synthesizer has it available). This is a form of stacking a number of the synthesizer's voices together to produce thicker and wider tones but it also reduces the polyphony available to the synthesizer, so you have to exercise caution as to the polyphony available to the synthesizer. Once a harmonically rich voice is created it can then be thinned, if required, with the filters or EQ and modulated with the envelopes and LFOs. These latter modulation options play a vital role in producing leads as they need some sonic movement to maintain interest.

Typically, these methods alone will not always provide a lead sound that is rich or deep enough; it's worth employing a number of methods to make it 'bigger' such as layering, doubling, splitting, hocketing or residual synthesis. We've already looked at some of the principles behind layering when looking at basses but with leads this can be stretched further as there is little need to keep the lead under any real control – its whole purpose is to sit above every other element in the mix! Alongside layering the sounds in other synthesizers, it's often worth using different amp and/or filter envelopes in each synthesizer. For example, one timbre could utilize a fast attack but a slow release or decay parameter, while the second layered sound could utilize a slow attack and a fast decay or release. When the two are layered together, the harmonic interaction between the two sounds produces very complex timbres that can then be mixed together in a desk and EQ'd or filtered externally to suit.

Doubling is similar to layering, but the two should not be confused as doubling does not use any additional synthesizer but the one you're using to programme the sounds. This involves copying the MIDI information of the lead motif to another track in the MIDI sequencer and transposing this up to produce a much richer sound. Usually transposing this copy up by a fifth or an octave will produce musically harmonious results, but it is worth experimenting by transposing it further and examining the effects it has on the sound. A variation on this theme is to make a copy of the original lead and then only transpose some notes of the copy rather than all of them so that some notes are accented.

Hocketing consists of sending the successive notes from the same musical phrase to different synthesizers or patches within the same module to give the impression of a complex lead. Usually, you determine which synthesizer receives the resulting note through velocity by setting one synthesizer or patch to not accept velocity values below, say, 64 while the second synthesizer or patch will only accept velocity values above 64. If this is not possible, then simply copying the MIDI file onto different tracks and deleting notes that should not be sent to the synthesizer will produce the same results.

Splitting and residual synthesis are the most difficult to implement but often produce the best results. Fundamentally, splitting is similar in some respects to layering but rather than produce the same timbre on two different synthesizers, a timbre is broken into its individual components which are then sent to different synthesizers. For example, you may have a sound that's constructed from a sine, sawtooth and triangle wave but rather than have one synthesizer do this, the sine may come from one synthesizer, the triangle from another and the saw from yet another. These are all modulated in different ways using the respective synthesis engines, but by listening carefully to the overall sound through a mixing desk, the sound is constructed and manipulated through the synthesizer's parameters and mixing desk as if it were from one synthesizer. Residual synthesis, on the other hand, involves creating a sound in one synthesizer and then using a band-pass or notch filter to remove some of the central harmonics from the timbre. These are then replaced using a different synthesizer or synthesis engine and recombined at the mixer.

Finally, effects also play a large part in creating a lead timbre. The most typical of these that are used on leads are reverb, delay, phasers, flangers, distortion and chorus, but experimentation with different effects and even the order of the effects can all produce great results.

As always the key to producing great leads, as with all other sounds, is through experimentation and familiarization with effects and the synthesis engines you use.

EUPHORIC TRANCE LEAD

The euphoria trance lead is probably the most elusive lead to programme properly but, in many cases, this is simply because it cannot be recreated on any synthesizer. To capture the sound properly requires an analogue synthesizer (emulated or real) with oscillators that have the right character for the genre. This means that you should use a synthesizer that employs methods to dynamically alter both the tone and the pitch of the oscillators in a slightly different manner every time you hit a key. This method, often referred to as phase initialization, produces the characteristics distinctive of any good analogue synthesizer. Most software or hardware analogue emulations will, or should, employ this but the amount of initialization depends entirely on the synthesizer and for trance leads, the higher this is, the better the results will be. As a side note, the most commonly used synthesizer to create the trance lead is the access virus (in fact, nearly all professional trance musicians will use the virus!) but the Novation SuperNova, Novation A station, Novation K station or the Novation V station (the software VST Instrument of the K station) can also produce the requisite timbres with a little extra work.

Alongside the 'unreliable' feature of analogue oscillators, the real secret behind creating any good trance lead is through clever use of effects and noise, not the bad type of noise, of course, but the noise produced by an oscillator in the synthesizer. As we've already touched upon, noise produces a vast number of harmonics, which is essential to creating the hands in the air vibe.

A basic trance lead timbre can be constructed with four oscillators. Two are set to pulse and detuned to produce a wide hollow sound while the third is a sawtooth to add further harmonics, and the fourth is set to create noise to add some 'top end' fizzy harmonics to the sound. The two pulse waves are detuned from one another by detuning one to −5 cents and the other to +5 cents to produce a wide sound while the saw is detuned from these by a full octave to add some bottom-end harmonics to the timbre. The final oscillator is left as it is but can use either pink or white noise depending on how fizzy you want the harmonics it produces to be. Generally, pink noise is the best choice since it contains a huge range of harmonics from low to high while white noise is more like radio static and tends to be a little too light and fizzy for trance leads.

Obviously the note needs to start as soon as the MIDI note is received so the amp's attack is set to zero but to create a small pluck use a medium decay with a small sustain and a fairly short release. This release can be lengthened or shortened further depending on the trance riff you've programmed. The filter envelope can use the same settings as the amplifier, although in some cases it may be worth shortening the decay to produce a better 'plucking' sound. This envelope should be applied from ¾ to full as positive modulation to the filters and the filter key follow should be on so that they track the pitch of the notes being played. The filter itself is set to a low-pass with a high cut-off but a low resonance to prevent the timbre from squealing and it should use a 4-pole filter rather than the usual 2-pole, so that the filter sweeps sharply, helping the lead become more prominent and allowing it to cut through the mix.

To add some energy and interest to the timbre, it's an idea to modulate the pulse width of both pulse oscillators and, if at all possible, you should use two LFOs so that each pulse can be modulated differently. The first LFO modulates the pulse width of the first oscillator with a sine wave set to a slow-to-medium rate and full depth, while the second modulates the pulse width of the second oscillator. This, however, is set to a triangle wave with a faster rate than the first LFO and at a full depth. The resulting effect is that both pulse waves beat against each other creating a more interesting timbre. Finally, the timbre will need to be washed in both reverb and delay to provide the required sound. The reverb should be applied quite heavily as a send effect but with 50 ms of pre-delay so that the transient pulls through undisturbed and the tail set quite short to prevent it from washing over the successive notes. It's also prudent to set a noise gate to remove the subsequent reverb tail as this will produce a heavier timbre which cuts through the mix and prevents it from becoming pushed into the background (an effect known as Gate-Tailing). Decay should also be applied but again this is best used as a send effect so that only a part of the timbre is sent to the delay unit. The settings to use will depend on the type of sound you require but the delays should be set to less than 30 ms to produce the granular delay effect to make the timbre appear big in the mix.

PLUCKED LEADS

Plucked leads are often used in most genres of music from house through trance but can be typified in most of Paul Van Dyk's work. This is not to say that they will all sound like Paul Van Dyk, however, as they can be constructed in numerous ways depending on the type of sound you want to accomplish and the genre of music you write. Thus, what follows are two of the most commonly used basic patches for plucked leads in dance which, as always, you should manipulate further to suit your own music.

Plucked Lead 1

The first plucked lead consists of three oscillators, two of which are set to saw-tooths to add plenty of harmonics and the third set to either a sine or triangle wave to add some bottom-end weight. The choice between whether to use a sine or triangle is up to you, but using a sine will add more bottom end and is useful if the bass used in the track is rather thin while if the bass is quite heavy, a triangle may be better as this will add less of a bottom end and introduce more harmonics. The sawtooth waves are detuned from each other by 3 cents (or further if you want a richer sound) and the third oscillator is transposed down by an octave to introduce some bottom-end weight into the timbre. Obviously, the amp's attack is set to zero as is the sustain portion but both decay and release are set to a medium depth depending on the amount of pluck you require and the motif playing the timbre. A low-pass filter is used to remove the higher harmonics of the sound and this should be set initially at midway while the resonance should be set quite low. This is controlled with a filter envelope set to a zero attack and sustain but the decay and release are set just short of the amplifier's decay and release settings so if you adjust the amp's decay and release, you'll also need to reduce the filter's decay and release. Finally, it's worthwhile applying some reverb to the timbre to help widen it a little further but if you take this approach you'll also have to employ a tailing gate to prevent the lead from being pushed too far back into the mix.

Plucked Lead 2

The second plucked lead is a little easier to assemble and consists of just two oscillators, a saw and a triangle. The saw wave is pitched down to produce a low end while the triangle wave is pitched up to produce a slight glistening effect. The amount that these are detuned by depends on the timbre you require so you'll need to experiment through detuning them by different amounts until the frequencies sit into the mix you have so far. As a general starting point, detune them from one another as far as you can without the sounding as two different timbres and then reduce this tuning until the timbre fits to the music. Due to the extremities of this detuning sync both oscillators together to prevent the two from beating too much and then apply a positive pitch envelope to the saw wave with a fast attack and medium decay so that it pitches up as it plays. The sound needs to start on key press so set the attack to zero with no attack, sustain or release and set the decay to halfway. This envelope is copied to the filter envelope but set the decay to zero and increase the sustain parameter to about halfway. Finally, using a low-pass filter set the cut-off to around three-quarters open and set the resonance to about a quarter. This will produce the basic pluck timbre but it may be worth using the synthesizer's unison feature to thicken the sound out further depending on the mix. If this isn't possible granular delay can be used in its place to widen the timbre.

TB303 LEADS

Although we've already seen the use of a TB303 when looking at programming basses, they are a versatile machine and are also equally at home creating lead sounds by simply pitching the bass frequencies up by a few octaves. The most not-able example of this was on Josh Wink's *Higher State of Consciousness* and this same effect can be recreated on analogue (or analogue emulated) synthesizer. Only one oscillator is required and this is a sawtooth due to the high number of harmonics required for it to cut through a mix. The sound should start immediately on key press, so it requires the amp's attack to be set to its fastest position but the decay can be set to suit the type of sound you require (as a starting point try a medium decay and no sustain or release). Using a low-pass filter set the cut-off quite low and the resonance so that it sits just below self-oscillation. The filter envelope is set to a fast attack with no sustain or release, but the decay needs to be set so that it's just short of the amp's decay stage. Filter key follow is often employed to create additional movement but it's also prudent to modulate the filter's cut-off with velocity so that the harder the key is struck the more it opens.

Finally the timbre is run through a distortion unit and the results are filtered with an external or plug-in filter to finally mould the sound. In some instances, it may also be worth increasing the amplifier's decay so that the notes overlap each other and then switch on portamento so that they slur into each other while the riff is played.

DISTORTED LEADS

There are literally hundreds of distorted leads but one of the most popular is the distorted/phased lead used in many house tracks and, in particular, by Daft Punk and Tomcraft. The basic patch is created using single sawtooth or if you need it slightly thicker, two saws detuned from one another by 5 cents. The amp envelope is set to a zero attack with a full sustain, a short release setting and the decay to around a quarter. As always, the decay will have the most influence over the sound so it's worth experimenting with to produce the sound you need. The filter cut-off should be set quite low, while the resonance should be pushed up pretty high to produce overtones which can be distorted with effects in a moment. The filter's envelope will need adjusting so that it reacts with the sound, so set the attack, sustain and release to zero and the decay so that it's slightly longer than the amp's decay setting. Ideally, the filter's envelope should fully affect the sound so as a starting point set this so that it fully affects the filter and then experiment by reducing the amount until the basic timbre you want appears.

Distortion and phaser are also applied to the sound, but as previously touched upon, the distortion should come first so that the subsequent phaser also works on the distortion. It is important, however, not to apply too much distortion otherwise it'll may overpower the mix and become difficult to mix. Preferably, you should aim for a subtle but noticeable effect before finally applying a phaser to the distorted signal. How much to apply will obviously depend on the sound you require but exercise caution that you do not apply too much otherwise the solidity of the timbre can be lost. In some cases it may also be worth applying portamento with a slow rate to enable the sound to slew into one another which, as a result, brings more attention to the phased effect.

THEREMINS

For those with no idea of what a theremin is, it consists of a vertical metal pole approximately 12–24″ in height which responds to movements of your hands and creates a warbling low-pitched whistle dependent on your hand's position. The sound was used heavily in the 1950s and 1960s sci-fi movies to provide a 'scary' atmosphere when the aliens appeared but has made many appearances in lo-fi music. The most notable example of this use in music can be heard on Portishead's *Mysteron*.

These are incredibly simple to create and can be reproduced on any synthesizer using a sawtooth wave with a low-pass filters cut-off parameter reduced until it produces a soft constant tone. On some digital synthesizers you may need to raise the resonance to produce the correct tone but most analogue synthesizers will produce the tone with no resonance at all. This is also the example of one lead where the amplifier's attack is not set at zero; instead this is set at halfway along with the release while the sustain is set to full. Theremins are renowned for their random variations in pitch, so you'll need to emulate this by holding down a key (around C3) and using the pitch wheel to introduce the variations from waving your hands around. If the synthesizer has access to portamento then it's also advisable to use this to recreate the slow shifting from note to note.

HOOVERS

The hoover sound originally appeared on the Roland Juno synthesizers and has been constantly used throughout all genres of dance music including techno, house, acid house, breakbeat, drum 'n' bass and big beat. In fact, these still remain one of the most popular sounds used in dance music today. Originally in the Juno they were aptly named *What the…* but due to their dissonant tonal qualities they were lovingly renamed 'Hoover' sounds by dance artists since they tend to share the same sound as vacuum cleaners… Honestly…

The sound is best constructed in an analogue/DSP synthesizer as the tonal qualities of the oscillators play a large role in creating the *right* sound. Two sawtooth oscillators are used to create the initial patch and these need to be detuned as far from each other as possible but not so far that they become two individual timbres.

The sound starts immediately, so the amp's attack is set at zero mixed with a short decay and release and sustain set to just below full so that there is a small pluck evident from the decay stage. The filter should be a low-pass with a low cut-off and a medium resonance setting which is modulated slightly with the filter's envelope. This latter envelope is set to a medium attack, decay and release but no sustain, but you need to exercise caution as to how much this envelope modulates the filter, since settings that are too high will result in a timbre that's nothing like a hoover. To add the typical dissonance feel that hoovers often exhibit, the pitch of both oscillators will also need modulating using a pitch envelope set to a fast attack and a short decay. This, however, should be applied negative rather than positive so that the pitch bends upwards into the note rather than downwards. Finally, depending on the synthesizer recreating the timbre it may need some widening, which can be accomplished by washing the sound in chorus or preferably stacking as many voices as possible using a unison mode.

'HOUSE' PIANOS

The typical house piano was drawn directly from the Yamaha DX range of synthesizers and is usually left unmodified (due to the complexity and painful experience that is programming with frequency modulation). In fact, if you want to use the house piano, the best option is to either purchase an original Yamaha DX7 synthesizer or alternatively invest in Native Instruments FM-8 VST instrument. This is a software emulation of the FM synthesizers produced by Yamaha which can also import the sounds from the original range of DX synthesizers.

Acquiring the FM piano is difficult on most analogue synthesizers and often impossible of most digital synthesizers (well apart from the DX range of course!) because of the quirks of FM. Nevertheless, if you want to give this a go it can be accomplished by using two oscillators set to sine waves with one of the oscillators detuned so that it's at a multiple of the second oscillator. These are then frequency modulated to produce the general tone. The amp envelope is then set to a fast attack, short decay and release and a medium sustain with the filter key-tracking switched on. To produce the initial transient for the note a third sine wave pitched high up on the keyboard and modulated by a one shot LFO (i.e. the LFO acts as an envelope – fast attack, short decay, no sustain or release) will produce the desired timbre. As a side note to this, if you want to produce the infamous bell-like or metallic tones made famous by FM synthesizers, use two sine oscillators with one detuned so that its frequency is at a non-related integer of the second and then use FM to produce the sound.

ORGANS

Organs are commonly used in the production of house and hip-hop, with the most frequent choice being the Hammond B-4 drawbar organ. This general timbre, however, can be emulated in any subtractive synthesizer by using a pulse and a sawtooth oscillator. As the sound starts on key press the amp uses a zero attack with a full sustain and medium release (note that there is no decay since the sustain parameter is at maximum). A low-pass filter is used to shape the timbre with the cut-off set to zero and the resonance increased to about halfway. A filter envelope is not employed since the sound should remain unmodulated; but if you require a 'click' at the beginning of the note you can turn the filter envelope to maximum but the attack, release and sustain parameters should remain at zero with a very, very short decay stage. Finally the filter key follow should be set so that the filter tracks the current pitch which will produce the typical organ timbre that can then be modified further.

GENERIC LEADS

So far, we've looked at some of the most popular timbres used for dance leads but as previously touched upon, there are literally thousands of combinations available, so what follows is a brief overview of how to produce the basic patches that can be used as the basic building blocks to further develop upon.

1. As a starting point to trance, techno, big beat and lo-fi leads, mix a sawtooth oscillator with a square wave and detune them by at least 3 to produce a wide timbre. The amp EG is set to a fast attack with no sustain and a medium release with a short decay. The filter's (usually a low-pass to keep the body of the sound) cut-off is set low with a medium resonance setting with the envelope using a fast attack, medium decay, low sustain and no release. The amount the filter envelope modulates the filters will determine much of the sound, so start by using a high value and reduce it as necessary. Employ filter key follow so that the filter tracks the pitch and then experiment with LFOs modulating the filters and pitch. If the sound still appears thin at this stage, make use of the unison feature or add another square or sawtooth to produce a thicker sound.

2. As a starting point for hip-hop (assuming that you don't want to sample it) and chill-out a triangle wave mixed with a square or saw wave detuned by 5 or 7 will produce the basic harmonic patch. Using a low-pass filter, set the cut-off and resonance quite low and set the key-tracking to full. For the amplifier envelope, use a fast attack with a medium decay and release and a sustain set just below the decay parameter. For the filter envelope, it's prudent to use an attack that's slightly longer than the amp's attack stage, along with a medium decay, low sustain and a release just short of the amp. Sync both oscillators together and use an LFO to add some vibrato to the filter and the oscillators.

3. As a starting point for UK garage, try starting with a square, sine and triangle wave each detuned from each other as much as possible without them appearing as distinct individual timbres. From this, use a low-pass filter with the cut-off set quite high and no resonance, and set the key-tracking to full. Use a zero attack, long decay, medium sustain and no release for both the filter and the amp envelopes and modulate the pitch of the triangle wave with either positive or negative values.

4. The basic timbres that the vintage techno and house sounds were based around can be easily created on any analogue/DSP synthesizer by detuning

saws, triangles or square waves by a 3rd, 5th or 7th. The choice of which of these oscillators to use obviously depends on the type of sound you require but once elected they should be detuned by an odd amount. The amplifier EG is generally set at a zero attack, sustain and release while the decay is used to shape the overall sound. The filter envelope is normally set to match the amp's EG, although if you require a pluck in the sound, the decay should be set slightly shorter than the amp's decay. The filter is always set at low-pass and a good starting point is to use a high resonance with a cut-off set about midway. Key follow is employed so the filter tracks the keyboards pitch and it's quite common to employ a pitch envelope on both oscillators so that they pitch either up or down while playing.

> The data CD contains audio examples of these timbres being programmed.

PROGRAMMING SOUND EFFECTS

Sound effects can play a fundamental role in the production of dance music as they have a multitude of uses from creating drops and adding to builds to sitting in the background and enhancing the overall mix. Indeed, their importance in dance music production should not be underestimated since without them a mix can sound dry, characterless or just plain insipid.

For creating sound effects in a synthesizer, the most important parameter on a synthesizer is the LFO as this can be used to modulate various parameters of any sound. For instance, simply modulating a sine wave's pitch and filter cut-off with an LFO set to S&H or noise waveform will produce strange burbling noises while a sine wave LFO modulating the pitch can produce siren-type effects. There are no limits and as such no real advice on how to create them as it comes from experimenting with different LFO waveforms modulating different parameters and oscillators. Generally, though, the waveform used by the LFO will contribute a great deal to the sound effect you receive. Triangle waves are great for creating bubbly, almost liquid sounds, while saws are suitable for zipping type noises. Square waveforms are good for short percussive snaps such as gunshots and random waves are particularly useful for creating burbling, shifting textures. All of these used at different modulation depths and rates, modulating different oscillators and parameters will create wildly differing effects.

Additionally, if you're using wave editors you shouldn't discount their uses in creating sound effects. Most wave editors today feature their own synthesis engines that allow you to create and mix raw waveforms that can then be affected and treated to any plug-in effects you may own. For example, Steinberg's WaveLab test signal generator is very similar to an additive synthesizer, allowing you to create sounds containing up to 64 different layers of waveforms and treat each individually to different frequency, tremolo and vibrato values. There's little need to use this many layers as this will often results in a sound that's far too complex to be used as a sound effect so you don't have to worry about typing in 64 different values for each.

Normally, most sound effects can be acquired from using just two or three waveforms and some thoughtful and creative thinking. The real solution to creating effects is to experiment with all the options at your disposal and not be afraid of making a mess. Even the most unsuitable noises can be tonally shaped with EQ and filters to be more suitable.

To help you along in your experiments, what follows is a small list of some of the most popular effects and how they're created. Unfortunately, though, there are no specific terms to describe sound effects so what follows is a rough description of the sound they produce, but from what we've covered so far in this chapter simply reading about how the effect is created should give you a good idea of what they will sound like. And, of course, if you're still not sure, try creating them to see how they sound…

SIREN FX

The siren is possibly the easiest sound effect to recreate. Set one oscillator to produce a sine wave and use an amp envelope with a fast attack, sustain and release and a medium decay. Finally, use a triangle wave or sine wave LFO to modulate the pitch of the oscillator at full depth. The faster the LFO rate is set, the faster the siren will become.

WHOOPING FX

To create a 'whooping effect' use one oscillator set to a sine wave with a fast attack, no sustain or release and a longish decay on the amp EG. Modulate the pitch of this sine wave with a fast attack and long decay set to a positive amount and then programme a series of staccato notes into a MIDI sequencer. Use these to trigger an arpeggiator set to one octave or less so that it constantly repeats the same notes in a fast succession to create a *whoop, whoop, whoop* effect.

ZAP FX

To create a zapping effect turn the filter cut-off down to zero and increase the resonance until it breaks into self-oscillation, creating a pure sine wave. Set the amplifier's attack to a fast attack, sustain and release and a medium decay, and use these same settings on the filter envelope. Set this latter envelope to fully modulate the filter and then use either a triangle or saw wave LFO set at a medium speed and full depth to modulate the filter's cut-off.

EXPLOSION FX

To create explosive type effects requires two oscillators. One oscillator should be set to a saw wave while the other should be set to a triangle wave. Detune the triangle from the saw by +3, +5 or +7 and set a low-pass filter to a high cut-off and resonance (but not so high that the filter self oscillates). Set the amp's envelope to a medium attack, sustain and release but with a long decay and copy these settings to the filter envelope but make the decay a little shorter than the amp's EG. Use a sawtooth LFO set to a negative amount and use this to control the pitch of the oscillators along with the filter's cut-off. Finally, use FM from the saw onto the triangle and play low down on the keyboard to produce the explosive effects. Notably, the quality of the results from this will depend on the synthesizer being used. Most digital and a few analogue synthesizers don't produce a good effect when FM is used so the effect will differ from synthesizer to synthesizer.

ZIPPING FX

This effect is quite popular in all forms of dance and is created with two oscillators and an LFO to modulate the filter frequency. Start by selecting a saw and triangle as the two oscillator waveforms and detune one from the other by +7. Use a fast attack and release along with a medium-to-long decay and medium sustain on the amp's EG and using a low-pass filter, set the cut-off quite low but use a high resonance. Finally, set a sawtooth LFO (using a saw waveform that moves from nothing to maximum) at full depth to slowly modulate the filter's cut-off and, if possible use an envelope to modulate the LFOs speed so that it gradually speeds up as the note plays. If this is not possible, you'll have to increase the speed of the LFO manually and record the results into a sampler or audio sequencer.

RISING SPEED/FILTER FX

Another popular effect is the rising filter, whereby its speed increases as it opens further. For the oscillator you can use a saw, pulse or triangle wave but a self-oscillating filter provides the best results. Set both the amp and filter EG to a fast decay and release but a long attack and high sustain. Use a triangle or sine LFO set to a positive mild depth and very slow rate (about 1 Hz) to modulate the filter's cut-off. Finally use the filter's envelope to also modulate the speed of the LFO so that as the filter opens the LFO also speeds up. If the synthesizer doesn't allow you to use multiple destinations, you can increase the speed of the LFO manually and record the results into a sampler or audio sequencer.

FALLING SPEED/FILTER FX

This is basically the opposite of the previously described effect so that rather than the filter rising and simultaneously speeding up, it falls while simultaneously slowing down. Again, set both the amp and filter EG to a fast decay and release with a long attack but don't use any sustain. Use a triangle or sine LFO set to a positive mild depth and fast rate to modulate the filter's cut-off. Finally use the filter's envelope to also modulate the speed of the LFO so that as the filter closes the LFO also slows down. If the synthesizer doesn't allow you to use multiple destinations, you can decrease the speed of the LFO manually and record the results into a sampler or audio sequencer.

DALEK VOICE

The Dalek voice isn't just for Dr Who fans as it can be used in place of a vocoder to produce more metallic style voices that are suitable for use in any dance genre. This can be accomplished by recording your voice into a sampler or sequencer and feeding it, along with a low-frequency sine wave, into a ring modulator. You can experiment with the results of this by changing or using cyclic modulation on the pitch of the sine wave entering the ring modulator.

SWEEPS

Sweeps are generally best created using sawtooth oscillators as their high harmonic content gives the filter plenty to work with. Start by using two oscillators both set to sawtooth waves and detune them as far as possible without them becoming individual timbres and feed the results into a ring modulator. On the amp's envelope, use a fast attack, decay and release but set the sustain parameter to maximum and then use a slow saw or triangle wave LFO to modulate the pitch of one of the oscillators and the filters. For the filters, a band-pass will produce the best results set to a medium cut-off but a very high resonance (but not so high that is self oscillates). Finally, use a second saw, sine or triangle LFO to modulate the filter's cut-off to produce the typical sweeping effect.

GHOSTLY NOISES

To create ghostly noises from a synthesizer use two oscillators both set to triangle waves. Using a low-pass filter set the cut-off quite low but employ a high resonance and set the filter to track the keyboards pitch (filter key follow). Adjust the amp's EG to a fast attack with a long decay, high sustain and medium release and set the filter's envelope to a fast attack, long decay but no sustain or release. Finally, using an LFO sine wave very slowly (about 1 Hz) modulate the pitch of the oscillators and play chords in the bass register to produce the timbres.

COMPUTER BURBLES

Computer burbling noises can be created in a number of ways but by far the most popular method is to use two oscillators both set to triangle waves. Detune one of these from the other by $+3$ cents and then set the filter envelope to a medium attack with a long decay with a medium release and no sustain. Do the same for the amp's EG but use a high sustain and set the filter's envelope to positively but mildly modulate a low-pass filter with a low cut-off and a high resonance. Finally ensure that filter key-tracking is switched on and modulate the pitch of one, or both, of the oscillators with a noise or sample and hold waveform. This should initially be set quite fast will a full depth but it's worth experimenting with the depth and speed to produce different results.

SWOOSH FX

To create the typical swoosh effect (often used behind a snare roll to help in creating a build-up) the synthesizer will need both filter and amp envelopes but also a third envelope that can be used to modulate the filter as well. To begin with, switch all the oscillators off, or reduce their volume to zero, and increase the resonance so that the filter breaks into self-oscillation. Use a filter envelope with no decay, a medium release and attack and a high sustain and set it to positively affect the filter by a small amount and then use a second envelope with these same setting to affect the filter again, but this time set it to negatively affect the filter by the same amount as before. Set the amp's EG to no attack or sustain but a small release time and a long decay and use a saw or triangle wave LFO set to a medium speed and depth to positively modulate the filter's cut-off. This will create the basic 'swoosh' effect but if possible employ a second LFO set to a different waveform from the previous one to negatively modulate the filter by the same amount.

The data CD contains audio examples of these timbres being programmed.

PROGRAMMING NEW TIMBRES

Of course, there will come a time when you want to construct the timbre that's in your head but for this it's important to note that you need to be experienced in identifying the various sounds produced by oscillators and the effects they have when mixed together. Also you need to be able to identify the effect positive and negative envelope can impart on a sound along with the effects produced by an LFO augmenting parameters. If you have this knowledge then constructing sounds isn't particularly complicated and just requires some careful consideration.

Firstly, when you're attempting to come up with sounds on your own, it's important to be able to conceptualize exactly what you want the instrument to be. This means that you need to imagine the completed sound (and this is much more difficult than you envisage!) and then take it apart in your mind by asking a series of questions such as:

- Does the timbre start immediately?
- How does it evolve in volume over time?
- Is the instrument synthetic, plucked, struck, blown or bowed?
- What happens to its pitch when it's sounded?
- Are there any pitches besides the fundamental that stand out enough to be important to the timbre?
- Does it continue to ring after the notes have been sounded?
- How bright is the sound?
- How much bass presence does the sound have?
- Does it sound hollow, rounded, gritty or bright and sparkly?
- What happens to this brightness over time?
- Is there any modulation present?
- What does the modulation do to the sound?

The answer to all these should be written down on paper otherwise you can easily loose track and wind up going in another direction altogether. As a more practical example of how this form of sound design is accomplished, we'll tear a sound apart and reconstruct it.

> The data CD contains audio examples of these timbres being programmed.

We need to begin by examining a sound and then break it down into its subsequent parts. For this we can use the following chart as a general guideline:

General Sound	Technical Term	Synthesizers Parameter
Type of sound	Harmonic content	Oscillators waveforms
Brightness	Amplitude of harmonics	Filter cut-off and resonance
Timbre changes over time	Dynamic filtering	Filter's envelope
Volume changes over time	Dynamic amplitude	Amplifier envelope
Pitch	Frequency	Oscillators pitch
Sound has a cyclic variation	LFO modulation	LFO waveform, depth and rate
Tremolo (cyclic variation in volume)	Amplitude modulation	LFO augments the amplifier
Vibrato (cyclic variation in pitch)	Pitch modulation	LFO augments the pitch
Sound is percussive	Transient	Fast attack and decay on the amplifier
Sound starts immediately or fades in	Attack time	Lengthen or shorten the attack and decay stages
Sound stops immediately or fades out	Release time	Lengthen or shorten the release on the amplifier
Sound gradually grows 'richer' in harmonics	Filter automation	Programmed CC messages or a slow rate LFO augmenting the filter cut-off

A useful step when first starting out the process of recreating timbres is to try and emulate the sounds properties with your voice and mouth and record the results. As madcap as this probably sounds (and to those around you while you try it), not only will you have a physical record of the sound but the movements of the mouth and tongue can often help you determine much about the sound you're trying to construct. The expansion and contraction of the mouth's muscles can often be related to the filter, while movement of the tongue can often be related to the LFO augmenting a parameter.

For this example, the sound is similar to saying *wwOOOwww* (listen to the timbre and replicate it with your voice and mouth and this will make a little more sense).

From this we can determine that there is some filter augmentation in the timbre since it opens and closes and that when it does, it allows higher harmonics through as it sweeps. This means that it's using a low-pass filter but as there is no evidence of the characteristics of using an LFO waveform to modulate the timbre (for instance a saw waveform would be evident by opening slowly and closing suddenly etc.) it must be the filter envelope that's modulating the filter's cut-off. What's more, it's quite easy to determine that the filter envelope is being applied positively rather than negatively otherwise it would sound entirely different.

Therefore, we've determined so far that

- The filter is augmented with an envelope and that it's positive.
- The filter is a low-pass.

Next, listening to the sound's overall timbre it's obviously quite rich in harmonics all of which change with the movements of the filter, so there is no sine wave present. Additionally the timbre doesn't exhibit a particularly hollow character or the bright 'fizzy' harmonics related with noise, so it's safe to assume that there is no square or noise waveform present either. This (generally) leaves two waveform options: a saw or a triangle. Finally, as we can hear the filter moving in and out at key press, and it most definitely has a pluck to it, we can also determine that the amp envelope is using a fast attack with a long decay, little sustain, if any and a short release.

Thus we have

- The timbre consists of either a saw or triangle wave, or both.
- The amplifier envelope utilizes a fast attack, a long decay, perhaps a small sustain and a short release.
- The filter is augmented with an envelope and that it's positive.
- The filter is a low-pass.

At this point, it's worth inputting these parameters into a synthesizer and experimenting with the oscillators and envelopes to see how close to the timbre you can get with the parameters you've theorized thus far. If you're not pitch perfect it may first be worth setting the synthesizers to a single sine oscillator with the aforementioned envelope settings and finding the key that the riff is in first as this will help things along no end. Further experimentation should also reveal that the sound consists of both a saw and a triangle wave that are detuned from each other to make the timbre appear as wide as it does (the saw is transposed down an octave from the triangle).

Now that we have the basic timbre and amplifier settings, we can listen back to the sound again. The sound isn't particularly resonant, so the resonance must be set quite low. Also the filter is sweeping the harmonic content quite rapidly, so the filter must be 24 dB rather than 12 dB. What's more, listening to the way that the filter's envelope augments the filter's cut-off we can determine that the envelope is using a long attack and decay with a medium sustain and release.

The final result is that to recreate the timbre two oscillators are used: a sine and a triangle wave with the saw wave transposed down by an octave. The amp envelope uses a fast attack with a long decay, no sustain and a fast release while the filter envelope has a longish attack and decay with a medium sustain and release.

GENERAL PROGRAMMING TIPS

- Never bang away at a single key on the keyboard while programming, it will not give you the full impression of the patch. Always play the motif to the synthesizer before any programming.
- Ears become accustomed to sounds very quickly and an unchanging sound can quickly become tedious and tiresome. Consequently, it's prudent to introduce sonic variation into long timbres through the use of envelopes or LFOs augmenting the pitch or filters.
- Generally speaking the simpler the motif, the more movement the sound should exhibit. So for basses and motifs that are quite simple assign the velocity to the filter cut-off, so the harder the key is hit the brighter the sound becomes.
- Don't underestimate the uses of keyboard tracking. When activated this can breathe new life into motifs that move up or down the range as the filter's cut-off will change.
- Although all synthesizer share the same parameters, they do not all sound the same. Simply copying the patch from one synthesizer to another can produce totally different results.
- Although the noise oscillator can seem useless in a synthesizer when compared to the saw, sine, triangle and square waves, it happens to be one of the most important oscillators when producing timbres for dance and is used for everything from trance leads to hi-hats.
- To learn more about the character of the synthesizers you use, dial up a patch you don't like, strip it down to the oscillators and then rebuild it using different modulation options.
- If you have no idea of what sound you require, set every synthesizer parameter to maximum and then begin lowering each parameter to sculpt the sound into something you like.

- Many bass sounds may become lost in the mix due to the lack of a sharp transient. In this instance, synthesize a click or use a woodblock timbre or similar to enhance the transient.
- The best sounds are created from just one or occasionally two oscillators modulated with no more than three envelopes: the filter, the amplifier and the pitch. Try not to overcomplicate matters by using all the oscillators and modulation options at your disposal.
- Try layering two sounds together but use different amplifier and filter settings on both. For example, using a slow attack but quick decay and release on one timbre and a fast attack and slow decay and release on another will *hocket* the sounds together to produce interesting textures.
- Never underestimate the uses of an LFO. A triangle wave set to modulate the filter cut-off on a sound can breathe new life into a dreary timbre.
- Remember that we determine a huge amount of information about a sound from the initial transient. Thus, if you replace the attack of a timbre with the attack portion of another it can create interesting timbres. For instance, try replacing the attack stage of a pad or string with a guitar's pluck.
- For more interesting transients, layer two oscillators together and at the beginning of the note use the pitch envelope to pitch the first oscillator up and the second one down (i.e. using positive and negative pitch modulation).
- For sounds that have a long release stage, set the filter's envelope attack longer than the attack and decay stage but set the release slightly shorter than the amp's release stage. After this, send the envelope fully to the filter so that as the sound dies away the filter begins to open.

In the end programming good timbres comes from a mix of experience, experimentation and serendipity. What's more, all of the preceding examples are just that – examples to start you on the path towards sound design and you should be willing to push the envelope (pun intended) much further. Try adding an extra oscillator, changing the filters to high pass or band select, adjust the amp and/or filter envelopes and try changing the LFOs modulation source, destination and/or waveform. The more time you set aside to experiment, the more you'll begin to understand not only the characteristics of your synthesizer but also the effects each controller can impart on a sound. This will ultimately be time well spent as professionally programmed sounds will make the difference between a great dance track and an average one.

CHAPTER 5
Digital Audio

'Why go digital? I come from a breed of musical technicians that couldn't live without getting the latest technology on the market...'

Anonymous

While programming synthesizers is obviously important, dance music relies on more than just this; today, it also relies on working directly with audio. Whether the audio is sampled from another record, from film, or from TV or recorded directly from the real world, the capability to record the results accurately can make the difference between a hit track and a middle-of-the-road demo track.

With digital audio now at the forefront of technology and being used by all dance musicians, it is imperative that you know how to perform digital recording accurately and precisely. Therefore, before we look at introducing real-world audio into your mix, be it through samplers or recording real-world instruments/vocals, it is imperative that you have a thorough understanding of how to capture the audio as perfectly as possible.

Indeed, when recording anything directly into a sampler, computer soundcard or any digital recording device, you will get the best results by combining high-level recordings with good sample and bit rates.

With any digital recording system, sound has to be converted from an analogue signal (i.e. the sound you hear) to a digital format that the device can work with. Any digital recording device accomplishes this by measuring the waveform of the incoming signal at specific intervals and converting these measurements into a series of numbers based on the amplitude of the waveform. Each of these numbers is known as an 'individual sample' and the total number of samples that are taken every second is called the 'sample rate' (Figure 5.1).

On this basis, the more the samples taken every second, the better the overall quality of the recording. For instance, if a waveform is sampled 2000 times/s,

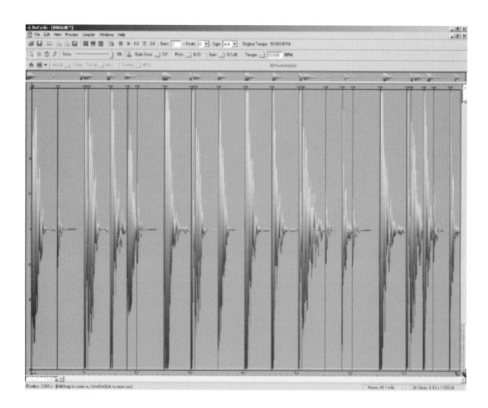

FIGURE 5.1
Frequency

FIGURE 5.2
Example of sample rate measurements

it will produce more accurate results than if it were sampled 1000 times/s (Figure 5.2).

While this may seem basic in principle, it becomes more complex when you consider that the sampling rate must be higher than the frequency of the waveform being recorded in order to produce accurate results. If it isn't, then the analogue-to-digital converter (ADC) could miss anything from half to entire cycles of the waveform, resulting in a side effect known as 'aliasing'.

This is the result of a real-world audio signal not being 'measured' in the correct places. For instance, a high-frequency waveform could confuse the ADC into believing it's actually of a lower frequency, which would effectively introduce a series of spurious low-frequency spikes into the audio file.

To avoid this problem, you must make sure the sampling rate is greater than twice the frequency of the waveform: a principle called the Nyquist–Shannon theorem. This states that to recreate any waveform accurately in digital form, at least two different points of a waveform's cycle must be sampled. Consequently, as the highest range of the human ear is a frequency of approximately 20 kHz, the sample rate should be just over double this range. This is the principle from which CD quality is derived.

That is, the sample rate of a typical domestic audio CD is 44.1 kHz which is derived from the calculation:

Human hearing limit = 20 000 Hz

20 000 Hz × 2 = 40 000 Hz + 4 100 Hz (to make the rate more than twice the optimum frequency and compensate for the anti-alias filter slope).

Although this frequency response has become the de facto standard used for all domestic audio CDs, there are higher sampling frequencies available consisting of 48 000, 88 200, 96 000 and 192 000 Hz.

Though these sampling rates are far beyond the frequency response of CD, it is quite usual to work at these higher rates while processing and editing. Although there has been no solid evidence to support the theory, it is believed that higher sample rates provide better spatial signal resolution and reduce phase problems at the high-frequency end of the spectrum as the anti-alias filters can be made much more transparent due to gentle cut-off slopes. Thus, if the ADC supports higher rates, theoretically, there is no harm in working at the highest available rate. Notably, many engineers argue that because the signal must be reduced to 44.1 kHz when the mix is put onto CD, the downward conversion may introduce errors, so working at higher sampling rates is pointless. If you do decide to work at a higher sample rate, it may be worthwhile using a rate of 88 200 kHz, as this simplifies the down-sampling process.

BIT RATES

In addition to the sample rate, the bit rate also determines the overall quality of the results of recording audio into a digital device. To comprehend the importance of this, we need to examine the binary language used by all computers.

All computers utilize the binary language, that consists of two values, 1 and 0. The computer can count up to a specific number depending on how many bits are used. For example, if an 8-bit system is used, it is possible to count to a maximum of 256 values (0–255), while in a 24-bit system the maximum value is 16 777 215 (Table 5.1).

Relating this to a digital audio recording system, the number of bits determines the number of analogue voltages that are used to measure the volume of a waveform, in effect increasing the overall dynamic range.

Table 5.1	Bit Rates						
8-Bit	**Binary**	**16-Bit**	**Binary**	**20-Bit**	**Binary**	**24-Bit**	**Binary**
0	1	0	1	0	1	0	1
1	1	1	1	1	1	1	1
2	1	2	1	2	1	2	1
4	1	4	1	4	1	4	1
8	1	8	1	8	1	8	1
16	1	16	1	16	1	16	1
32	1	32	1	32	1	32	1
64	1	64	1	64	1	64	1
128	1	128	1	128	1	128	1
		256	1	256	1	256	1
		512	1	512	1	512	1
		1024	1	1024	1	1024	1
		2048	1	2048	1	2048	1
		4096	1	4096	1	4096	1
		8192	1	8192	1	8192	1
		16384	1	16384	1	16384	1
		32768	1	32768	1	32768	1
				65536	1	65536	1
				131072	1	131072	1
				262144	1	262144	1
				524288	1	524288	1
						1048576	1
						2097152	1
						4194304	1
255		65535		1048575		16777215	

FIGURE 5.3
Example of bit depth

In technical applications the dynamic range is the ratio between the residual noise (known as the noise floor) created by all audio equipment and the maximum allowable volume before a specific amount of distortion is introduced.[1]

In relation to music, the dynamic range is the difference between the quietest and loudest parts. For instance, classical music utilizes a huge dynamic range by suddenly moving from very low to very high volume, the cannons in the *1812 Overture* being a prime example. Most dance and pop music, on the other hand, has a deliberately limited dynamic range so that it remains permanently up front and 'in your face'. Essentially, this means that if a low bit rate is used for a recording, only a small dynamic range will be achieved. The inevitable result is that the ratio between the noise floor and the loudest part of the audio will be small, so background noise will be more evident. When the bit rate is increased, each additional bit introduces another analogue voltage, which adds another 6 dB to the dynamic range.

For example, if only one bit is used to record a sound, the recorder will produce a square wave at the same frequency as the original signal and at fixed amplitude. This is because only one voltage is used to measure the volume throughout the sampling frequency. However, if an 8-bit system is used, the signal's dynamic range will be represented by 256 different analogue voltages and a more accurate representation of the waveform will result. It's clear, then, that a 24-bit audio signal will have a higher dynamic range than a 16-bit signal (the bit rate used by CDs) (Figure 5.3).

At the time of writing, although 24-bit is the highest resolution available to samplers' and soundcards' ADCs, a proportionate amount of audio-capable software utilizes internal 32- or 64-bit processing. The reasoning behind using bit rates this high is that whenever any form of processing is applied to a digitized audio signal, quantization noise is introduced. This is a result of the hardware rounding up to the nearest measurement level. The less rounding up the hardware has to do (i.e. the more bits used), the less noise is introduced.

[1]According to the IEC 268 standard, the dynamic range of any professional audio equipment is measured by the difference between the total noise floor and the equivalent sound pressure level where a certain amount of total harmonic distortion (measured in THD) appears.

This is a cumulative effect, meaning that as more digital processing is applied, more quantization noise is introduced, and quantization noise needs to be kept to a minimum. In more practical terms, this means that while a CD may only accept a 16-bit recording, if a 24-bit process is used throughout digital mixing, editing and processing, when the final sound is dropped to 16-bit resolution for burning to CD the quantization noise will be less apparent.

The process of 'dropping out' bits from a recording to reduce the bit rate is known as 'dithering'. Understanding how this actually works is not vital; what is important is that the best available dithering algorithms are used. Poor-quality algorithms will have a detrimental effect on the music as a whole, resulting in clicks, hiss or noise. As a reference, Apogee is well known and respected for producing excellent dithering algorithms.

It isn't always necessary to work at such a high bit rate; some genres of dance music benefit from using a much lower rate. For instance, 12-bit samples are often used in hip-hop to obtain the typical hip-hop sound. This is because the original artists used old samplers that could only sample at 12-bit resolution; thus, to write music in this genre it's quite usual to limit the maximum bit rate in order to reproduce these timbres. Similarly, with trip-hop and lo-fi, the sample rate is often lowered to 22 or 11 kHz, as this reproduces the gritty timbres that are characteristic of the genre.

Ultimately, no matter what sample or bit rate is used, it's important that the converters on the soundcard or digital recorder are of a good standard and that the amplitude of the signal for recording is as loud as possible (but without clipping the recorder). Although all digital editors allow the gain of a recorded signal to be increased after recording, the signal should ideally be recorded as loud as possible so as to avoid having to artificially increase a signal's gain using software algorithms. This is because all electrical circuitry, no matter how good or expensive, will have some residual noise associated with it due to the random movement of electrons. Of course, the better the overall design the less random movement there will be, but there will always be some noise present so the ratio between the signal and this noise should be as high as possible. If not, when you artificially increase the gain after it has been recorded, it will increase any residual noise by the same amount.

For instance, suppose a melodic riff is sampled from a record at a relatively low level, then the gain is artificially increased to make the sound audible; the noise floor will increase in direct proportion to the audio signal. This produces a loud signal with a loud background noise. If, however, the signal is recorded as loud as possible, the ratio between the noise floor and the signal is greatly increased. There is a fine line between capturing a recording at optimum gain and actually recording the signal too loud so that it clips the recorder's inputs.

This isn't necessarily a problem with analogue recorders as they don't immediately distort when the input level gets a little too high – they have 'headroom' in case the odd transient hit pushes it too hard – but digital recorders will 'clip' immediately. Unlike analogue distortion, which can often be quite pleasant, digital distortion cuts off the top of the waveform, resulting in an ear-piercing clicking sound. This type of distortion obviously should be avoided, so you need to set a recording level that is not loud enough to cause distortion, yet not low enough to introduce too much noise.

All recording software or hardware, including samplers, will display a level metre informing you how loud the input signal is, to help you determine the appropriate levels. Before beginning to record, it's vital to set this up correctly. This is accomplished by adjusting the gain of the source so that the signal overload peaks light on only the most powerful parts and then backing off slightly so that they are just below the clipping level. This approach is suitable only if there isn't too much volume fluctuation throughout the entire recording, though. If a sample is taken from a record, CD or electronic instrument there's a good chance that the volume will be constant, but if a live source, such as vocals or bass guitars, are recorded there can be a huge difference in the dynamics. It's doubtful that any vocalist, no matter how well trained, will perform at a constant level; in the music itself it's quite common for the vocals to be softer in the verse sections than they are in the chorus sections.

If the recording level is set so that the loudest sections don't clip the recorder, any quieter section will have a smaller signal-to-noise ratio, while if the recording levels are set so that the quieter sections are captured at a high level, any louder parts will send the metres into the red. To prevent this, it's necessary to use a device that will reduce the dynamic range of the performance by automatically reducing the gain on the loud parts, while leaving the quieter parts unaffected, thus allowing the recording levels to be increased on the quietest parts.

This dynamic restriction is accomplished by strapping a compressor between the signal source and the input of the sampler or recording device. When the compressor is set to squash any signal from the source above a certain threshold, any peaks that could cause clipping in the recorder are reduced and the recording can be made at a substantially higher volume.

CHAPTER 6

Sampling and Sample Manipulation

'Culture, like science and technology, grows by accretion, each new creator building on the works of those who came before. Over-protection stifles the very creative forces it's supposed to nurture.'

Alex Kozinski
(Chief Judge of the Federal Claims Court)

The introduction of samplers in the 1980s changed the face of music forever. In fact, they have made such a significant impact on music that all commercially produced music today, not just dance music, has been at least part realized with the aid of one. This isn't solely due to the numerous genre-specific sample CDs that are available, although these do often play a role, but because samplers are one of the most creative tools you can have at your disposal.

With some creative thought, a recording of a ping-pong ball being thrown at a garage door can become a tom drum, hitting a rolled up newspaper against a wall can make an excellent house drum kick and the hissing from a compressed air freshener can produce great hi-hats. As a way of making music, sampling can be quick, easy and an immensely creative resource.

At a very basic level a sampler is the digital equivalent of an analogue tape recorder, but rather than recording the audio signal onto a magnetic tape, it is recorded digitally into random access memory (RAM) or directly onto a hard disk drive. After a sound has been recorded, it can then be manipulated using a series of editing parameters, very similar to those on synthesizers, and also played back at varying pitches from any attached controller keyboard. The variations in pitch are created by the sampler, artificially increasing or decreasing the original frequency of the sampled sound according to the key that is pressed. To raise the pitch the frequency is increased while the pitch is lowered as the frequency is decreased.

This principle is analogous to the way that analogue tape behaves, whereby the speed that the tape is played alters the pitch. However, as with analogue tape, if the sample's speed is increased or decreased by a significant amount it will no longer sound anything like the original source. For instance, if a single-key strike from a piano is sampled at C3 and is played back at this same pitch from a controller keyboard, the sample will play back perfectly. If, however, this same sample were played back at C4 the sampler would increase the frequency of the original sound by 12 semitones (from 130.81 to 523.25 Hz) and the result would sound more like two spoons knocking together than the original recording.

These extreme pitch adjustments aren't necessarily a bad thing – particularly for a creative endeavour such as dance music – but to recreate a real instrument, sampling a single note doesn't provide a realistic sounding instrument throughout the octaves. Indeed, most samplers only reproduce an acceptable instrument sound four or five keys from the original root key, so if an original instrument is needed throughout the keyrange, samples should be taken every few keys of the original source instrument. This is called 'multi-sampling' and when the source is sampled at every couple of keys and the samples assigned to the same keys in the sampler. For example, with a piano it is prudent to sample the keys at C0, E0, G0 and B0, then C1, E1, G1, B1 and so forth until the entire range has been sampled.

Naturally, recording a sound in this way would equate to over 16 samples and the more samples that are taken, the more memory the sampler must have available. Because most hardware (and some software) samplers hold the sampled sounds in their onboard RAM, the maximum sampling time is limited by the amount of available memory. At full audio bandwidth (20 Hz–20 kHz) and a 44.1 kHz sampling rate, 1 min of mono recording will use approximately 5 megabytes (MB) of RAM. In the case of sampling a piano, this would equate to 80 MB of memory, and you can double this if you wanted it in stereo! Consequently, over the years samplers have adopted various techniques to make the most of the available memory, the first of which is to loop the samples.

As the overall length of a sample determines the amount of RAM that is required, reducing the sample's length means more samples will fit into the memory space. Because most sounds have a distinctive attack and decay period but the sustain element remains consistent, the sustain portion can be continually looped for as long as the key is held, moving to the release part after the key is released. This means that only a short burst of the sustain period must be sampled, helping to conserve memory. Well, that's the theory anyway. In practice sustain looping can prove particularly difficult.

The difficulty arises from the fact that what appears to be a consistent sustain period of a sound is, in most cases, rarely static due to slight yet continual changes in the harmonic structure. If only a small segment of this harmonic movement is looped, the results would sound unnatural. Conversely, if too long a section is looped in an effort to capture the harmonic movements, the decay or some of the release period may also be captured and again, when looped, the final sound will still be unusual. In addition, any looping points must start and end at the same phase and level during the waveform. If not, the difference in phase or volume could result in an audible click as the waveform reaches the end of its looped section and jumps back to the beginning.

Some samplers offer a work around to this latter problem and automatically locate the nearest zero crossing to the position you choose. While this increases the likelihood that a smoother crossover is achieved, if the waveform's level is different at the two loop points there will still be a glitch. These glitches can, however, be avoided if the sampler has a cross-fading feature.

Cross-fading works by fading out the end of the looped section and overlapping this fade-out with the start of the loop that's fading in. This creates a smooth crossover between the two looping points, reducing the possibility that glitches are introduced. Although this goes some way to resolve the problems associated with looped samples, this is not necessarily the best solution because if the start and end of the looped points are at different frequencies there will be an apparent change in the overall timbre during the cross-fade. Unfortunately, there is no quick fix for avoiding the pitfalls associated with successfully creating a looped segment, so success can only be accredited to patience, experimentation and experience.

Having said that, some samplers now utilize 'sample streaming', this eliminates the need to squeeze all the samples into available RAM. Often referred to as 'ROMplers', samples are stored on a local hard disk and only the start of the sample is held in the RAM. By using a small buffer, the RAM begins playback after the key is pressed and is constantly updated with the new data directly from the hard disk, allowing samples of any length to be played back. With today's disk drives capable of storing up to and above 200 gigabytes (GB) of data, it's possible to store massive multi-samples. Indeed, it's quite usual for ROMplers to use sample sets made up of the source instrument sampled at every note, at different velocities, and with the foot pedals (if applicable) in different positions.

It'll come as no surprise to learn that this form of key multi-sampling is incredibly time consuming to undertake yourself and without access to the right equipment, impossible to accomplish accurately. Even most professional dance artists do not have the experience or equipment required to successfully multi-sample real instruments. With this in mind, if real instruments are required it's much easier to buy a collection of well-recorded sounds on a sample CD than to attempt to build them yourself.

SAMPLE CDS

Sample CDs are produced in several formats to suit different samplers and essentially fall into four categories:

- Audio CD,
- Wave/Aiff CD (WAV),
- CD-ROM and,
- Sample-based instruments.

FIGURE 6.1
East West's Goliath
Instrument

Due to the popularity of using PCs and Macs for music, Wave CD formats are generally the most popular format as these can be copied digitally from the CD into a compatible software sampler where they can then be key mapped and edited further. Sample-based instruments are beginning to take over from the standard Wave style CDs. Although these cannot be directly compared to a sampler, they are software instruments that contain a number of samples already placed along the keyboard. For example, East West's *Goliath* is a software instrument containing over 35 GB (Gigabytes!) of sampled instruments, each instrument is held in a bank (similar to a sampler or other instrument) and the instruments are pitched right across the keyboard meaning you do not have to sample any instruments yourself (Figure 6.1).

Audio CDs can be played by any conventional CD player and can be recorded into any sampler hardware or software. Alternatively, if a software sampler is used, the audio can be 'ripped' into the computer and used directly with no loss in audio quality and no need to constantly check the recording levels to ensure that you have the best signal-to-noise ratio.

The disadvantage, however, is that after the sounds have been sampled, they must be named, loops created if needed, cross-fades created and key ranges defined. This can be incredibly time-consuming, taking anything from a few hours to over a few days to set up a multi-sampled instrument, depending on the sampler's interface.

This is where sample CD-ROMs offer the advantage, though they are considerably more expensive than both Wave and audio CDs. Because the samples are all stored on a CD in a data format the samples are already named, key mapped and looped (if necessary) to suit the sampler. This means that it's not necessary to spend time arranging the samples across the keyboard, setting up loops and so on. The data files are simply loaded into the sampler ready for work.

While this has the obvious advantage of easing the sampling process, typically, there are also some disadvantages. The data on the CD must be compatible with the operating system used by the sampler. This isn't so much of a problem with the most established sampler manufacturers such as AKAI and E-MU since these are cross-compatible with each other to a certain extent. For instance, the AKAI S6000 sampler will import CDs that are designed for use with E-MU samplers (and vice versa), but keep in mind that there may be slight discrepancies when using a format-specific CD across different samplers in this way.

Even though the basic operating systems of most samplers are quite similar, each has slightly different parameters on offer. If these have been exploited in the CD-ROM then they may not interpret well on a different sampler. The most common consequence is a difference in the timbre, simply because of different filter algorithms and settings employed by different samplers but in more severe cases, the key-mapping may not come out quite as expected, the sustain samples may not loop properly, or velocity modulation may not work correctly.

Although sample CDs have some quite obvious advantages (and are quite heavily used in all dance music genres), it's important to understand the various clauses involved in their commercial use.

When a sample CD is purchased, the cost includes a licence to use the sounds in your own projects. This licence does not include the copyright to the sounds as these belong to the manufacturer. You are simply granted a licence to use them. It's important to understand the content of this licence because if a mistake is made and a commercial track using them is released, you could be in for a heavy ride for copyright infringement.

Most sample CDs specifically mention the 'musical context' for their use, meaning that they can be used in musical productions without having to notify the original artist or pay royalties but you cannot use them for library music or other sample CDs. It is vitally important that you read all the terms, though, since some sample CDs have quite unusual terms ranging from acknowledging the sample's source on the documentation accompanying a commercial releases, asking the copyright owner for permission or paying them a share of the royalties from every sale! These latter clauses are rare but always check first.

BASIC SAMPLING PRACTICES

Nearly all the sample CDs that are aimed at dance musicians consist of single hits rather than multi-sampled instruments. This is because the bass and melodies in dance music are usually simple with very little movement throughout the octave. Also, single hit samples of synthesized basses and leads, along with most woodwind and brass instruments, can be successfully stretched over more keys than string instruments such as pianos, violins, harps or guitars.

Nevertheless, before any of these are imported into a sampler, the 'root' key must be set. This is the key that the sampler will use as a reference point for stretching the subsequent notes up or down along the length of the keyboard. For instance, bass samples are typically set up with the root key at C1, meaning that if this key is pressed the sample will play back at its original frequency. This can then be stretched across the keyboard's range as much as required. In most hardware samples, the root key is set in the keyrange or keyzone page along with the lowest and highest keys available for the sample. Using this information the sampler automatically spreads the root note across the defined range and the bass can be played within the specified range of notes. To gain the best results from the subsequent pitch adjustments on each key, it is preferable that the sampler's keyrange is set to 6 notes above and below the root. This allows the sample to be played over the octave, if required, and will produce much more natural results than if the root note were stretched 12 semitones up or down. Even if you're not planning to play the instrument over an octave, it's still prudent to set the keyrange to span an octave anyway, as this makes more pitches available if they're needed later.

When setting a keyzone for bass sounds, it may be possible to set the range much lower than 6 semitones, as pitch is determined by the fundamental frequency and the lower this is the more difficult it is to perceive pitch. Thus, for bass samples it may be possible to set the lowest key of the keyzone to 12 semitones (an octave) below the root without introducing any unwanted artefacts. In fact, pretty much anything sounds good if it's pitched down low enough.

Most competent samplers will also allow a number of keyzones to be set across the keyboard. This makes it possible to have, say, a bass sample occupying F0 to F1 and a lead sample occupying F2 to F3. If set up is in this way then both the bass and lead can be played simultaneously from an attached controller keyboard. In taking this approach, it is worth checking that each keyzone can access different parameters of the sampler's modulation engine. If this is not possible, settings applied to the bass will also apply to the lead. To avoid this it's prudent to allocate each keyzone to an individual MIDI channel. Because the majority of samplers are multitimbral, it's usually possible to set different synthesis parameters on each MIDI channel.

After the samples are arranged into keyzones, you can set how the sampler will behave depending on how hard the key is hit on a controller keyboard. This uses 'velocity modulation', which is useful for recreating the movement within sounds. The immediate use for this is to set the velocity to react with a low-pass filter so that the harder the key is hit the more the filter opens. This can be used to add expression to any riffs or melodies and prevent the static feel that is often a consequence of using a series of sampled sounds.

'Velocity cross-fading' and 'switching' are also worth experimenting with, if the sampler has these facilities. Switching involves taking two samples and using velocity values to determine which one should play. The two samples are imported into the same keyrange and hitting the key harder (or softer) changes between the two samples. Velocity cross-fading uses this same principal but morphs the two samples together creating a (hopefully) seamless blend rather than an immediate switch between one and the other. Although velocity cross-fading produces more convincing results it also reduces the polyphony of the sampler as it has to play two samples together while it cross-fades between them. Both these velocity-related parameters are typically used to emulate the nuances of real-world instruments such as a piano where the harder a note is struck the more the tonal content changes.

WORKING WITH LOOPS

Alongside single synthesizer hits, the majority of sample CDs also feature a number of programmed or recorded drum loops to suit the genre of the CD. These can be imported/recorded into a sampler and subsequently re-triggered continually to create the effect of a constant drum track to play throughout the length of the music. More interestingly, because samplers store the audio in RAM (or on a hard drive) once a loop is recorded the sound will start over from the beginning every time a key is pressed. Thus, if a key on a controller keyboard is tapped continually the loop will start repeatedly, producing a stuttering effect. This technique can be used to create breakdowns in dance music. Alternatively, the sampler can be set to 'one-shot trigger mode', so that a quick tap on a controller keyboard plays the sample in its entirety even if the key is released before the sample has finished playback. This is useful if you want the original sample to play through to its normal conclusion while triggering the same sample again to play over the top. This technique can be used to create complex drum loops and break beats; a few instances of the same loop are playing simultaneously, each out of time and perhaps pitch with one another.

If either of these methods are used to trigger a drum loop or motif to play throughout the length of the track, it is important to ensure that the sample is not looped from within the sampler but is, instead, re-triggered every couple of bars by the sequencer. Bear in mind that whenever a sample is triggered, the only part that will be in sync with the rest of the music is the start of the sample. After this, you're relying entirely on the sampler's internal clock to keep in time with the sequencer. This may not necessarily be an issue if the sampler can be locked to a MIDI clock, but if this is not possible the two could drift apart. For example, if a 2-bar drum loop is sampled and programmed to constantly repeat for as long as a MIDI note-on is present, unless it has been looped with sample accuracy the timing will begin to drift as the track continues. While a couple of milliseconds drift near the beginning of the track may be unnoticeable, over the length of a 5 or 6 min track the timing could drift out by a couple of seconds.

This form of looping, however, shouldn't be confused with the term '*phrase sampling*' as this consists of sampling a short musical phrase that is only used a couple of times throughout a song. This method is commonly used in dance music for sampling a short vocal phrase or hook that is then triggered occasionally to fit with the music. This phrase sampling technique can also be developed further to test new musical ideas. By slicing your own song into a series of 4-bar sections and assigning each loop to a particular key on the keyboard you can trigger the loops in different orders so you can determine the order that works best. What's more, if 4 bars of each individual track of the mix (drums, bass, lead, etc.) are sampled, each one can be played back from the keyboard, allowing you to quickly change the timing of each track in relation to the others. Many club records are constructed this way, but with the sampler's one-shot triggering deactivated so that either the entire track or the groove of the record can be 'stuttered' to signify a break or build in the music.

So far, we've assumed that the phrase or loop that has been sampled is at the same tempo and/or pitch as the rest of the mix but more often than not this isn't the case. Accordingly, all samplers provide pitch-shifting and time-stretching functions so that the pitch and tempo of the phrase or loop can be adjusted to fit with the music. Both these functions are pretty much self-explanatory: the pitch-shift function changes the pitch of notes or an entire riff without changing the tempo and time stretching adjusts the tempo without affecting the pitch.

These are, of course, both useful tools to have but it's important to note that the quality of the results is proportionate to how far they are pushed. While most pitch-shifting algorithms remain musical when moving notes up or down the range by a few semitones, if the pitch is shifted by more than 5 or 6 semitones the sound may begin to exhibit an unnatural quality. Similarly, while it should be possible to adjust the tempo by 25 BPM without introducing any digital artefacts, any adjustments above this may introduce noise or crunching that will compromise the audio.

Due to the limitations of time stretching, it's often more sensible to 'comp' parts together to increase or decrease their speed. This involves importing the audio loop into the sequencer of choice and cutting each constituent hit to create a number of single hits. These can then be spaced apart to fit appropriately with the tempo and small amounts of time stretching/compressing can be applied to each hit to make the loop fit together again. This method of comping is particularly important when working with vocals. By cutting them into separate words and sometimes even syllables, it's possible to correct small nuances in timing which produces a better end result.

It can also be advantageous to sample some of your own synthesizer's sounds, although the reason for taking this approach may not be immediately obvious.

For example, if you want to play a timbre polyphonically but the synthesizer is monophonic (can only play one sound at any one time) or monotimbral (can only play one type of sound at one time) and you want to use more than one type of sound from the same synthesizer, the sampler will allow these possibilities. Multi-sampling the synthesizer at every key essentially recreates the synthesizer in the sampler allowing for polyphonic or multitimbral operation. In some instances the sampler may also even offer more synthesis parameters than are available on the synthesizer enabling you to, say, synchronize the LFO to the tempo, use numerous LFOs or access numerous different filter types.

There's more to sampling a synthesizer than simply recording the timbre into a sampler, though, and creating an acceptable multi-sampled (or even single-keyed) instrument requires careful consideration. For starters, it's prudent to expose the filter (by setting the cut-off fully open and resonance fully closed so that they have no effect on the timbre), deactivate any LFO modulation, and set the amplifier envelope to its fastest attack and longest release. By doing so, a relatively static sound will be easier to sustain in a loop, and the envelope, filter and LFO movements can be replicated using the sampler's own synthesis engine.

While this approach is generally the best, there will undoubtedly be occasions where it isn't possible to remove the modulation (such as sampling from previous 'hit' records) in which case you'll have to exercise care. Bear in mind that any timbre that evolves over time will be difficult to map across a keyboard because as the pitch is adjusted by playing over the keyboard, the modulation rate will increase or decrease. This can be particularly apparent when an LFO has been used to augment the timbre, because the timing may be slightly out making any sustain looping impossible to achieve. Unfortunately, there is no direct way of overcoming these limitations short of applying a series of effects such as reverb or distortion in an attempt to mask some of the modulation.

CREATING PROFESSIONAL LOOPS FROM SINGLE HITS

Many sample CDs not only contain drum loops but also single hits. Often when you cannot find a loop appropriate for the music, you may prefer to use the individual sampled hits to create your own loops. This, however, can be more difficult than it sounds.

The big 'secret' behind getting a good, professional sounding loop is to keep the length of the hits short to add a dynamic edge to the rhythm. Keep in mind that long drum sounds cover the gaps between each of the drum hits and this lessens the impact of the rhythm. Jumping from silence to sound then back again to silence will have a much more dramatic impact than sounds occurring directly one after the other. What's more, long samples can also cause problems with the bass line as a deep kick will tend to merge with the bass line resulting in a muddy sounding bottom end, so it's prudent to keep the kick drum snappy as this will introduce space between the hits resulting in a more dynamic rhythm.

Generally speaking, if the bass line is quite heavy, the drums are tighter and more controlled whereas if the bass is quite bright and breezy the drums are more 'sloppy' with less space between hits (i.e. by increasing the decay).

This can be accomplished by reducing the decay of the less important percussion (such as toms, claves, tambourines, etc.) and then moving onto the more important elements such as the kick, snare and hi-hats. It's also worthwhile sampling/bouncing down the loop and then run a noise gate across the loop.

By setting the threshold so that most of the percussive elements lie above it and experimenting with the hold and decay times it's possible to introduce character to loops that were initially quite superficial. Heavier beats or rhythms can also be made by applying light reverb over the loop as a whole and then employing a noise gate to remove the reverbs tail. The gate can also be used as a distortion tool for drums provided that it's set up correctly.

If you play back a loop into a noise gate and set the attack, hold and release parameters to zero the gate will open and close in quick succession often resulting in the gate following the individual cycles of the low-frequency kick. As a result, the waveforms that fall below the threshold become a series of square waves while the peaks remain unmolested, resulting in a distorted loop. The amount of distortion can then be controlled by lowering or raising the threshold.

Compression can also be used to produce a crunchy distortion that is particularly suited towards big beat, hip-hop and techno drum loops. For this, two compressors are used in serial with the drum loop fed into the first compressor set to a high ratio and low threshold mixed with a fast attack and release. If the output gain of this compressor is pushed high enough it results in distortion of the mid-range which adds the typical vintage character of these genres. By feeding this output into a second compressor, the distortion can be controlled to prevent it from clipping the inputs of the recording device (or the outputs of a wave editor).

Generally, an opto compressor, such as the Waves Renaissance Compressor or the Sonalksis TBK3, produces the best results but it is worth experimenting with other vintage-style compressors. It's also worth noting that hip-hop, lo-fi and big beat will also often use 'crunchy', dirty timbres which is best accomplished by lowering the bit rate to 12-bit or the sample rate to 22 kHz prior to compression or distortion. This replicates the 'feel' of these rhythms as many are commonly sampled from other records.

On that point, although I can't condone lifting samples from previous records because of the legal consequences, it would be incredibly naive to suggest that some house and hip-hop artists, in particular, do not sample drum loops and sometimes entire grooves from other vinyl records. In fact, in many instances this sampling is absolutely paramount in attaining the 'feel' of the music.

Although there are plenty of sample CDs dedicated to both these genres, they are generally best avoided as everyone will have access to the very same CDs. The trick is to find old records that have preferably only been pressed in a small amount due to their obscurity. Indeed, the more obscure the better as it's unlikely that any other producers will have access to the same records and if required, you'll be able to get copyright permission much more easily. These records can be sourced in the majority second-hand and charity shops but try to ensure that the records are over 20 years old as it's the character of the sound that matters the most.

Once you have a good collection (over 100 records is often classed as an average collection) you'll need to listen to each for an exposed segment of the drums to sample. Although you can sample these straight off the record and use them as it is, it's much better to be a little more creative and drop the loop into a sequencer to comp the parts together. Once this is accomplished the rhythm can be shifted around and the timing can be adjusted to produce new variations.

Of course, comping loops in this fashion can be incredibly time-consuming and, in some instances, may not be possible if the loop is quite complex or played live as it's likely that hi-hats may occur partway through a snare or kick. If you were to cut these and move them the rhythm may lose its cohesion and sound dreadful so as an alternative it may be worthwhile layering other hits over the top of the original loop or write a new pattern to sit over the original. If you take this latter approach, however, you need to exercise caution as it's a fine line between a great loop and a poor one. The most common problem experienced is that the loop becomes too complex and if this is the case it's prudent to place the loop in a wave editor or sequencer and reduce the volume of the offending parts.

CREATIVE SAMPLING AND EFFECTS

The main key with sampling is to be creative. While many musicians simply rely on the sampler to record a drum loop (which can be accomplished with any sequencer) they are not using them to their full extent and experimentation is the real key to any sampler. Therefore, what follows are some general ideas to get you started in experimental sampling.

Creative Sampling

Although cheap microphones are worth avoiding if you want to record good vocals they do have other creative uses. Simply plugging one into a sampler and striking the top of the mic with a newspaper or your hand can be used as the starting point of kick drums. Additionally scratching the top of the microphone can be used as the starting point to Guiro's. You should also listen for sounds in the real world to sample and contort – FSOL have made most of their music using only samples of the real world. Todd Terry acquired snare samples by bouncing a golf ball off a wall, and Mark Moore used an aerosol as an open hi-hat sound. Hitting a plastic bin with a wet newspaper can be used as a thick slurring kick drum, and scraping a key down the strings of a piano or guitar can be used as the basis for whirring string effects (it worked for Dr Who's TARDIS anyway). Once sampled these sounds can be pitched up or down, or effected as you see fit. Even subtle pitch changes can produce effective results. For example, pitching up an analogue snare by just a few semitones results in the snare used on drum 'n' bass, while pitching it down gives you the snare typical of lo-fi.

Sample Reversing

This is probably the most immediate effect to try but while it's incredibly simple to implement it can produce great results. The simplest use of this is to sample a cymbal hit and then reverse it in the sampler (or wave editor) to produce a reverse cymbal that can be used to signify the introduction of a new instrument. More creative options appear when you consider that reversing any sound with a fast attack but long decay or release will create a sound with long attack and an immediate release.

For example, a guitar pluck can be reversed and mixed with a lead that isn't. The two attack stages will meet if they're placed together and these can be cross-faded together. Alternatively, the attack of the guitar could be removed so that the timbre begins with a reverse guitar which then moves into the sharp attack phase of a lead sound or pad. On the other hand, if a timbre is recorded in stereo, the left channel could be reversed while the right channels stays as is to produce a mix of the two timbres. You could then sum these to mono and apply EQ, filters or effects to shape the sound further.

Pitch Shifting

A popular method used by dance producers is to pitch-shift a vocal phrase, pad or lead sound by a fifth as this can often create impressive harmonies. If pads or leads are shifted further than this and mixed in with the original it can introduce a pleasant phasing effect. Despite the advice from some, however, this will not work well on drum loops. Keep in mind that these form a crucial part of the track and should be as dynamic as possible.

Time Stretching

Time stretching adjusts the length of a sample while also leaving the pitch unchanged. It does this by cutting or adding samples at various intervals during the course of the sample so that it reaches the desired length, while, to a certain extent, smoothing out the side-effects of this process on the quality and timbre of the sound. This is a complex, processor-intensive, process and is not usually suitable for extreme stretching. For instance, stretching a 67 BPM loop into 150 BPM introduces unpleasant digital noise into sounds but this isn't something that you should always want to avoid. Continually stretching and shortening loops, vocals, motifs or even single notes will introduce more and more digital noise which is great for dirtying up sounds for use in hip-hop, lo-fi and big beat. If a timbre is programmed on a synthesizer, sampled and stretched numerous times the resulting noisy timbre can be sampled and used to play the motif.

Perceptual Encoding

Similarly, any perceptual encoding devices (such as minidisk) can be used to compress and mangle loops further. Fundamentally, these work by analysing the incoming data and removing anything that the device deems irrelevant. In other words, data representing sounds that are considered to be inaudible in the presence of the other elements are removed, which can sometimes be beneficial on loops.

Physical Flanging

Most flanger effects are designed to have a low noise floor which isn't of much use if you need a dirty flanging effect for use in some genres of dance. This can, however, be created if you own an old analogue cassette recorder. If you record the sound to cassette and applying a small amount of pressure on the drive spool the sound will begin to flange in a dirty uncontrollable manner, this can then be re-recorded into the sampler.

Gritty Sounds

Most dance musicians will not only use the latest samplers but also own a couple of old samplers. Due to the low bit and sample rate quality of these samples, any sounds that are recorded using them take on a gritty, dirty nature. This same effect can be accomplished in most of today's samplers and wave editors by reducing the bit and sample rates of the audio. This type of effect is especially used in hip-hop and lo-fi to create the gritty drum loops.

Transient Slicing

A popular technique for house tracks is to design a pad with a slow resonant filter sweep and then sample the results. Once in a sampler, the slow attack phase of the pad is cut-off so that the sound begins suddenly and sharply. As the transient is the most important part of the timbre, this creates an interesting side-effect that can be particularly striking when placed in a mix.

In the end creative use of samplers comes from a mix of experience, experimentation and serendipity. What's more, all of the preceding examples are just that – examples to start you on the path towards creative sampling and you should be willing to push the envelope much further. The more time you set aside to experiment, the more you'll learn and the better results you'll achieve.

SAMPLES AND CLEARANCE

I couldn't possibly end a chapter on sampling without mentioning copyright law. Ever since dance music broke onto the scene, motifs, vocal hooks, drum loops, basses and even entire sections of music from countless records have been sampled, manipulated and otherwise mangled in the name of art. Hits by James Brown, Chicago, Donna Summer, Chic, Chaka Khan, Sylvester, Lolita Holloway, Locksmith and Michael Jackson along with innumerable others have come under the dance musicians' sample knife and been remodelled for the dance floor.

Although in the earlier years of dance artists managed to get away with releasing records without clearing the samples, this was because it was a new trend and neither the original artists nor record companies were aware of a way that they could prevent it. This changed in the early 1990s, when the sampling of original records proved it was more than just a passing fad. Today, companies know exactly what to do if they hear that one of their artists' hits has been sampled and, in most cases, they come down particularly hard on those responsible.

In other words, before you begin sampling another artist's motif, drum, vocals or entire verse/chorus consider that by law you cannot sample anything from anybody else and use it commercially without clearance.

To help clear up some of the myths that still circulate – about how small initial pressings or samples under a certain length are not covered by law – I spoke to John Mitchell, a music solicitor who has successfully cleared innumerable samples for me and many other artists.

What is copyright?
'Copyright exists in numerous forms and if I were to explain the entire copyright law it would probably use up most of your publication. To summarise, it belongs to the creator and is protected from the moment of conception, exists for the lifetime of the creator and another 70 years after their death. Once out of copyright, the work becomes 'public domain' and any new versions of that work can be copyrighted again.'

So it would be okay to sample Beethoven or another classical composer?
'In theory Yes, but most probably No. The question arises as to where did you sample the recording from because it certainly wouldn't have been from the original composer. Although copyright belongs to the creator, the performance is also copyrighted. When a company releases a compilation of classical records they will have been re-recorded and the company will own the copyright to that particular performance. If you sample it you're breaching the performance copyright.'

What if you've transcribed it and recorded your own performance?
'This is a legal wrangle that is far too complex to discuss here as it depends on the original composer and the piece of music. When the original copyright expires and it's re-recorded the subsequent company may own the copyright to both the performance and the creation. My advice is to check first before you attempt to sample or transcribe anything.'

So what can you get away with when sampling copyright music or speech?
'Nothing, there seems to be a rumour circulating that if the sample is less than 30s in length you don't need any clearance but this simply isn't true. If the sample is only a second in length and its copyright protected you are breaking the law if you commercially release your music containing it.'

What if the sample were heavily disguised with effects and EQ?
'You do stand a better chance of getting way with it but that's not to say you will. Many artists spend a surmountable amount of time creating their sounds and melodies and it isn't too difficult to spot them if they're used on other tracks. It really isn't worth the risk!'

Does this same copyright law apply to sampling snippets of vocals from TV, radio, DVD, etc.?

'Yes, sampling vocal snippets or music from films breaches a number of copyrights, including the script writer (the creator) and the actor who performed the words (the performer).'

Is there an organization that monitors all the releases?

'Not as far as I'm aware but I do know that the Mechanical Copyright Protection Society employs a number of DJs who actively listen out for records containing illegal samples. Plus, if the record becomes a big hit more people will hear it and the chances are that someone will recognise it eventually.'

So what action can be taken if clearance hasn't been obtained?

'It all depends on the artist and the record company who own the recording. If the record is selling well then they may approach you and work a deal, but be forewarned that the record company and artist will have the upper hand when negotiating monies. Alternatively, they can take out an injunction to have all copies of the record destroyed. They also have the right to sue the samplist.'

Does it make any difference how many pressings are being produced for retail with the uncleared samples on them?

'This reminds me of another urban myth: if you produce and sell fewer than 5000 copies then there is no need to obtain sample clearance. This couldn't be further from the truth. If just one record were released commercially containing an illegal sample the original artist has the legal rights to prevent any more pressings being released.'

What if the record were released to DJs only?

'It makes no difference. The recording is being aired to a public audience.'

Can you go to prison for illegal use of samples?

'It hasn't happened yet but that's not to say that it never will. At the moment, breaching copyright law is viewed as a civil, not criminal offence, but times change and I would fully recommend acquiring the clearance before using any samples. You don't want to be the first exception to the rule.'

How much does it cost to get sample clearance?

'It depends on a number of factors. How well known the original artist is, the sampled record's previous chart history, how much of the sample you've used in your track and how many pressings you're planning to make. Some artists are all too willing to give sample clearance because it earns them royalties while other will want ridiculous sums of money.'

How do you clear a sample?

'Due to the popularity of sampling from other records there are companies appearing every week who will work on your behalf to clear a sample. However, my personal advice would be to talk to the Mechanical Copyright Protection Society. These guys handle the licensing for the recording of musical works and started a Sample Clearance Department in 1994. Although they will not personally handle the clearance itself – for that you need someone like myself – they can certainly put you in touch with the artist or record company involved.'

CHAPTER 7
Recording Vocals

'I recorded one girl down the phone from America straight to my Walkman and we used it on the album. If we'd taken her into the studio it would have sounded too clean...'

Mark Moore

Although originally vocals played a particularly small role in the dance music scene, over the years the vocal content has grown significantly to the point that, unless you can gain permission to use another artist's vocal performance, you need to be proficient at recording them.

This is much more difficult than it first appears and there's certainly much more to it than simply plugging in a microphone, pressing record and hoping for the best. In fact, if you take this approach hope is about the closest you'll get because you certainly won't record a good performance.

We've been exposed to the human voice since the day we were born so we instinctively know how it's supposed to sound. As a result, any poor recording equipment or technique will be immediately noticeable and as the vocals can form the centrepiece of a track, if they're wrong, the rest of the track, no matter how well produced, will appear just as bad.

To record vocals proficiently requires not only a good vocalist but a good understanding of the effects the surrounding environment can have on a recording, and more importantly, the technology used to record them. This latter knowledge can be particularly significant since it would be easy to simply say which microphone you should use and where to place it; with a good understanding of the equipment involved you can make an informed decision yourself as to which microphone and positioning is best suited for the current project.

FIGURE 7.1
A dynamic microphone

MICROPHONE TECHNOLOGY

Starting with the basics, a microphone simply converts sound into an electrical current that is then transformed into an audio signal at the end of the chain. As straightforward as this appears, though, there are different ways of accomplishing it and the way sound is captured determines the quality of the overall results. Of course, all quality microphones will invariably produce a good sound, but the tonal quality between each is very different, so it is important to choose the right microphone for the vocalist and genre of music you produce.

Despite the number of different microphones available, ultimately for vocals, there are only two real choices: electromagnetic and electrostatic. Both of these use the principle of a moving diaphragm to capture sound, but they use different methods for converting it into an electrical signal and so provide entirely different tonality from one another.

Possibly the most instantly recognizable of these is the electromagnetic which is also referred to as a dynamic microphone. They're always used in live situations and music TV programmes and are best described as a small stick with a gauze ball sitting on top (Figure 7.1).

Dynamic microphones work under the principle of a moving coil which is very similar to how a typical loudspeaker operates (although loudspeakers operate in reverse). The microphones consist of a very thin plastic film, known as the diaphragm, which is coupled to a coil of wire that's suspended in a magnetic field. When any sound strikes this diaphragm, it begins to vibrate sympathetically, which causes the coil to vibrate too. As this coil vibrates in the magnetic field, it creates an alternating current which the microphone then outputs and is subsequently converted into sound.

Generally speaking, this type of assembly is particularly hard wearing and is the reason why it's so often used in live performances. If you drop the microphone, there's a good chance that it will still work. On the downside, though, since the vibrations have to move a relatively heavy coil assembly to produce the sound, they're not very sensitive to changes in air pressure and so are relatively slow to respond to transients. Consequently, the higher frequencies are not captured reliably so sounds recorded using them will often exhibit a 'nasal' quality. While this generally makes them unsuitable for recording the delicate vocals that are required for trip-hop and trance, they are perfectly suited towards rap as the 'nasal' sound provides more mid-range frequencies helping the sound to remain up front and 'in your face', an effect that can sometimes be vital for the genre.

It's because of this nasal quality that many artists prefer to use electrostatic (otherwise known as capacitor or condenser) microphones. These can be seen in use in all professional studios and are best described as 'flat looking' with angled gauze at either side. These are much more sensitive than electromagnetic microphones as the design doesn't rely on generating a signal by moving a coil in a magnetic field but is based on capturing sound using a varying capacitance. This means that the diaphragm can be much lighter and thus captures sound much more accurately (Figure 7.2).

Typically in these microphones the diaphragm consists of a gold-plated light Mylar plastic which is separated from a back plate by a gap of a few microns. A small electrical charge is imposed onto the back plate or the diaphragm or both (depending on the model) which creates a capacitance charge between the two. When sound strikes the diaphragm, the distance between the mylar plastic and the back plate varies, which in turn creates a fluctuation in the capacitance resulting in small changes in the electrical current. These changes in the current produce the audio signal at the end of the chain. While this means that less air pressure is required to create the signal (and so produces a more accurate response), it is important to note that different capacitor microphones will use differently sized diaphragms, and this also has a direct effect on the frequency response.

A larger diaphragm will obviously have a heavier mass and therefore will react relatively slowly to changes in air pressure when compared to a smaller diaphragm. Indeed, due to the smaller overall mass of these, they respond faster to changes in air pressure which results in a sharper and more defined sound. From a theoretical point of view, this would mean that to capture a perfect performance you would be better using a small diaphragm microphone since it would produce more accurate results, but in practice this isn't always required. To better understand this, we need to examine the relationship between the wavelength of a sound and the size of the diaphragm.

Typically, large diaphragm microphones will have a diameter of approximately 25 mm and if this is exposed to a wavelength of equal size (10 kHz is circa 34 mm, wavelength of 344 ms), there is a culmination of frequencies at this point. Effectively, this results in more directional sensitivity at higher frequencies. If we compare this reaction to a smaller diaphragm microphone that has a 12 mm diameter, this build-up would occur much higher in the frequency range (20 kHz is circa 17 mm, wavelength of 344 ms) and is therefore not noticeable to the human ear.

FIGURE 7.2
A condenser microphone

While this means that a larger diaphragm microphone doesn't have a frequency response as accurate or as flat as a smaller diaphragm, the subsequent colouration results in a rounder smoother timbre that pulls through a mix much more readily. What's more, a large diaphragm microphone will also have a lower noise floor than a smaller diaphragm because it has a larger surface area to conduct so the electrons can distribute more readily.

If you decide to use a condenser microphone rather than electromagnetic, an electrical charge is required to provide the capacitance between the diaphragm and back plate. This can be provided by batteries, phantom power or it may even already have a charge retained in the diaphragm or back plate from the manufacturer. In most instances, though, this charge is +48V D.C. phantom power which is received through the microphone cable itself. The term phantom power is used since you can't 'physically' see where it receives its power from and is commonly supplied from a mixing desk or a microphone pre-amplifier.

Whichever microphone you choose the signal produced is incredibly small (−60 dBu), so it needs to be amplified to an appropriate volume for recording using a pre-amp. These are relatively simple in design and will perform just two functions: supply the microphone with power, if required, and amplify the incoming signal. Unsurprisingly, there is a huge range of pre-amps to choose from and can range in price from £100 to £6000 and above, depending on the design and parameters they have on offer.

If at all possible, though, you should aim for a pre-amp that uses valves in its design as these add warmth to the signal that's archetypal of all genres of dance, but if this isn't feasible (due to the higher price tag) then you can record using a solid-state design and feed the results into a valve compressor or valve emulation plug-in.

In an ideal situation you should use a microphone pre-amp rather than a mixing desk because the circuitry in a pre-amp is specifically manufactured for the purpose rather than an 'added extra' as it is in most mixing desks (which as a consequence are prone to a much higher noise floor).

RECORDING PREPARATION

While it is important to have a good microphone and pre-amp combination the most common problem doesn't actually come from the equipment used but from a poor vocal technique. Indeed, as long as the equipment isn't truly 'bargain basement' and the microphone is placed thoughtfully, it's perfectly possible to attain respectable results. The only real secret to capturing a great performance is to ensure that you have nothing but a great performance going into the microphone. If you settle for anything less at this stage, no amount of effects or editing you apply later are going to make it sound any better.

With this in mind perhaps the most obvious start is to ensure that the vocalist has attended at least six to eight singing lessons. Anyone can sing, but not everyone can sing well! While very few vocalists will admit that they need any form of voice training – their gift is natural of course – every professional vocalist on the circuit has attended lessons with a voice trainer or a singing teacher, or both. Which lessons they should attend will depend on the vocalist, but if they have trouble singing in the correct key, breathing properly or holding notes, then they should see a voice trainer. If, however, they can sing fairly well then they may benefit from attending lessons with a singing teacher as they'll teach them how to perform a song with depth, emotion and feeling.

It's doubtful that you'll be able to find a teacher who specializes in house or rap or any other dance-related genres (in fact, you may have difficulty finding one who has actually ever heard of them) as most will specialize in classical performances or musicals, but the genre is unimportant. Simply attending singing lessons will help the vocalist learn the techniques every professional vocalist will use and this can then be applied to any form of music.

Provided that the vocalist is capable of delivering a good performance (which is much more difficult than it sounds), the location can also have a significant effect on the quality of the end result. No one is going to perform well in a clinical environment with bright fluorescent lighting. Apart from the fact that fluorescent lighting is prone to introducing hum into a recording, it's difficult for a performer to concentrate in such a 'scientific' environment. Ask yourself if you could perform your best sitting in a dentist's chair. Instead, vocalists produce a much more energetic and passionate performance if the room has a welcoming feel, and this means using bulbs that give off a warm glow.

The temperature of the room is also important and ideally you should look towards maintaining a temperature of around 22°C (approximately 72°F). Any colder and the performer's vocal chords may tense up.

If the environment is suitable, the next consideration is the foldback mix. Essentially, this is the *complete* mix plus the vocalist's microphone signal returned to a pair of headphones for the performer to perform along to. It is vital that this foldback mix is complete and not simply a 'rough draft', though, since if you play an unfinished or badly mixed track back for them to perform to, they may find it difficult to determine the pitch, the vocal tone they should use or to move closer/further away from the microphone so that they can balance their voice with the current mix. If, on the other hand, the track has been well mixed and engineered, the performer will not only find it much easier to pitch but will also want to make an extra effort to match the quality of what's already down.

Usually, the foldback mix is fed from tape, minidisk, DAT or CD to headphones with the computer or dedicated hard disk recording the resulting vocals, but whatever the configuration, the balance between the instruments and the performer's voice is critical. More often than not, a mix that is either too loud or quiet in respect to the microphone's output is responsible for poor timing, dynamics and intonation, so you should spend plenty of time ensuring that both mix and vocals are at the appropriate volume.

The headphones used to play the foldback mix should offer plenty of isolation to prevent any spill encroaching onto the vocal recording. Consequently, open headphones are not suitable as they leak sound, but equally closed headphones aren't suitable either. Even though these will not leak any sound, the closed design surrounds the ears and creates a rise in bass frequencies. Because of this, headphones with a semi-open design are the best, but you should also ensure that they are tight enough on the head so that they don't fall off!

Some professional vocalists will not want to listen to the full mix and instead prefer to listen to just the drum track with a single-pitched instrument sitting on top as this can help them to keep better timing and intonation. Always take the effort to check exactly what the vocalists want and no matter how difficult some of their requests or suggestions may be, try to comply. If they are happy, they're more likely to produce a great performance, but if the vocalists don't ask for reverb on the foldback vocals, apply it anyway.

Reverb can play a key role in helping a singer to pitch correctly plus it also invariably makes them sound much better, which in turn, helps them to feel better about their own voice. This only needs to be lightly applied and, typically, a square room setting with a short tail of 5 ms is usually enough. Alongside this reverb, some engineers also choose to compress the vocal track for the performer's foldback, although whether this is a good idea or not is under constant debate. From a personal point of view, I find that an uncompressed foldback mix tends to persuade the vocalists to control their own dynamics which invariably sounds much more natural than having a compressor do it.

RECORDING TECHNIQUES

Typically, for most genres of dance, with the possible exception of rap, the preferred choice of microphone is a condenser with a large diaphragm, such as the AKG C3000. This is generally accepted as giving the best overall warm response that's characteristic of most genres but any large diaphragm will suffice provided that the frequency response is not too flat. Note that some large diaphragm models will deliberately employ peaks in the mid-range to make vocals appear clearer but this can sometimes accentuate sibilance with some female vocalists. Sibilance is the result of over emphasized 'S' sounds which can create unwanted hissing in the vocal track.

Additionally, some mics may also roll off any frequencies below 170 Hz to prevent the microphone from picking up any rumble that may work its way up the stand, but rolling off this high can sometimes result in the vocals appearing too weak or 'thin' which can be difficult to widen later. It's therefore preferable to use a microphone that doesn't roll off above 120–150 Hz and to use it in conjunction with a quality pre-amp. Usually these microphones employ a filter in the signal path allowing you to manually define the frequency to roll-off.

The positioning of the microphone depends entirely on the style of performance and the performer but generally it should be positioned away from any walls, but not in the centre of a room. Any condenser microphone picks up both the direct sound from the vocalist and the sound from the surrounding ambience of the room. Thus, if the microphone is positioned in the centre of the room or too close to a wall, the reflections can create phase cancellations making the vocals appear thin, phased or too heavy. It should also be mounted on a fixed stand and situated so that the bottom of the housing is in level with the performer's nose (with their back straight!). This will ensure a good posture while also helping to avoid heavy plosive popping or breath noises.

Plosive popping is perhaps the most common problem when recording vocals and is a result of the high-velocity bursts of air that are created when we say or sing words containing the letters P, B or T. These create short, sharp bursts of air that force the microphone's diaphragm to its extremes of movement, which results in a popping sound. These can be incredibly difficult to remove at a later stage so you need to remove the possibility of them appearing during the recording by using a 'pop shield' positioned between the vocalist and mic. These consist of a circular frame approximately 6–8″ in diameter that is covered with a fine nylon mesh and are available as free standing units or with a short gooseneck for attaching to the microphone stand. By placing these pop shields approximately 5 cm (about 2″) from the microphone, any sudden blasts of air are diffused, preventing them from reaching the diaphragm.

Although pop shields are available commercially, many artists make their own from metal coat hangers or embroiders' hoops and their girlfriend's/wive's nylon stockings. Provided that the frame is large enough to act as a pop shield, the nylon can be stretched over the ring which can then be held in position next to the microphone with some stiff wire.

If you are experiencing constant plosive problems, it is better to attend a series of lessons with a professional singing teacher. This will not only help in learning how to reduce the emphasis on plosive sounds but also contribute to preventing any sibilance. This latter effect can sometimes be removed by placing a thin EQ cut somewhere between 5 and 7 kHz or by using a professional de-esser. These are frequency-dependent compressors which operate at frequencies between 2 and 16 kHz. Using these, it's possible to boost the EQ on vocals to make them appear brighter and the de-esser will jump in to prevent them from becoming too harsh.

As for the distance between mouth and microphone, ideally it should be placed approximately 20–30 cm (8–12″) away from the mouth of the vocalist. At this distance, the voice tends to sound more natural, plus any small movements on the vocalist behalf will not create any serious changes in the recording level. Naturally, keeping this microphone to mouth distance can be difficult to achieve, especially if the vocalist is used to performing in live conditions or is a little over excited.

In these situations, vocalists tend to edge even closer to the microphone as they continue the performance, and if they get too close, it will not only boost the lower frequencies but also create deposits of condensation on the diaphragm, which reduces its sensitivity. Since microphones are high-impedance devices, using them in a cold room or having the vocalist stand too close to them can introduce condensation onto the diaphragm. If this happens, the sensitivity of the mic will fall sharply and crackles or pops can start to appear in the recording. Subsequently, it's prudent to ensure that any capacitor mics are at room temperature before they're used, and it's also worthwhile making sure that the room isn't too cold. If condensation does form on the capsule, the only solution is to put them in a warm place for a few hours.

Obviously, prevention is always better than cure, so to prevent the vocalist from getting too close to the mic it's sensible to position a pop shield 10 cm (approximately 4″) away from the microphone. This is the preferred approach to placing the microphone above the singer's mouth and angling the mic downwards. While some engineers believe that this approach ensures a 'fixed' distance from the mouth to the mic while also helping the vocalist straighten their back to improve breathing and projection; if vocalists are forced to raise their chins, the throat becomes tense which can severely affect the performance.

If, however, the genre demands a more up-front vocal sound that's typical of commercial pop (and trance, house), then it's worthwhile reducing the pop shield to mic distance to just 2″ and have the performer almost pressing their lips to the shield while singing. You may have to reinforce the pop shield with this approach by stretching another stocking over it but this close positioning will take advantage of the proximity effect. This creates an artificial rise in the bass frequencies which can make the vocals appear warmer, more up front and in your face. Occasionally, this approach can also increase the chance of plosives, but if this happens, turning the microphone slightly off axis from the performer can help to prevent them from occurring.

Once the mic has been set up and positioned correctly, it's prudent to insert a compressor between the pre-amp and recording device. Some pre-amps have a built-in compressor dedicated for this function, but if not, you will need to use an additional hardware unit to control the dynamics. The human voice has an incredibly wide range of dynamics and these will need to be put under some control to prevent any prominent parts clipping the recording.

Notably, some professional vocalists will use their own form of compression by backing away from the microphone on louder sections and moving closer in quieter parts, but nevertheless, even in these circumstances it is still prudent to use a compressor just in case. A single uncontrolled clip can easily destroy an otherwise great performance! This must be set carefully, however, as once recorded you cannot undo the results later, so you should set the compressor to squash only the very loudest parts of the signal.

Ideally, the compressor should be analogue and employ valves as the second-order harmonic distortion they introduce is typical of the vocal sound of the genres, but if this is not possible, then you can use a solid-state compressor and compress them after recording with a valve emulation plug-in. A good starting point for recording compression is to set a threshold of $-12\,dB$ with a 3:1 ratio and a fast attack and moderately fast release. Then, once the vocalist has started to practice, reduce the threshold so that the reduction meters are only lit on the strongest part of the performance.

These compression settings can also be applied to rap music, but generally speaking, this genre benefits more from using a dynamic microphone such as the Shure SM58, to capture the performance. Also, rather than mounting the mic on a stand better results can occur if the vocalist holds the mic. This not only captures the slight nasal quality synonymous to most rap music but also allows the rapper to move freely. Much of rap music relies on the different vocal tones that are produced by the rapper changing their position from upright to bowing down, in effect 'compressing' their abdomen to produce lower tones. If this technique is used, though, ensure that the vocalists keep their hand clear of the top of the microphone since obstructing this will severely change the tonal character of the vocals. Also try to ensure that the microphone is not held too far away from the mouth. Most dynamic microphones will roll off all frequencies below 150 Hz to prevent the proximity effect, but if the vocalist has it too far away, then this can result in a severe reduction in bass frequencies. In general, a dynamic microphone should be approximately 10–20 cm (approximately 4–8″) away from the mouth.

This is the usual distance that most vocalists will use naturally, but if you have a loud foldback mix with a bass cut, they'll tend to hold it closer to their mouths to compensate for the lack in mix presence so it's vital that the mix is bass accurate before adding vocals. It's also worth noting that, due to the design, these mics have a diminished frequency response and can be quite easy to overload if placed too close to the mouth. This can result in light distortion in the higher mid-frequencies, so to prevent this it may be prudent to feed the pre-amps output into an EQ unit and then feed the results into the side chain of the compressor at the problematic frequencies.

With the equipment set-up to record, the vocalist/rapper will, or should, need an hour to warm up their voice to peak performance. Most professional vocalists will have their own routine to warm up, and many prefer to do this in private, so it may be necessary to arrange some private space for them. If they are inexperienced they may insist that they do not need to warm up, and in this instance, you should charm them into it by asking them to sing a number of popular songs that they enjoy while you 'set up the recording levels…'

During this time, you should listen out for any signs that the performer has poor intonation or is straining and if so, ask them to take a break and relax. If they need a drink, fruit juices are better than water, tea or coffee as these can dry the throat, but you should avoid any citrus drinks. From personal experience, two teaspoons of honey with one teaspoon of glycerine mixed with a 100 ml of warm water provides excellent lubrication and helps to keep the throat moist, but persuading the vocalist to drink it can be another matter altogether. Alternatively, Sprite (the soft drink) seems to help, but avoid letting them drink too much as the sugar can settle at the back of the throat. Failing that, sucking lozenges such as Lockets with Honey before the performance can help with the lubrication.

Once the vocalists have warmed up, their voice should be at peak performance and they should be able to achieve depth and fullness of sound. It's at this point that you want them to belt out the track with as much emotion as possible, from beginning to end. As difficult as it may be, you should avoid stopping the vocalists at any mistakes apart from the most severe intonation problems and ask them to continue the song despite any mistakes as you'll be recording more takes later.

Unequivocally, there is no music today that is completed in just one pass and even the most professional artists will make three or four passes, so keep things light and friendly and do not put any pressure on the performer to get the take right in one.

It's also worthwhile ensuring that the vocalists know the lines off hand without having to read them off a sheet of paper. Encourage them to learn the lines and if they forget them, ask them to have fun and improvise. The occasional 'lah' or 'ooh' can help to enhance the performance and may actually sound better than the original line, plus if they know the words, they are likely to sound more energetic and lively than a droll reading. More importantly, though, lyric sheets can have an adverse effect on the recording as the vocals will reflect from the paper, which can create a slight phase in the recording. This has often been responsible for numerous scratched heads as the engineer tries to figure out why the vocals sound so strange.

In some situations the performer may insist they have lyric sheets to hand. If this is the case, position the lyric sheets behind the microphone and slightly off axis so that the possibility of phase problems is reduced.

Once this first performance has been recorded, listen back to the recording and try to identify any problems such as very poor intonation, the inability to hold high notes, uneasiness and tenseness, phasing effects or too much bass presence. If the microphone has been placed correctly, there shouldn't be any phasing or bass problems, but if there are, then you'll need to reconsider the microphone's placement. If there is too much bass presence, ask the performer to move further away from the microphone, while if there are phasing problems, move the microphone closer, or further away from the wall.

Severe intonation problems, the inability to hold high notes, uneasiness or tenseness, on the other hand, are a little more difficult to resolve as they reside in the psychological state of the performer and no amount of equipment, no matter how good, will turn a poor or uninspiring performance into a great one. Instead, you'll need to look at the monitoring environment and your own communication skills.

Assuming that the vocalist can actually sing in tune, the most common reason for poor intonation comes from the performer's monitoring environment or a poor overall mix. As previously touched upon, they will most probably monitor the mix through headphones and this can result in an increase in bass frequencies or difficulty in comprehending the psychological acoustic coupling. If the headphones are tight fitting on the performer's head, there will be an increase in bass frequencies and it is difficult to perceive pitch in low-frequency waveforms which can force the performer to loose intonation. This problem can usually be identified by the vocalist 'involuntarily' moving further away from the microphone and singing louder as this reduces the bass presence in their voice on the feedback mix. To prevent this, it's worth placing a wide cut of a few dB at 100–300 Hz (where the bass usually resides) or increasing the volume of a higher pitched instrument that the vocalist can pitch to.

Psychological acoustic coupling is less of a problem with experienced performers but is a result of the singer's vocal chords vibrating their ear drums along with all the bones in the ear cavity – the reason why you sound different when you listen to a recording of yourself. When you place a pair of headphones on the vocalists, their 'natural' environment is upset, which can make it difficult for them to pitch correctly. In this instance, it's worth muting the vocals returned in the foldback mix to see if this makes a difference, but if not, it may be worth muting one side of the headphone feed so that they can hear themselves naturally in one ear. Some of the higher priced mixing desks and headphone distribution amps offer a kill switch to terminate one of the headphones stereo channels for this purpose, but if this option isn't available, you'll need to purchase a single channel headphone, manufacture a mono converter or ask the vocalist to move one of the phones to uncover their ear.

Of course, if the monitoring volume is quite high, this latter solution can result in the foldback mix spilling out of the unused earpiece and working its way onto the recording. However, as long as this isn't ridiculously loud, a little overspill isn't too much of a problem as it will be masked by the other instruments. Indeed, even professional artists such as Moby, Britney Spears, Madonna and Michael Jackson have overspill present on some of their individual vocal tracks and it's something you'll undoubtedly have to deal with if you have the chance to remix a professional artist.

Some vocalists simply can't perform wearing headphones so you need a way of playing the mix back without recording any spill. The best way to accomplish this it to place both monitors 90 cm apart and 90 cm from the rear of the microphone. Following this, reverse the wires (i.e. the polarity) of the left speaker and then feed both monitors with a mono signal. This will allow the performer to hear the mix but the phase of the speakers will match at the microphone, in effect, cancelling out the sound.

If the problem is an inability to hold high notes or a tense or uneasy voice, then this is the result of a nervous performer not breathing deeply enough and this can only be resolved through tact, plenty of encouragement and excellent communication skills. Vocalists feel more exposed than anyone else in a studio so they constantly require plenty of positive support and plenty of compliments after they've completed a take. This also means that eye contact throughout is absolutely essential, but refrain from staring at the vocalist with a blank expression – try to keep an encouraging smile on your face. This can give a vocalist a massive confidence boost which will pull through into the recording. More importantly, though, do not swamp the vocalist with techno babble, or try to impress them by giving them a technical run through of what you plan to do. If the vocalist has too many technical comments to think about it can ruin the performance.

EDITING VOCALS

Once you have the first 'tonally' correct, emotionally charged take recorded, there will undoubtedly be a few mistakes throughout that will need repairing or replacing. At this stage, before you send the vocalist home, you should ask them to perform the 'dodgy' parts again. It isn't advisable to ask them to perform the whole take again as it may exhaust their energy before they reach the part that needs to be replaced, so it's a better idea to punch in and out on the parts that need it.

Essentially this means that you need to make it clear to the performer which parts you want to re-record and these should be entire lines, not single words. Also, the vocalist should perform the lines before and after the faulty part to prevent a cold start and ensure that the same breathing pattern as the original take is more or less maintained.

Above all, remember that very, very few vocalists today can perform a perfect take in one go and it certainly isn't unusual for a producer to record as many as eight takes of different parts to later 'comp' (engineer speak for compiling not compressing) various takes together to produce the final, perfect, vocal track. This is where you need to be willing to spend as much time as required to ensure that the vocal track is the best it can be. In some cases, you may even have to recall the vocalist to re-perform some lines and this can often result in a different tone of voice, but it is better to have a slightly different vocal tone than have a poorly sung line. Although ideally you should have ensured that you had enough to work with originally, re-recording some vocal parts at a different time isn't unusual and even professional recording studios do this from time to time.

If problems do occur later and it isn't possible to recall the vocalist, then you'll have no choice but to use a series of studio processors. This approach should *always* be viewed as the very last option but can range from cross-fading individual words together, using pitch correction software such as Antares Auto-Tune, harmonizing, double tracking, applying chorus, flange or reverb to thicken out thin vocals, time stretching some parts while time compressing others to reconstructing the entire vocals word for word in an audio sequencer or editor. This may all seem a little excessive but it is quite normal for a producer to spend days and even weeks editing a vocal track to achieve perfection.

If you are experiencing severe problems with intonation, you should look at using a different vocalist and try not to rely too much on pitch correction software. While these are adept at correcting the odd out of tune words they do not work as well when applied to an entire vocal line, phrase or chorus section. Although many like to believe that 'manufactured' bands are incapable of holding a single note and rely entirely on pitch correction software, disappointingly, they can actually sing quite well (if chosen for looks alone they'll certainly have attended a series of singing lessons!). Any 'good' singer will naturally bend and modulate notes with vibrat and if a vocal line, phrase or chorus section is particularly slow during the bend, the intonation unit can become confused and attempt to correct it. This can result in it bending the note when it doesn't require it, bend it in the wrong direction altogether or quickly jump to the next note rather than keep the bend. All of these side-effects can produce a synthetic unnatural sound that you would usually want to avoid, although of late the 'fake' pitching has come to the fore in music since it was used by Cher's producer on 'Believe'.[1]

VOCAL EFFECTS

When it comes to dance music, it isn't extraordinary to apply various processing to the vocals to make them different. As previously touched upon, we've all heard the human voice naturally since the day we were born so any effects that make them different will inevitably attract attention – so long as the effect isn't cliché. In fact, with dance music it's quite usual to treat vocals as any other instrument and heavily EQ or affect them to make them prominent and hopefully grab attention. Naturally, there is no definitive method to affecting vocals and it all depends on the overall effect you want to achieve. As Cher's producer has proved, experimentation is the real key to affecting the human voice and you certainly shouldn't be afraid of trying out new ideas to see what you can come up with. Having said that, there are some vocal effects that are still in wide use throughout the genres and knowing how to achieve these may help aid you to further experimentation.

COMMERCIAL VOCALS

Possibly the first effect to achieve is the archetypal contemporary up-front sound that's typical of most commercial pop songs and of late, trance. Although much of this is acquired at the recording stage by positioning the vocalist closer to the microphone, effects can be used to help widen it across the image and add some depth if it's required. Stereo-widening effects are not used for this as you're working with a mono source to begin with and recording the vocals in stereo is not only difficult, requiring two microphones, but it will rarely produce the right results. Instead a much better way is to employ a very subtle chorus or flange as a send, not insert effect. By then setting the effect rate as slow as possible you can send the vocals to the effect by the required amount, thus keeping the stability of the vocals while also applying enough of an effect to spread it across the stereo image. There are no universal settings for this as it depends entirely on the vocals that have been recorded, but a good starting point is 80% unaffected with a 20% effect.

An alternative method to this is to double track the vocals by copying them onto a second track before applying a pitch-shift effect to both. This, obviously, should be applied as an insert effect, otherwise you will experience a strange phased effect, but by setting one channel to pitch up by 2–5 cents and the other

[1]Cher's 'Believe' track did not use an intonation programme but the Digitech Talk-Box to acquire the effect.

down by 2–5 cents it can produce a wide, thick up-front sound. Additionally, if you move the second vocal track forward by a few ticks so that it occurs slightly later than the original, this can help to thicken the overall sound.

Many engineers will also recommend applying reverb to a vocal track to help thicken them in a mix but if it is not applied cautiously and sparingly it will push them into the background producing a less defined, muddied sound. Nevertheless, if you do feel the need to apply reverb to vocals, use it as a send effect rather than insert as this will allow you to keep a majority of the vocal track dry. Also, if reverb is being used on any other instruments within the mix, use a reverb with a different tonal character for the vocals as this will help to draw a listener's attention to them.

The trick behind using any reverb is to set the effect so that it's only notice-able when it's removed, but keep the decay time short for a fast tempo sing and lengthen it for those of a slower tempo. As a very general starting point, try setting the reverb's pre-delay to 1 ms, with a square room and 7 ms decay. By then increasing the pre-delay you can separate the effect from the dry vocal to enable the voice to stand out more.

The most widespread processing used to obtain the up-front sound is compres-sion. If you really want that big, in your face, up-front vocal, then compres-sion is the real key to acquiring it, but for this it is imperative that you use compressors with the best sound possible. As every dance genre will use valve compressors, this means you will need to use them too and whether hardware or software emulations they should ideally exhibit 0.2% total harmonic distor-tion (THD) or more for a good sound. Notably, most musicians use more than one compressor for this and will usually employ two or three. The first is com-monly a solid-state compressor that's used to smooth out the overall level with a ratio of 4:1, a fast attack and release and threshold set so that the gain reduc-tion meter reads 4 dB. The results of this are then fed into a valve compressor and this is generally set with a ratio of 7:1, a fast attack and release along with a threshold set so that the gain reduction meter reads 8 dB. Naturally, these are general settings and should be adjusted depending on the vocals and the microphone used to capture them. As we've touched upon in earlier chapters, applying compression will reduce the higher frequency as it squashes down on the transients of the attack, but if you find that the high-end detail becomes lost, it can often be brought back up again using a sonic exciter or enhancer.

'TELEPHONE' VOCALS

The second effect that's often used in the dance genre is the 'telephone' vocal, called so because it sounds as though it's being played through a restricted bandwidth (i.e. a telephone or transistor radio). This effect is quite easy to achieve by inserting (not sending) the vocals into a band-pass filter so that frequencies above 5 kHz and below 1 kHz are removed from the signal. If you don't have access to a band-pass filter, the same effect can be acquired by first inserting the vocals into a low-pass filter set to remove all frequencies above 5 kHz with the results of this inserted into a high-pass filter set to remove all frequencies below 1 kHz. If you can adjust the poles on these filters then, generally speaking, a 2-pole (12 dB) slope will produce the best results but depending on the mix that the vocals are sitting in, it may be worth using a 4-pole (24 dB) or, if possible, a single pole (6 dB) slope. Alternatively an EQ unit can be used to achieve much the same results by cutting all frequencies below 1 kHz and above 5 kHz. Once you have this general effect, it can be expanded upon by automating the filters (or EQ) to gradually sweep through the range to add movement and interest.

PITCHED VOCALS

Another, more recent, effect to apply to vocals is the unnatural pitch-shifting that's acquired by using an intonation unit to alter the pitch. How this is accomplished depends entirely on the effects unit itself, but fundamentally they all work on the principle that you can analyse the incoming signal and then select the key signature that it should be in. By using this as an insert effect so that it affects the entire vocal line and setting it to a key that's at least an octave out of the range of the original input, the processor will attempt to correct the vocals to the selected scale and, as a side-effect, produce a strange moving pitch very similar to the effect used on Cher's 'Believe'. On top of this, it is also worth experimenting by running the vocals through the intonation unit a number of times, each time using a different key setting before 'comping' the different results together.

VOCODERS

One final effect that's particularly useful if the vocalist is incapable of singing in key is the vocoder. Of all the vocal effects, these are not only the most instantly recognizable but also the most susceptible to changes in fashion. The robotic voices and talking synth effects they generate can be incredibly cliché unless they're used both carefully and creatively, but the way in which they operate opens up a whole host of creative opportunities.

Fundamentally vocoders are simple in design and allow you to use one sound – usually your voice (known as the modulator) – to control the tonal characteristics of a second sound (known as the carrier), which is usually a synthesizer's sustained timbre. However, as simple as this may initially appear actually producing musically useable results is a little more difficult since simply dialling up a synth preset and talking, or singing, over it will more often than not produce unusable results. Indeed, to use a vocoder in a musically useful way, it's important to have a good understanding of exactly how they work and to do this we need to begin by examining human speech.

A vocoder works on the principle that we can divide the human voice into a number of distinct frequency bands. For instance, plosive sounds such as 'P' or 'B' consist mostly of low frequencies, 'S' or 'T' sounds consist mostly of high frequencies, vowels consist mostly of mid-range frequencies and so forth. When a vocal signal enters the vocoder, a spectral analyser measures the signal's properties and subsequently uses a number of filters to divide the signal into a number of different frequency bands. Once divided, each frequency band is sent to an envelope follower which produces a series of control voltages[2] based on the frequency content and volume of the vocal part. This exactly same principle is also used on the carrier signals and these are tuned to the same frequency bands as the modulator's input. However, rather than generate a series of control voltages, they are connected to a series of voltage-controlled amplifiers. Thus, as you speak into the microphone the subsequent frequencies and volume act upon the carrier's voltage-controlled amplifiers which either attenuate or amplify the carrier signal, in effect, superimposing your voice onto the instrument's timbre. Consequently, since the vocoder analyses the spectral content and not the pitch of the modulator it isn't necessary to sing in tune as it wouldn't make any difference.

From this, we can also determine that the more the filters that are contained in the vocoders bank, the more accurately it will be able to analyse and divide the modulating signal, and if this happens to be a voice, it will be much more comprehensible. Typically, a vocoder should have a minimum of six frequency bands to make speech understandable but it's important to note that the number of bands available isn't the only factor when using a vocoder on vocals.

The intelligibility of natural speech is centred between 2.5 and 5 kHz, higher or lower than this and we find it difficult to determine what's being said. This means that when using a vocoder, the carrier signal must be rich in harmonics around these frequencies since if it's any higher or lower then some frequencies of speech may be missed altogether. To prevent this, it's prudent to use a couple of shelving filters to remove all frequencies below 2 kHz and above 5 kHz before feeding them into the vocoder. Similarly, for best results the carrier signal's sustain portion should remain fairly constant to help maintain some intelligibility. For instance, if the sustain portion is subject to an LFO modulating the pitch or filter the frequency content will be subject to a cyclic change which may push it in and out of the boundaries of speech resulting in some words being comprehensible while others become unintelligible. Plus, it should also go without saying that if you plan on using your voice to act as a modulator it's essential that what you have to say, or sing, is intelligible in the first place. This means you should ensure that all the words are pronounced coherently and clearly.

More importantly, vocal tracks will unquestionably change in amplitude throughout the phrases which will create huge differences in the control voltages generated by the vocoder. This results in the VCA levels that are imposed onto the

---

[2]Control voltages (often shortened to CV) measure the signal at predetermined points and convert each of these measurements into differing voltages to represent the volume and frequency of the waveform.

carrier signal to follow this change in level, producing an uneven vocoded effect which can distort the results. Subsequently, it's an idea to compress the vocals before they enter the vocoder and if the carrier wave uses an LFO to modulate the volume, compress this too. The settings to use will depend entirely on the vocals themselves and the impact you want them to have in the mix (bear in mind that dynamics can affect the emotional impact), but as a very general starting point set the ratio on both carrier and modulator to 3:1 with a fast attack and release, and then reduce the threshold so that the quietest parts only just register on the gain reduction meter. Additionally, remember that it isn't just vocals that will trigger the vocoder and breath noises, rumble from the microphone stand and any extraneous background noises will also trigger it. Thus, along with a compressor you should also consider employing a noise gate to remove the possibility of any superfluous noises being introduced.

With both carrier and modulator under control there's a much better chance of producing a musically useful effect, and the first stop for any vocoder is to recreate the robotic voice. To produce this effect, the vocoder needs to be used as an insert effect, not send, as all of the vocal line should go through the vocoder. Once this modulator is entering the vocoder you'll need to programme a suitable carrier wave. Obviously, it's the tone of this carrier wave that will produce the overall effect and two sawtooth waves detuned from each other by + and −4 with a short attack, decay and release but a very long sustain should provide the required timbre. If, however, this makes the vocals appear a little too bright, sharp, thin or 'edgy' it may be worthwhile replacing one of the sawtooth waves with a square or sine wave to add some bottom-end weight.

Though this effect is undoubtedly great fun for the first couple of minutes, after the typical *Luke, I am you're father* it can wear thin, and if used as it is in a dance track, it will probably sound a little too cliché, so it's worthwhile experimenting further. Unsurprisingly, much of the experimentation with a vocoder comes from modulating the carrier wave in one way or another and the simplest place to start is by adjusting the pitch in time with the vocals. This can be accomplished easily in any audio/MIDI sequencer by importing the vocal track and programming a series of MIDI notes to play out to the carrier synth, in effect creating a vocal melody. Similarly, an arpeggio sequence used as a carrier wave can create a strange gated, pitch-shifting effect while an LFO modulating the pitch can create an unusual cyclic pitch-shifted vocal effect. Filter cut-off and resonance can also impart an interesting effect on vocals and in many sequencers, this can be automated so that it slowly opens during the verses, creating a build-up to a chorus section. Also, note that the carrier does not necessarily have to be created with saw waves, and a sine wave played around C3 or C4 can be used to recreate a more tonally natural vocal melody that will have some peculiarity surrounding it.

Vocoders do not always have to be used on vocals and you can produce great results by using them to impose one instrument onto another. For instance using a drum loop as a modulator and a pad as the carrier the pad will create a gating effect between the kicks of the loop. Alternatively using the pad as a modulator and the drums as the carrier wave the drum loops will turn into a loop created by a pad!

Ultimately these have only been simple suggestions to point you in a more creative direction and you should be willing to try out any effects you can lay your hands on to hear the effect it can have on a vocal. Bear in mind that due to the very nature of dance music it's always open to experimentation and it's much better to initiate a new trend than simply follow one set by another artist.

CHAPTER 8
Recording Real Instruments

'In the late Sixties a record producer was a more hands-on person. A record producer was also a skilled music arranger or an engineer or both. ...'

Tony Visconte

Although many of the timbres within dance music are created with synthesis and many of the real-world instruments are often sampled from a sample CD – or more commonly, another record – if you have access to the appropriate instruments, microphones, direct injection (DI) boxes and pre-amps, recording your own instruments can be incredibly rewarding. Not only will they be guaranteed to fit with the tempo of your music, they're also pretty much guaranteed to be in the right key too.

Recording real-world instruments is not something that should be approached lightly, however, since to gain the best results you need to not only have the appropriate environment for the instrument you wish to record but also good-quality microphones, a good pre-amp and – depending on what you are recording – a good amplifier and DI Box.

It must also be stressed that different engineers will use different techniques to record any instrument, and the judgement on how to record the instrument will depend on a large number of factors including the instrument's character, the room it's recorded in, its tuning and, most important of all, the player's style. Therefore, what follows in this chapter should only be viewed as general guidelines and you should adapt these to ensure you receive the best recording possible.

RECORDING PREPARATION

The first priority to record any instrument should be the room's acoustics. It is not necessary to have a recording room with a totally flat frequency response for a majority of real-world instruments; in fact, if you do record in a room such as this, you may find that the instrument is lacking in character. This is because with any real instrument the sound you hear is a result of a number of factors. For example, the sound of an acoustic guitar is not a result of only the sound hole but from the entire guitar vibrating sympathetically along with the reflected frequencies within the room. The same applies for brass instruments; the sound does not come directly from the bell but from the bell's vibrations and the reflected frequencies of the room it is played in. Because of this, it is vital that you choose a room that has the best acoustics for the instrument you wish to record. While you can add reverb at mix-down for example, you cannot remove it if it was captured at the recording stage!

From a general point of view, a typically decorated and carpeted room will prove suitable for recording most real-world instruments. Garages, basements and rooms with wooden floors should normally be avoided since these create brighter reflections which can result in a recorded sound that lacks any bottom end. However, that is not to say you shouldn't try playing the instrument in these rooms and listening to the results. In some instances, they may produce the sound you're after, particularly if you're recording brass instruments.

The best way to know if a room is suitable for recording is to actually play the instrument in that room. The principle to work with is that if it sounds good to the performer, then it stands to reason that it will sound good to the microphone, provided of course, that you use a good-quality microphone.

For many instruments, either ribbon or large diaphragm capacitor microphones are regarded as the best suited for the job. As touched upon in the previous chapter, larger diaphragm microphones have a heavier mass which produces a rounder, warmer timbre that is often suited for many instruments.

Ribbon microphones, although not as popular or widely used as large diaphragm microphones, are often used since these have a bidirectional polar pattern in that they capture sound from both sides of the microphone. By doing so, they cannot only capture the direct sound but also the ambience of the room. Naturally, this can be accomplished by increasing the distance from the instrument with a cardioids microphone, but for some engineers, the sound of a ribbon microphone produces the best results.

Ribbon microphones work similar to a dynamic microphone; however, whereas the diaphragm in a dynamic microphone is coupled to a coil of wire that's suspended in a magnetic field, with a ribbon microphone there is an extremely thin 'ribbon' of aluminium suspended at both ends which is vibrated by sound pressure within a magnetic field. This means that the ribbon can vibrate from both sides but not from its edges, allowing it to record from both sides of the microphone and producing what's known as a figure 8 pick-up pattern.

For this reason, ribbon microphones are considered the most natural sounding microphones and perfect for capturing instruments, but they are also *extremely* delicate. Blowing into a ribbon microphone can damage the thin aluminium ribbon and even carelessly slamming down the lid of the microphone box can create a surge of air that will damage the microphone's ribbon. Unfortunately, due to this delicate nature, ribbon microphones are few and far between and those that are in production are expensive (Figure 8.1).

It should also be noted that due to the low output of a typical ribbon microphone, you need a powerful and quiet pre-amp to bring the gain up to an acceptable level (60 dB of amplification to bring it up to 0 VU!). Therefore, if you plan on using a ribbon microphone, you'll need to ensure that you have a powerful and high quality pre-amp that has a very low noise floor.

As previously touched upon, when recording instruments, each engineer will approach the project differently and this includes the choice of microphones for the job at hand. Consequently, recommending microphones can be a minefield since whereas one engineer may choose one particular microphone for a particular instrument, a different engineer will have a totally different opinion. On top of this, the room's frequencies and the player's personal style will also affect the choice.

Nevertheless, at the risk of leaving myself open, I would personally recommend the *Neumann KM184* capacitor microphone for some instruments since it produces particularly pleasing results and rolls off at 200 Hz keeping the instruments free of any low end rumble. Alternatively, the *B&K 4011*, the *Shure SM81* and the *AKG C1000s* can all produce excellent results if set up correctly.

FIGURE 8.1
An early ribbon microphone

If I were to recommend ribbon microphones (and if you have a large wallet), *Audio Engineering Associates* produce some outstanding microphones such as the *R44* or alternatively *Golden Age Companies R1 Active* and *Coles 4038*. It is doubtful that you will find a ribbon microphone for sale in the second-hand market since they're often snapped up by studios the moment they appear.

RECORDING ACOUSTIC GUITARS

Before recording an acoustic guitar, it's often preferable to tune the guitar to the Nashville System. This consists of using the octave strings from a 12-string guitar, fitting them and tuning E and B to their regular pitch but tuning G, D, A and (the lower) E an octave higher – as they would be on a 12-string guitar. By doing this, the guitar produces an incredibly bright and 'jingly' timbre, replicating the typical pop music guitar sound. This also reduces a proportionate amount of low-end boom which helps the instrument fit perfectly into a busy dance mix.

Ideally, acoustic guitars should be recorded in stereo. While you can record in mono and apply stereo effects afterwards, although the results are not as impressive as recording in stereo. Obviously, this means that to record in stereo requires two microphones and you should use a paired set of microphones.

In other words, they are paired to one another directly from the manufacturer and sold as a pair. This ensures that they both have the same response which makes it much easier at mix-down. If this isn't possible, you should try to use two of the same make and model of microphone to ensure you receive a consistent recording from the guitar. Of course, you should feel free to use different microphones if that is all you have; experimentation may provide the best results for you.

The microphones should be placed with the first pointing towards the bridge of the guitar and the second pointing towards where the fret meets the body of the guitar. The distance of the microphones from one another and from the guitar must be carefully considered, however, in order to reduce any phase cancellation.

Phase cancellation is a result of the sound waves from the guitar reaching the two microphones at differing times. As discussed in Chapter 1, if the waveforms are captured at different phases of its cycle by the microphones, when they are combined at the mixing desk it can create phase cancellations between the two waveforms. To avoid this, you should adopt the 3:1 rule. This means that the spacing between each microphone should be three times the distance they are from the guitar. For example, typically a microphone should be placed approximately 300 mm away from an acoustic guitar; therefore, the microphones should be placed at least 900 mm apart.

Of course, if you don't have access to two microphones you'll be left with no option but to record the guitar with one microphone. One of the most common mistakes when recording with just one microphone is to place it directed entirely at the sound hole. The problem with this approach is that rather than capturing the entire guitar's sound, it only captures the lower end of the frequencies produced by the guitar.

A preferential approach is to place the microphone around 300 mm away from the guitar aiming towards where the neck and body meet. This captures both the sound emanating from the sound hole and the natural vibrations from the guitar's body.

This microphone position shouldn't be considered de facto and you should always experiment by moving the microphone further away since the room will have an influence on the sound that you may want to capture. It is important, however, that you monitor through headphones so you only hear what the microphone is picking up and not frequencies from the room that the microphone may not be picking up.

Depending on the player's individual style, you may record fret squeals as the guitarist slides his fingers up and down the fret. While some engineers will attempt to avoid capturing any fret noise whatsoever, I find it removes the realism of the guitar since fret squeal – in limited amounts – is a natural occurrence from playing the instrument and often enhances the sound of the guitar.

During recording, you should attempt to keep processing to a minimum since it cannot be removed afterwards. This means compression should be kept to a minimum and only used to prevent any louder transients from clipping the recording device. A good starting point for compression is to use a 5:1 ratio, with a 5 ms attack, a 40 ms release and a hard knee setting. Once set up, lower the threshold so that the compressor only captures any hard transients that may clip recording. The idea is to set the compressors threshold to ensure that the guitar is not compressed whatsoever, and it only compresses when absolutely necessary. If it is not set up correctly, you may find that the compressor pumps the recording which ruins the end result.

ELECTRIC GUITARS

Electric guitars require a slightly different approach to recording acoustic and for many the best approach is to simply employ a DI box. A DI box is a device with – most commonly – a jack connection that allows any electric instrument with a high impedance (such as an electric guitar) to be connected directly into it. The unit then converts this to a low-impedance signal (like a microphone's output) which can then be connected directly into the microphone input on a mixing desk. More recently, however, DI boxes have started to appear that feature a USB connection that directly connects the DI box into a computer, allowing the user to record the instrument directly without having to connect the DI box output into a mixing desk. Many of these 'computer DI boxes' also feature amp simulations, allowing you to set a specific amp type which is then applied to the instrument before it is recorded into the computer. One of the most celebrated of these is the *Line 6 POD*, a DI box used by most professional engineers and very highly recommended (Figure 8.2).

FIGURE 8.2
The Line 6 POD

However, as good as the Line 6 POD does sound, there are still benefits of using a microphone to record the guitar. In fact, you will find that in many cases the best results come from using a DI box and a microphone set-up. With the DI box set-up to record the guitar, also use a dynamic microphone to record the sound directly from the amplifier cabinet.

The microphone placement needs to be considered carefully, so you should listen to the sound of the DI box and then determine what type of sound you require from the amp cabinet. If the microphone is placed directly facing the centre of the amp cabinet, it will capture more of the higher frequencies, whereas if it is placed further to the side it will capture more of the lower end of the spectrum.

As a starting point, try placing the microphone approximately 300 mm above the cabinet pointing downwards towards the speaker and roughly 400 mm away. Again, using this set-up, monitor the resulting sound through headphones or in a control room before recording and move the positioning around until you gain the sound you require. By recording in this way, the DI box records a clean guitar signal, while the real microphone records the ambience of the room, alongside the tonal effects of the amplification. The two signals can then be mixed to produce the timbre required at the mixing desk.

If you do not have access to a DI box, then you will have no choice but to set a microphone up to record the amp cabinet. Since you are recording with only a microphone and no DI box to back the timbre up, it is unadvisable to place the microphone too close to the amplifier since it will tend to capture a significant amount of low-end rumble from the cabinet, rather place the microphone on a stand and place it approximately 600 mm above the cabinet pointing downwards towards the amp and roughly 600 mm away.

Using this set-up, monitor the resulting sound through headphones or in a control room before recording and move the positioning around until you gain the sound you require. As a general rule of thumb, if you place the microphone close to the centre of the amp, you'll receive more of the high end of the guitar, further to the side and you'll receive the lower end energy, so experiment before committing to a recording.

While the amp cabinet and DI box will tend to add their own type of compression, it's prudent to still apply some during recording due to the playing styles of electric guitarists; it should be applied heavier than on standard acoustic guitars.

A good starting point for compression is to set the ratio at 10:1 with a 7 ms attack, a 50 ms release and a hard knee setting. From here, reduce the threshold so that it captures the transients and prevents clipping. Note that you shouldn't aim to crush the guitar with the compressor since more compression can be applied at the mixing stage, but too much compression applied at the recording stage cannot be removed later.

BASS

Bass is perhaps one of the most difficult instruments to record faithfully, since by its very nature it's a bass instrument and as discussed in Chapter 1, bass has a long waveform. While all frequencies will reflect off walls, due to the length of a low-frequency waveform, if you are recording in a small room there will be a culmination of low-end frequencies at the microphone, even if the microphone is pointed directly at the amp cabinet. This results in more ambience and colouration being captured by the microphone which can result in a muddy sound. On top of this, the bass is an absolutely vital element in a dance mix, it is part of the groove that you dance too, so it has to be recorded perfectly.

The first step to recording a bass instrument is to ensure that the bassist can play the instrument well and that it's tuned correctly. A poorly tuned bass instrument will result in a nightmare at the mixing stage. Also, while fret noise is acceptable on an acoustic guitar it should be avoided on a bass guitar; while this can sometimes be avoided by setting the string height from personal experience, the majority of the cause is a result of an inexperienced thumb-slapping dickhead attempting to show off his talents rather than concentrate on achieving a good take.

Similar to electric guitars, you can use DI box to record the bass but it is of paramount importance that you use a good DI unit such as the *Ridge Farm Gas Cooker*. However, while this type of DI box will result in a clean focused timbre, do not underestimate the room ambience that can be captured by capturing the sound through a microphone. Consequently, it is advisable when recording bass to use both a DI box and a microphone to capture the sound; both can then be mixed together during the mixing stage to attain the best results.

Recording directly from a bass amp is difficult since you should have a fairly dead room to record in. This is because both reflections and reverberations from the room will be captured and, as discussed in earlier chapters, reverb should be avoided on bass since it will severely affect the position of the bass when it comes to mix-down. You need to ensure that you record in the biggest room available to prevent the reflected waveforms returning to the microphone and adding more bass end than you wish to capture.

Generally speaking, microphones with a good low-end response should be used to record the amp cabinet; if pushed to recommend a microphone, the Sennheiser E602 produces very good results but some engineers will use a microphone designed to record kick drums since they're designed to capture the lower end of the frequency range. Again, experimentation is the key and you should experiment with as many different microphones that you have at your disposal.

The microphone placement is generally the same as electric guitars in that if the microphone is placed directly facing the centre of the amp cabinet it will capture more of the higher frequencies, whereas if it is placed further to the side it will capture more of the lower end of the spectrum. As a starting point, try placing the microphone approximately 600 mm above the cabinet, pointing downwards towards the amp and roughly 400 mm away.

Again, using this set-up, monitor the resulting sound through headphones or in a control room before recording and move the positioning around until you gain the sound you require.

Since a bass instrument is an incredibly dynamic instrument, compression is a must. The settings to use on a compressor will depend entirely on the player, since a player's technique will vary widely from one to another. A good starting point on compression is to set the ratio to 12:1, a 10 ms attack, 20 ms release and a hard knee. Set the threshold to -13 dB and then slowly increase it until all the transients are captured. This should give a good recording with a limited dynamic range that will fit into a dance mix.

BRASS

While bass is considered a difficult instrument to record, brass has to take the award for being the most difficult. Firstly, with brass instruments the room's acoustics have a huge influence over the sound, and since they are blown instruments, there is an astounding amount of pressure leaving the bell section which can easily destroy most microphones if they're not carefully positioned. Also, brass is incredibly loud – often reaching a sound pressure level of 135–140 dB. On top of this, the sound of the instrument changes as the player becomes tired and the instrument fills up with spit.

Generally, ribbon microphones are widely regarded as the best to record brass instruments since they cannot only capture the instrument directly but also the surrounding ambience. However, due to the fragile nature of the microphones, a pop shield is an absolute must. In fact, regardless of the microphone you choose to use, always employ a pop shield since the sheer velocity of the air leaving the bell can easily pop the diaphragm.

It is also important to note that the sound of a brass instrument does not come directly from the end of the bell; rather it is from the vibration of the bell itself. Therefore, it is prudent to place the microphone approximately 2 m in front of the instrument and around 400 mm above the bell's position. This allows the microphone to pick up the acoustics of the room which is where a majority of the sound's character will come from.

If the performer is mostly going to be performing lower notes, you can place the microphone closer to the instrument but it is inadvisable to go closer than 500 mm since you won't capture much of the sound's character. In some instances you can place the microphone just above the performer's shoulder since if it sounds good to them, then it'll probably sound good to the microphone.

Obviously, the room should be appropriate for the sound, so you should try recording in a number of rooms to see which gives the best result. Tiled rooms such as a kitchen or bathroom (provided the room is big enough for the performer to perform in) can provide plenty of character to the timbre, but be cautious since too much reverb captured in the recording stage cannot be removed afterwards. Experimentation is the key here and will as always yield the best results. I've achieved great results by having the performer (in this case a trumpet) play facing a window with a dynamic microphone positioned 1 m away facing the window to record the reflections.

Before recording you should wear headphones to double check how the microphone is picking up the signal and not be afraid to move the performer's and microphone's positions around until you have the best signal you could possibly achieve.

While setting up for recording it is particularly important not to tire out the performer; so it is inadvisable to have them continually recite a performance until you have found the perfect positioning. Playing any brass (or woodwind) instrument is particularly demanding on a performer, they're taking sharp intakes of breath and blowing hard continually. Try it yourself for 5 min and you'll understand.

With this in mind, once you have the correct positioning, get the most energetic parts down first and as quickly as possible. The most demanding parts will be the higher and hardest notes, so try to get all these down first and always listen carefully to what is being played. Few performers like to admit when they're tired, so listen out for missed notes, lower volumes, missing attack stages on the instrument and also keep an eye on the player. A player turning red, then blue, followed by collapsing to the floor provides a good indication of just how tired the player is becoming.

If you're after recording a brass ensemble but only have one performer, do not rely on chorus effects to reproduce an ensemble artificially since it very rarely works well. Instead, ask the performer to play the part again and record as many takes as possible. Once you have them all, you can layer them down on different tracks of the sequencer to recreate an ensemble performance.

If you only have one recording and there is no chance of repeating the performance, you can copy the performance onto another sequencer track and change the timing between the two tracks by a couple of ticks, but the result will not be as impressive as multiple recordings.

As with most instruments, compression is absolutely vital on brass instruments due to the sheer volume of the instruments. Start by setting the compressor to a ratio of 10:1 with a 7 ms attack and a 30 ms release, and then set the threshold to −15 dB with a hard knee setting. Ideally, the compressor should be set so that it captures all the notes to keep the volume level throughout the recording. A fluctuating volume once recorded can be compressed further, but for brass it's generally a good idea to keep the volume steady throughout the performance while initially recording.

GENERAL TIPS

As with the rest of the book, these techniques are not set in stone and you should develop and experiment on the principles discussed. The most important factor to bear in mind, however, is to remain patient.

Since real instrumentalists are prone to nerves, emotion and exhaustion, you must exhibit some patience; pressuring a performer will only result in a rushed and poor performance.

What follows are some guidelines and general tips to receive the best possible performance:

- Use as many microphones you have at your disposal to record instruments.
- If possible, experiment with different pre-amps to see which gives the best results for the instrument you are recording.
- Record in as many rooms as possible and choose the room that gives the best response.
- Allow the performers to take regular breaks, even if they suggest that they do not need one; ask them to take one anyway. Tired performers will not give their best.
- Always monitor the microphone's response using headphones. This ensures you are only hearing what the microphone is picking up.
- If you have a portable system, try recording instruments in small halls, etc., to see if it produces the results you need. Real reverb, in small amounts, is much better than artificial on some instruments, especially brass.
- Be prepared to regularly punch in and out of recordings.
- Do not expect to receive a perfect one-shot performance. It is not unusual to 'comp' together a number of recordings to create a perfect performance.

CHAPTER 9

Sequencers

'Some great records are being made with today's technology and there are still great artists among us. Likewise there are artists today who are so reliant on modern technology; they wouldn't have emerged when recording was more organic.'

Unknown

With a thorough understanding of the technology behind producing dance music, we now need to look at how it all ties together, and in nearly all instances this is accomplished via a sequencer. The sequencer should be considered the most important aspect of any dance musician's studio. Whether hardware or software, it ties together all the technology, and over recent years, the sequencer has become the central hub of just about every studio in the world.

With the increased power in computers and the microchip, today's sequencer can work with both MIDI and audio, permitting the musician to not only pro-gramme synthesizer to play specific notes at specific times but also record and edit audio and apply effects and processing – all in real time. Indeed, processing power in computers has now reached the stage where it is entirely possible to cre-ate, edit, mix and master an entire piece of music entirely on a laptop (notebook) computer. It should be noted that what follows is only a quick rundown of a sequencer's capabilities; to discuss every detail would take another book in itself.

Nearly all of today's sequencers can be broken down into working with two main elements; MIDI and audio. While MIDI is often seen as an outdated for-mat, a basic understanding of its principles is still important, especially in the current age of software synthesizers.

The main page in any sequencer is often referred to as the arrangement page. This consists of a series of tracks that can be created and listed down the left-hand side. These tracks can consist of MIDI tracks, audio tracks or both. When playback is initiated (usually by pressing the space bar, or clicking play on the

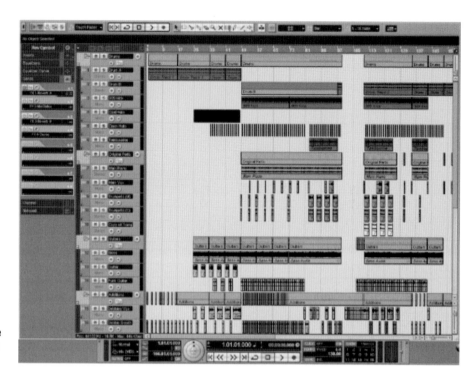

FIGURE 9.1
The arrangement page in Steinberg's Nuendo Sequencer

transport control), a timeline moves from left to right, playing back all the tracks simultaneously in the arrangement page (Figure 9.1).

In this arrangement page, it is possible to cut the tracks into segments/blocks using a scissor tool, and copy, repeat and move them around to produce an arrangement. For instance, if you have a 4 bar drum loop, you could continually copy this 4 bar 'block' over and over so that it plays for the full length of the arrangement. The same could be done with the bass and lead etc.

While you can organize all the individual parts in the arrange page, sequencers offer separate windows for performing more detailed editing. For MIDI, these commonly consist of a Piano Roll and CC editor. To further understand these, how they work and how they can be utilized to produce a record, we need to begin by examining the basics of MIDI.

In its most basic form a MIDI sequencer can be viewed as an advanced musical conductor, sending instructions to any number of synthesizer informing them which notes to play and when. Naturally, to achieve this, the synthesizer must be connected to the sequencer and the synthesizer must also be able to understand the messages the sequencer transmits to them. This connection and intercommunication is accomplished using a format known as Musical Instrument Digital Interface. Introduced in 1983, MIDI consists of a standardized set of instructions that all MIDI compatible synthesizer understand and using a simple 5-pin DIN plug allows them to connect to one another and a sequencer.

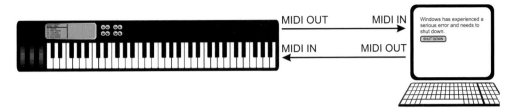

FIGURE 9.2
A simple MIDI set-up

In the early days of MIDI, a very simple set-up would consist of the MIDI OUT of a sequencer connected to the MIDI IN of a MIDI compatible synthesizer via a suitable MIDI cable. Connected in this way, any instructions sent from the sequencer would be received and performed by the synthesizer. The instructions sent by the sequencer consist only of a simple set of instructions that the synthesizer performs and do not consist of the sound itself. For example, a typical MIDI instruction consists of pitch with a note-on message, followed by a note-off message, for which the synthesizer would play a specific note and then stop (Figure 9.2).

Whereas this set-up may appear incredibly simple it does offer many benefits. You could programme the sequencer note by note at your own pace and then instruct the sequencer to play back the performance at any speed you desire. Or you could instruct the sequencer to transmit instructions to numerous synthesizers for all of them to play different parts of the pre-programmed performance. To accomplish this, you need to understand the Piano Roll Editor. Although different manufacturers of sequencers use different approaches to perform specific tasks, nearly all have adopted the 'piano roll' interface that was first introduced by Steinberg in their Cubase sequencer (Figure 9.3).

As you can see from Figure 9.2, down the left-hand side is a series of piano keys denoting the pitch. The higher up towards the top the notes are, the higher the pitch. To the right is the programming grid; when playback starts, any notes in this field are played back from left to right at a speed determined by the tempo of the piece. Drawing in notes is a simple as choosing the 'Pencil' tool from the sequencers toolbox, clicking in the programming grid and holding down the mouse button while you draw in the length of the note required. If numerous notes are drawn in at the same position in time, the notes will be chorded as the synthesizer plays back all the notes simultaneously. Once a series of notes have been drawn in, they can be lengthened, shortened, deleted or moved to perfect the performance.

Note, however, that the piano roll editor is split into a series of grids. These grid positions play an incredibly vital part of any music written on a sequencer, including drum sequencers, so it is important to have a thorough understanding of why they are present.

The main purpose of any sequencer is to transmit the appropriate message – such as a note-on message – to a sound-generating device at the specific times defined by the user. However, no computer is capable of rendering events at just

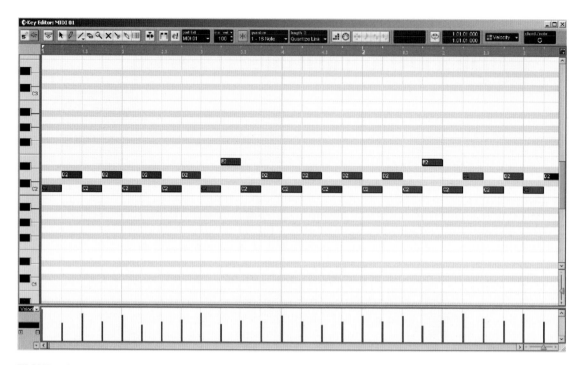

FIGURE 9.3
The piano roll editor in Steinberg's Nuendo Sequencer

any time and they are limited by the clock rate of the software or hardware circuit performing the process. Every time this clock pulses, the computer can perform another operation; the maximum number of clock pulses, also called clock 'ticks', that can occur within a given time period is known as the 'resolution'.

All MIDI sequencers specify this resolution in terms of pulses per quarter note (PPQN), and the higher this value the more useful the sequencer will generally be. As an example, if a sequencer has a PPQN of 4, each quarter note would contain four pulses, while each 8th note would contain two pulses and each 16th note would contain one.

What does this mean? If you're using a sequencer with a PPQN of 4, you couldn't draw in any notes that are less than a 16th because there isn't a high enough resolution to yield an integer number of clock pulses. Thus, not only would you be unable to create notes smaller than a 16th, it would also be impossible for any note timing to be adjusted to less than this either. This resolution is typical of some of the early analogue drum machines that were used to create dance rhythms and are responsible for the metronomic nature of the drum patterns.

While you could determine that you would only really need a PPQN of 16 to write a dance drum loop, the elements that sit upon this basic rhythm need a higher PPQN, otherwise the groove of the record will suffer because the sequencer will have no option but to keep the notes on the pulse of the PPQN. While on first impression, dance records may seem to remain very metronomic,

Table 9.1	Musical Notation and the Related Ticks	
Traditional Notation	**Length of the Note**	**Number of Ticks (Pulses)**
Whole note	One bar	1920
Half note	Half bar	960
Quarter note	Quarter bar	480
Eighth note	Eighth bar	240
Sixteenth note	Sixteenth bar	120
Thirty-second note	Thirty-second bar	60

that is far from the case. Very small sonic nuances give a record its groove, and if you can't accomplish small sonic nuances, your record will have no groove!

In the early 1990s, Roland came to the conclusion that a PPQN of 96 would be suitable for capturing most human nuances, while a PPQN of 192 would be able to capture the most delicate human 'feel'. Subsequently, most drum sequencers today use this latter resolution, while software sequencers have adopted a much higher PPQN of 480 and above. A sequencer that boasts a PPQN of 480 would work as shown in Table 9.1.

Each of these clock ticks can be referenced to a grid position within the piano roll editor (and indeed all of the sequencer), with the number of grids displayed on the screen determined by the quantize value. The quantize value will directly affect whereabouts the notes can be positioned, alongside the length of the note itself.

For instance, if the quantize value were set to 16, the grid would divide into 16th, and it would be impossible to place notes at smaller integers. The quantize value can obviously be used to affect rhythm, phrasing and embellishments, but its flexibility depends on the options offered by the sequencer.

Although quantizing features are helpful in creating the strict rhythms required in house drum loops for example, overuse of these techniques can produce anti-musical results. While there are some forms of music that, in theory at least, benefit from strict timing of each note, in reality even the most stringent forms of techno introduce subtle timing differences throughout to add interest and tension. Indeed, the importance of subtle timing differences shouldn't be underestimated when creating music that grooves.

Quantizing also plays an important part if you're more musically minded and prefer to actually play the instrument rather than programme it. Many MIDI instruments offer a MIDI OUT, and by connecting this to the MIDI IN of the sequencer, two-way communication becomes available. Using this, it's possible to play back a performance on the synthesizer itself, at the same time using the sequencer to record your performance. In many sequencers this is accomplished by creating a MIDI track on the sequencer, pressing record and playing the instrument.

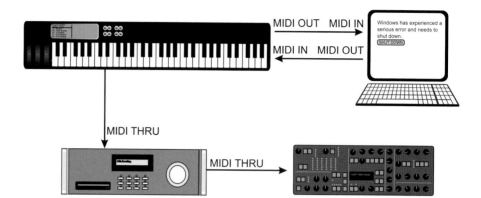

MIDI OUT   MIDI IN

MIDI IN   MIDI OUT

Windows has experienced a serious error and needs to shut down.
SHUT DOWN

MIDI THRU

FIGURE 9.4
A more elaborate MIDI set-up

MIDI THRU

All notes that are played are recorded direct onto the piano roll editor, whereby they can then be edited within the sequencer. For instance, badly played notes or poor timing can be corrected and then transmitted back to the synthesizer to recreate a perfect performance. Remember, however, that the quantize value is set too low, everything you record will be moved onto the next pulse and therefore you could end up losing the human feel you're trying to inject!

For many music projects one synthesizer will rarely be enough and you may wish to add samplers, drum machines and further sound-generating devices to the sequencer. This can be accomplished in one of two ways, utilize an instruments THRU port or purchase a multi-MIDI interface.

In addition to the IN and OUT MIDI ports, some of the more substantial synthesizer also features a THRU port. With this, information received at the IN port of the synthesizer can be transmitted back out (i.e. repeated) at the device's THRU port to the IN connector of the next device. Connecting synthesizer in this manner is often referred to as a 'daisy chain' and allows much more elaborate set-ups (Figure 9.4).

Using this method it is possible to connect, for example, a synthesizer, a drum machine and a sampler and have all these instruments playing together to recreate a performance. There are, however, a few problems with daisy chaining a number of synthesizers together.

Firstly, any message transmitted from a sequencer takes a finite amount of time to travel down the MIDI cable to the synthesizer. While this is certainly not noticeable on the first few synthesizers, if the information from the sequencer is destined for the ninth synthesizer down the daisy chain, it has to travel through the eight previous synthesizer first which can result in the information arriving 8 ms late resulting in a delay, an effect otherwise known as *latency*.

While 8 ms may not appear to be too drastic on reflection, the human ear is capable of detecting incredibly subtle rhythmic variations in music and even slight timing inconsistencies can severely change the entire feel of the music. Indeed, a timing difference of only 8 ms can destroy the strict rhythms that are required in some dance music genres and becomes especially evident with complex break beat or drum and bass loops.

The second problem arises when we consider how does each synthesizer know which information is destined for it. Simply sending an instruction such as a pitch and note-on/off message from the sequencer would result in every synthesizer in the chain playing that particular note. This latter problem can be solved through the use of MIDI channels.

All of today's sequencers can transmit MIDI information on different channels. Much like the different channels on a television, MIDI signals can be transmitted to a synthesizer on a number of different MIDI channels. The foremost reason for this is because many synthesizers are multitimbral, in other words, they can play back a number of individual instruments simultaneously. Typically, most of today's synthesizers are 16-part multitimbral but some of the more recent are 32-part multitimbral. This allows the sequencer to transmit, say, a piano part on channel 1 and a bass part on channel 2 and – provided the synthesizer was set up correctly – it would play back the piano part in a piano sound and the bass part in a bass sound.

This channel information, however, can also be used to communicate with specific synthesizer within a daisy chain. By specifying in the synthesizer to ignore (or pass through) channel 5, for example, any signal reaching that synthesizer on channel 5 would be passed to the MIDI THRU port and into the next synthesizer in the chain.

While this does provide a solution for connecting numerous MIDI devices together, it does come at the expense of losing a number of channels in each device so that the information can pass through it. It also doesn't solve the problem of possible latency further down the daisy chain. The only solution to circumvent these problems is to employ a multi-MIDI output device.

Multi-MIDI interfaces are most commonly external hardware interfaces connected to the sequencer, and offer a number of separate MIDI IN and OUT ports. Using these you can use different MIDI outputs to feed each different device rather than having to daisy chain them together. Notably, though, when looking to purchase a multi-MIDI interface, you should ensure that it utilizes multiple busses and does not operate on a single buss.

A single-buss interface will offer a number of MIDI outputs but they are all connected to one MIDI buss. This means that the 16 channels will be divided between the available outputs on the MIDI interface. In other words, these basically act as a substitute for devices that do not feature a MIDI THRU port.

Conversely, a multi-buss MIDI interface will provide 16 MIDI channels *per* MIDI output on the interface. Therefore, if a multi-buss interface offers four MIDI OUT channels, you'll have 64 MIDI channels to transmit over. Many of these multi-buss interfaces also feature a number of MIDI inputs too, saving you from having to change over cables if you want to record information from different synthesizers (Figure 9.5).

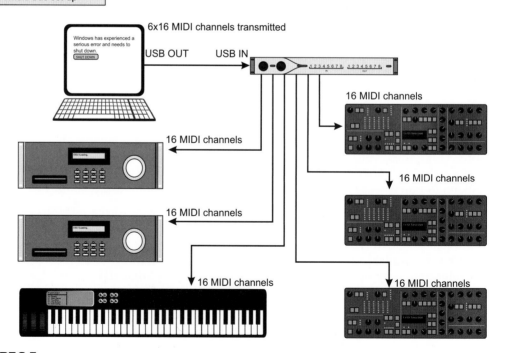

FIGURE 9.5
A multi-MIDI set-up

Of course, having multiple MIDI inputs begs the questions as to why would you want to record into a sequencer from numerous synthesizer. After all, if the sequencer only records simple data such as note-on/off and not the actual sounds, and the MIDI data received from all synthesizer keyboards is essentially the same, why not just one synthesizer as the main keyboard for recording?

If MIDI was only capable of simple pitch, note-on and note-off messages, anything performed via MIDI would sound incredibly dull and metronomic. While some dance music protagonists will argue that there is no human element behind any electronic dance music – it all sounds as though a machine has created them – this couldn't be further from the truth. As already touched upon, under close scrutiny you'll find that even the most metronomic sounding techno is crammed full of tiny sonic nuances to prevent it from becoming tedious. Indeed, this is the very reason why some songs appear to ooze energy and soul while others seem motionless and drab.

To give any music energy, drive and that indefinable character it requires the human element. Every 'real' musician will perform with expression by consciously or subconsciously using a series of different playing styles. For instance, a pianist may hit some notes harder than others; a guitarist may slide their fingers up their guitar's fret board bending the sound, or may sustain some notes while cutting others short. All of these playing nuances contribute to the human element behind the music and without it music would sound dull and lifeless.

Indeed, much of the energy and flow of any music can be attributed to slight or aggressive development of the sounds throughout the length of an arrangement. While any 'real' musician playing the piece naturally injects this into a performance, capturing this feel over MIDI requires you to be able to send more than simple on/off messages through MIDI.

With MIDI, sonic nuances can be either programmed or recorded live using a series of MIDI CC messages. As an example, when you play a note and move the pitch bend wheel, the wheel sends a continual stream of messages to the synthesizer's engine, which in turn bends the pitch either upwards or downwards. This message consists of a number informing the synthesizer what parameter is being adjusted, followed by another number informing it by how much it is being adjusted. This stream of messages can also be transmitted to the MIDI OUT and recorded into the sequencer as a series of MIDI control change messages (abbreviated to CC).

Of course, unless you happen to have three hands – two to play the instrument and one to move a controller on the facia of the instrument – it is much easier to record a performance and then create another MIDI track to then record any movements of the controller. While you can always record the movements onto the same track as the MIDI notes, it is a better practice to record them on a different track since it allows you to mute the CC movements without having to mute the MIDI notes too.

Once recorded, these CC messages can be edited within the sequencer to perfect the performance if required. This form of 'automation' can easily be accomplished in most sequencers using the sequencers toolbox. You can draw over the CC messages with a pencil, or straighten the movements into a gentle slope using line tools. Generally speaking, all sequencers will offer a graphical interface that permits you to do this.

CC MESSAGES

All synthesizer will (or should!) feature a pitch bend wheel but what if you want to control an element of the synthesizer that has no real-time control on its facia? There are two solutions to this. The first is to record pitch bend movements into a sequencer and then re-assign the pitch bend to control another aspect of the synthesizer, or you could just draw in automation using the sequencers toolbox.

Re-assigning CC controllers is dependent on the sequencer in question, but most will offer a method of doing so, typically, in the form of a drop-down box. You click on the drop-down box to choose a different CC number and the recorded (or drawn) automation will automatically be re-assigned to the new CC number.

However, it is important to note that not all synthesizers will understand the same CC messages. For example, you may re-assign pitch bend to control panning but if the synthesizer does not understand the CC number for panning it won't do anything. Alternatively, the CC number that equates to panning on one synthesizer may be set to control the resonance on another, so you wouldn't receive the desired effect.

In an effort to avoid this and encompass some kind of uniformity between different synthesizer, the general MIDI (GM) standard was developed. This is a generalized list of requirements that any synthesizer must adhere to for it to feature the GM symbol. Apart from determining what sounds are assigned to which numbers and how percussion sounds should be mapped across a keyboard, it also deals with the type of MIDI messages that should be recognized.

In total, there are 128 possible CC messages on any MIDI capable device but since sequencers all rely on computer chips in one form or another, 0 is classed as a number; therefore, they are numbered from 0 to 127. In the GM standard, many of these controllers are hard-wired to control particular functions on the synthesizer, so provided that the CC number is encompassed by the GM standard, your guaranteed that it will control the correct parameter no matter what synthesizer the messages are transmitted to, provided that the synthesizer is GM compatible.

Appendix D Features the Full List of CC Messages

As touched upon previously, each of these controllers must have variable parameters associated with it to permit you to set how much the specified controller should be moved by. Indeed, *every* CC controller message must be followed by a second variable that informs the receiving device how much that particular CC controller is to be adjusted. Some CC controllers will offer up to 128 variables (remember that's 0–127), while others will feature only two variables (0 or 1) – essentially on/off switches. It is also important to note that some CC messages will have both positive and negative values. A typical example of this is the pan controller, which takes the value 64 to indicate central position. Hence, when panning a sound, any values lower than this will send the sound to the left of the spectrum while values greater than this will send the sound to the right of the stereo spectrum.

While the GM standard is certainly useful in allowing synthesizer and sequencers to communicate effectively, it is not perfect since many synthesizer have outgrown the original specification of MIDI and offer many more user-accessible parameters. Consequently, two major synthesizer manufacturers developed their own MIDI standards that act as an extension to the GM standard. The first is Roland's GS system.

Contrary to popular belief, GS does not stand for general standard but is simply the name of the microprocessor used to power the engine (only Roland seems to know what it stands for). Nevertheless, this protocol is compatible with the GM standard while also adding variations to the original sound set, access to two drum kits simultaneously and a collection of new CC messages.

The second is Yamaha's XG system, which is also fully compatible with GM but offers more advanced functions than both the GM and GS formats. This system allows you to use three drum kits simultaneously, along with three simultaneous effects and is also partly compatible with the Roland's GS system. This is only partly compatible since Roland constantly updates the GS system with the release of new synthesizer.

As if these formats were not enough, the developers of the GM standard released the GM2 standard. This is yet again fully compatible with the original GM standard, but also adds a series of extensions to the original, including MIDI tuning and even more controllers.

Nevertheless, even with these new specifications and additional controllers, synthesizers are continually released which go beyond the standards set by the manufacturers and there will be instances where you need to access or control parameters that are not covered by these standards, therefore you will have to use System Exclusive messages.

Universal System Exclusive Messages

System Exclusive (often abbreviated to SysEx) is overlooked by an incredible amount of musicians since it can initially appear to be incredibly complicated. After all, musicians rarely want to start working with hexadecimal numbers they just want to make music.

Regardless, it is an art well worth learning since it can be used to great effect on *every* MIDI compatible unit. If you've programmed a synthesizer with your own set of personalized sounds, you could dump them all to one easy-to-manage file on the sequencer, allowing you to re-install then whenever required.

Alternatively it can be used to control or adjust parameters on a MIDI unit that has very few or no controls on its facia. It can even be used to adjust CC's that are listed on the MIDI specification but are not recognized by the MIDI unit. Furthermore, by inserting a string of SysEx messages at the beginning of a MIDI arrangement, you could take your own MIDI file to a friend's studio and, provided that they have the same synthesizer, the SysEx at the header of the file would not only set the sounds for each channel but also reprogramme the synthesizer to all the sounds you used.

To understand SysEx first requires an understanding of hexadecimal, or at the very least, how to count in hexadecimal. The normal way in which we count is known as Base 10, based around the logic that we have 10 fingers, therefore every time we count above 9, we have to move over one decimal point.

0…1…2…3…4…5…6…7…8…9…

Then we move over one decimal point and start again:

$1 \times 10 + 0$…1…2…3…4…5…6…7…8…9…

Take the number 1650 as an example. We can break this down into the following calculation:

$1 \times 1000 + 6 \times 100 + 5 \times 10 = 1650$

This could also be viewed as:

1000	100	10	0
1	6	5	0

With the number 1650 there are four decimal places; in other words, every decimal place in a number to the left of the decimal point is ten greater than the previous one.

Hexadecimal works at Base 16, not Base 10, we cannot move over 1 decimal place yet, so we need to continue by using letters in the place of 10, 11, 12, 13, 14 and 15. These are:

A = 10

B = 11

C = 12

D = 13

E = 14

F = 15

Therefore we count like this:

0...1...2...3...4...5...6...7...8...9...A...B...C...D...E...F

When we count to 16, we then move over one decimal place which would equate to 10, in other words F has moved over a decimal place to 16. Following this, 17 would equate to 11 (16 + 1), 18 would equate to 12 (16 + 2), 19 would equate to 13 (16 + 3) and so forth. When we reach 26, we then have to start using our letters again which would equal 1A.

That's $1 \times F(16) + 1 \times A(10) = 26$ (in other words 16 + 10 = 26).

In a more practical situation, we'll say that part way through an arrangement you want to change the waveform of a LFO. This is obviously not covered in the MIDI CC list but can be enabled using SysEx, usually listed in the final pages of a synthesizer manual.

To programme a certain type of behaviour using SysEx messages, you need to know what SysEx message to send to the synthesizer. SysEx messages typically consist of a hexadecimal address, which pinpoints the function within the synthesizer that you want to adjust, along with a hexadecimal value, telling the function how you want to adjust it.

Thus, by sending a SysEx message to the synthesizer reading 00 00 11 04 (derived in this case from the values shown in Table 9.2), the LFO rate would be set to waveform number 4.

It's all a little more difficult than this, however, because you can't just send the address and SysEx information. You also need to provide basic 'identification information' about the synthesizer that you are transmitting the message to, even if there is only one synthesizer in the entire set. If this additional information is not also sent, the message will simply be ignored. Thus, all synthesizers require that the full SysEx message is either eight or nine strings long, combining all the elements described, to complete a message. A typical message is F0, 41H, 10H, 42H, 12H, 00 00 11H, 04H, 10H, F7. To better understand what this means, the message can be divided up into nine distinct parts, as shown in Table 9.3.

Looking at Figure 9.3, the SysEx message we want to send, comprising the Address and the SysEx value, can be recognized in parts 6 (00 00 11) and 7 (04). The other seven parts of the message provide the required header and synthesizer identification information.

Table 9.2	SysEx Message Structure		
Address	**Parameter**	**SysEx Value**	**Adjustable Amount**
00 00 11	LFO waveform	00–04	0–4

Table 9.3		Example SysEx Message						
1	2	3	4	5	6	7	8	9
F0	41H	10H	42H	12H	00 00 11	7F	10H	F7

Parts 1, 2 and 9 are defined by the MIDI specification and are required in all SysEx messages. The first part simply informs the synthesizer that a SysEx message is on its way and sends the value F0, while the last part, part 9, informs the synthesizer that it is the end of the message with the value F7.

The second part of the message is specific to the manufacturer of the synthesizer. Each manufacturer employs a unique number, which is quoted in the back of the synthesizer's manual. In this example, 41H is the identification tag used by Roland synthesizers; thus, only a Roland synthesizer would prepare to receive this particular message and any other manufacturer's synthesizers will ignore it.

The third part of the message is used to identify a particular synthesizer even if it is from the same manufacturer. This can be changed in the parameter pages of the synthesizer in case another device in the set-up shares the same number, so that you can send SysEx messages to the one that you want to control rather than to all of them.

The fourth part of the message contains the manufacturer's model identification code for that particular synthesizer to ensure that the message is received and processed only by synthesizers with this particular model identifier (ID).

The fifth part can be one of two variables – 12H or 11H – used to specify whether the message is sending (12H) or requesting (11H) information. If a synthesizer receives a SysEx message it recognizes, it will look at this part to determine whether it needs to change an internal setting or reply with its own SysEx message.

An 11H message is usually employed to dump the entire synthesizer's patch settings into a connected sequencer. By sending this, along with the appropriate follow-on messages and pressing record on a sequencer, it's possible to save any user-constructed patches. This is useful if the synthesizer has no user patch memories of its own.

As already observed, sixth part contains the address of the function on which the SysEx message is to act, which in this case is the LFO rate. Most synthesizer manuals provide an address map of the various functions and the corresponding addresses to use.

The seventh part can serve two different functions. If the fifth part contains the value 12H (indicating that the message is being sent), then part 7 will contain the data being sent, as shown in the example we're using (where the SysEx value 04 is sent to set the LFO rate). If the fifth part contains the value 11H, indicating that information is being requested, the value of this part indicates the number of bytes you want the synthesizer to return in its reply message.

Whether eighth part is included will depend on the type of synthesizer you use. Some manufacturers employ this 'extra' byte as a checksum, which is used to validate the message and to ensure that the receiving synthesizer does only what the hexadecimal code asks it to. Although MIDI is a fairly stable platform, there will be the odd occasion when notes stick and messages become corrupted during transmission. If this happens, it's possible that the original message could end up asking the synthesizer to do something entirely different, such as erasing all user presets. Errors like this are avoided with a checksum because if the checksum and message do not match, the SysEx message will be ignored. This is why you need to understand how to calculate and count in SysEx.

A checksum is calculated using the following formula, taking the SysEx message shown in Figure 9.3 as the basis for the calculation.

For example, convert the address and SysEx hex values from parts 6 and 7 of the message to decimal values and add them together to give the value H. Based on the values from Figure 9.3:

Part 6 _ 00 00 11 (hex) and converts to a decimal value of 17.

Part 7 _ 04 (hex) and converts to a decimal value 4. Therefore; H _ 4 _ 17 _ 22.

If the value of H is greater than 127, then we need to minus 128 (H _ 128) from it to produce the value X. In this case H _ 22, which is less than 127, so X is also 22.

H _ 22, which is less than 128; therefore X _ 22.

Finally, we convert the value X to hexadecimal to derive the checksum value for the message. The decimal value 22 in hexadecimal is 16. Therefore, the checksum value is 16.

Although creating your own messages can be boring and time-consuming, it does give you complete control over every aspect of a synthesizer, allowing you access to parameters that would otherwise remain dormant.

It should, however, be noted that only one SysEx message can be transmitted at any one time and additional messages cannot be received while the synthesizer is using that particular function. Furthermore, any synthesizer will pause for a few microseconds while it receives and processes the command, so any two messages must not be sent too close together so as to give the synthesizer time to process the previous command. If a second message is transmitted while the current command is being processed, the synthesizer may 'lock up' and refuse to respond and a 'hardware boot' (switch off then on again) must be performed to reset it. Consequently, the timing between each SysEx message should be considered when transmitting any messages.

These timing problems make SysEx pretty useless if you want to continually increase or decrease a parameter while it is playing back. For instance, if you wanted to emulate the archetypal dance music 'filter sweep' where the sound becomes gradually brighter, you would need to send a continual stream of messages to the filter cut-off to steadily open it. As this is often considered to play an important part of music, it is included in the list of CC messages (CC74 brightness), but if you wanted to progressively adjust the rate of an LFO during playback this is not covered, so many manufacturers designate these types of non-registered messages to non-registered parameter numbers (NRPN).

Most dance music relies heavily on cleverly programmed sounds as well as samples. By accessing and controlling these 'hidden' elements of a synthesizer from a MIDI arrangement using registered parameter numbers (RPN) and USEM messages, it is possible to switch the functions of a synthesizer part way through a performance or develop sounds that are not accessible through the CC list over the length of the arrangement.

NRPN AND RPN CONTROLLERS

NRPN, or 'Nerpins', are similar to regular CC messages but can be set to control elements that are not available in the MIDI protocol or that need to be finely controlled. Nerpins are similar to SysEx and are high-resolution controllers that address up to 16384 separate parameters on a synthesizer. They were introduced into the MIDI format as a means of allowing access to a wider range of parameters that are specific to any one particular synthesizer.

For instance, while the rate of an LFO is not accessible through CC messages, the manufacturer could assign any NRPN number to select this function in the synthesizer, which could then be followed by two other messages (Data Button Increment and Data Button Decrement controllers) to adjust the parameters' values.

This negates the need to use any SysEx messages and makes it possible to adjust any parameter on a synthesizer, provided that the manufacturer has previously assigned it. Unfortunately, very few synthesizers actually utilize NRPN controllers, so it is always worth checking the back of a manual to see if they are supported and, if so, what they can control.

It is important to note that if a manufacturer has assigned Nerpins you can only access and adjust one at once. For instance, if you were using them to adjust the LFO rate and then wanted to adjust, say, the amplifier's attack, you would need to redefine the current NRPN so that the Data Button Increment and Data Button Decrement adjust this rather than the LFO rate. There may be over 16 000 NRPNs but there are only three controllers to adjust their properties.

Fortunately, there are some settings that most manufacturers now define as a standard, such as the Bend Range of the pitch wheel and the master tuning. It's for this reason that we also have a pair of controllers for these defined parameter numbers. These are referred to as RPN (generally called 'RePins'). These are universal to all MIDI devices but at the moment only six of them are actually used: Pitch Bend, Fine Tuning, Coarse Tuning, Tuning Programme Select, Tuning Bank Select and Null. As the MIDI specification continues to evolve this list will undoubtedly continue to grow and should eventually provide universal access to the basic sound editing and effect processing parameters that are currently the non-standard domain of Nerpins.

AUDIO SEQUENCING

Sequencers have moved on from simple MIDI sequencing and now offer a variety of audio sequencing parameters. Similar to MIDI these allow you to record, import and edit audio in much the same way. The audio can be displayed graphically and edited with a mouse, and it is also possible to apply a number of real-time effects to the audio such as delay, reverb and compression, negating the need to purchase additional hardware units to accomplish the same job. Additionally, all the audio is mixed together using a virtual software mixer, so there is little need to purchase a hardware alternative. All the audio can be mixed and EQ'd inside the computer and the resulting stereo signal can be output from the soundcard directly into the loudspeakers.

This has the obvious advantage of being much cheaper than having to purchase the hardware equivalents but the disadvantage that working with audio requires a lot more power than simply working with MIDI hardware. You're no longer simply sending small bytes of information to attached synthesizers or samplers because you're storing and manipulating audio on the computer itself, and unless the computer is of a high enough spec, this may not be possible. While all PCs and Macs available today are capable of acting as a digital audio workstation, there are a number of important factors that determine how many audio tracks you can run at once and how well they will perform when a number of effects are applied to the audio tracks.

Most of today's software sequencers are more than capable of playing back more than 200 audio tracks at a time but this does not necessarily mean they can; it depends entirely on the access speed of the hard drive, the processor speed and the amount of memory installed. All audio used by a sequencer is stored on the hard drive and this is read directly from the drive on playback; thus, the speed at which data can be read (or written while recording) forms a fundamental part of how many audio tracks can be played simultaneously.

Typically most computers come equipped with a minimum of 100 GB hard disk space available, and while this may seem plenty, in many cases this will be used up incredibly quickly. If you want to record long digital audio tracks, for example a live performance, then a typical CD quality stereo file will require 10 MB/min, so a single 3 min track will need 30 MB of free hard disk space, and as some music can consist of 10–15 tracks played simultaneously, it would require 300–450 MB of free space. Add to this a number of sample-based instruments; these are virtual instruments containing samples and often come with over 10 GB of samples that must also be stored on a hard drive and 100 GB drives can soon be filled!

An important consideration, if a PC is to be used as the sequencer, is the motherboard itself. For any audio applications a board that uses an Intel-based chipset will perform much more reliably than any other. All audio software manufacturers test their equipment using Intel chipsets, but not as many will test them on others and there have been innumerable reports from users of AMD-based systems of incompatibilities between soundcards, SCSI adaptors and USB devices (such as USB-based soundcards and multi-MIDI interfaces). Indeed there have been some serious issues reported with PCI to USB host controller chips with non-Intel-based chipsets, such as problems that prevent any attached USB devices from working at all.

Though Intel-based chipsets are more expensive, the reliability offered by them is worth it in the long run, especially if you want to make music rather than spend most of the time chasing incompatibility issues. Notably, many sequencers and other audio software packages are also specifically programmed to utilize the SSE II instruction sets used on Pentium 4 processors and some companies will not offer product support if there are incompatibilities with their software and non-Pentium-based systems. As the computer is the basis on which the entire studio will be based, money should be no object.

The most important link in the chain, however, is the soundcard as this defines the ultimate quality of the music you produce. Essentially, a soundcard performs two distinct functions; providing a basic MIDI interface and enabling audio playback and recording. There is a huge variety of audio cards available, some manufactured specifically for audio work while others aimed more towards computer gamers. However, although there are a number of factors to bear in mind when looking for any suitable audio card, for music use the three most important questions to ask are as follows.

HOW GOOD IS THE AUDIO QUALITY?

All audio cards will offer much the same audio quality provided that the audio remains inside the computer (digitized audio simply consists of numbers) but the quality of the audio being recorded from the soundcard's inputs is entirely dependent on the quality of the card's analogue to digital converter. Cheaper cards will have a poor signal-to-noise ratio resulting in an audible hiss on any recordings taken using them and many of these cards also remove some of the higher and lower frequencies of a sound. Also it is likely that they will use 16-bit converters rather than 24-bit, so they cannot capture the dynamic range as well (don't panic we'll be looking at all this in Chapter 4), That said, while 24-bit soundcards offer superior recording quality when recording real instruments or vocals, to record this high requires more memory and CPU power and, generally speaking, 16-bit soundcards can still achieve excellent results.

ARE THE SOUNDCARD'S DRIVERS 'MULTI-CLIENT'?

As soon as any audio sequencer is opened it captures the audio cards inputs and outputs to use within the programme. If the drivers on the card are not multi-client, the sequencer will be the only programme that can access the card's connections, thus it is not possible for any further audio applications to access the card while the sequencer is running. A number of companies may promise that multi-client capabilities will be included in later driver updates but this shouldn't be taken literally, and if the card does not have them, you should look for one that does. Keep in mind that many companies will promise updated drivers but instead release newer cards in their place. Visiting the manufacturer's website and checking the currently available drivers and their frequency is probably the best way to discover whether they are willing to update them to keep up with moving technology.

HOW MANY INPUTS AND OUTPUTS (I/O) ARE ON OFFER?

All soundcards will offer a stereo input and output but this may not be enough to suit your needs. In some situations you may wish to output each audio channel of the sequencer to a different output so that you can mix down on a hardware mixing desk. Additionally, if you have plans to record more than one instrumentalist at a time then you'll need an input for each performer. Although you may not feel that you need more than one stereo input or output, it is wise to plan for the future.

Provided that you have a capable computer and a good soundcard, possibly the largest benefit from using an audio sequencer is the opportunity to use a series of plug-ins. These are small auxiliary programmes that add features to the host programme, and for digital audio software, they can give you access to additional effects such as EQ units, filters, vocoder's and reverb algorithms.

Plug-ins for 'home' sequencing platforms usually appear in a number of formats, Audio Units, Universal Binary, Direct X and VST 2. Audio Units and Universal Binary are Mac only interfaces for use on their OSX and later operating systems, while Direct X is a Windows only application and VST 2 is available for both platforms.

Nevertheless, AU, UB, Direct X and VST are essentially one and the same as they allow access to real-time effects but whereas Direct X and AU can be used in a wider range of sequencing programmes, VST 2 and UB is limited to sequencers that support the standard. This limited support is advantageous, however, as the programming is more integrated providing more stability while also using less processor power.

More recently, Steinberg developed the Virtual Instrument 2 plug-in interface. This works on a principle similar to the aforementioned plug-ins but allows you to use software emulations of real-world instruments within any VST 2 compatible sequencers.

Additionally, software samplers have also started to appear, offering all the benefits of sampling but from within the sequencer, negating the need to use any external MIDI devices whatsoever. More significantly, as their audio output can be directed to a number of channels in the audio sequencer's mixer, the signal can be mixed with the other audio channels contained in the virtual mixer while allowing you to EQ and apply any plug-in effects to their output.

To further explain the advantages of a software-based studio, Ronan Macdonald, an accomplished musician and editor agreed to be interviewed.

Q: What are the benefits of a studio based entirely on software?

'The benefits of the software-only studio are many, and so compelling that the music technology hardware industry is genuinely suffering for it (as, indeed, are recording studios the world over), as more and more musicians get their home studios up to professional standards thanks to the power and price of software.'

'First, then, price. Music software, generally, is much more cost-efficient than hardware. For the price of a decent hard disk recorder, you can now buy a very powerful computer AND a MIDI/audio sequencer that leaves the hardware equivalent standing.'

'Second, convenience and mobility. The advantage of having your entire studio running inside your Mac or PC is self-evident in terms of the space it takes up and the lack of need for cables, racks and all the rest of it. And if that Mac or PC happens to be a laptop, the freedom to take your entire studio wherever you want is nothing short of a modern miracle.'

'Third, total recall. Generally, with a hardware-based studio, it's only possible to work on one mix at a time, since all mixer settings, patch bays and instrument and effects settings have to be cleared for each new project. A totally software-based studio doesn't have this problem, as all settings are saved with the song, meaning you can work on as many projects at once as your brain can handle.'

'Fourth, sheer power. Today's computers are really, really powerful. It's now possible to run well over 60 tracks of plug-in-processed audio alongside numerous virtual instruments in real time. Sequencers also make MIDI and audio editing highly user-friendly, very quick and very easy, and in terms of sound quality, since nothing need ever leave the digital domain, low noise and crystal clarity can be taken for granted with a software-only setup.'

Q: What are the advantages of using VST instruments over the hardware alternative?

*'Again, price is a major factor here. Hardware synth's can cost hundreds or even thousands of pounds; soft synth's very rarely cost more than £400 and usually a lot less. These days it's possible to buy some incredibly powerful virtual instruments on the net for as little as £30 and that's not to mention the countless free ones out there! On top of that is the fact that you can run multiple instances of most virtual instruments at the same time. As if having a totally realistic software Prophet-5 (a staple instru-*ment in the creation of dance music) *wasn't cool enough, how about having eight of them all running at once, and all for the price of just one?'*

'When it comes to samplers in particular, hardware simply can't compete with computer-based systems. Editing a big sample is infinitely easier on a computer monitor than it is on a sampler's LCD screen, and being able to store gigabytes of samples on a computer hard drive makes maintaining and archiving a large sample library a mundane task rather than a constantly fraught one.'

Q: But do virtual instruments sound as good as hardware instruments?

'Absolutely. While early VST synth's were limited in terms of quality by the speed of computers a few years ago, these days there are no such problems. The latest generation of analogue emulations, such as GMedia's Oddity and Arturia's Moog Modular V, have shown that we now have enough power at our disposal to be able to process exceptionally realistic oscillator algorithms, leading to sound quality that simply can't be distinguished from 'the real thing.'

Q: So you can play these in real time like any hardware instrument?

'While there is always a certain amount of latency involved in playing soft synth's (the delay between pressing a key and actually hearing the sound come out of the computer), with a high quality (which doesn't necessarily mean expensive) soundcard, this can be easily brought down to less than 10 ms, which is all but unnoticeable. And once you've actually recorded the performance into the computer, there's no latency at all on playback.'

Q: What would you recommend as the basic requirements to those starting out writing dance music on a computer?

'While as with everything in computing, the more money you spend, the more power you'll get, any off-the-shelf PC will be capable of running even the most demanding of today's music software. The soundcard is more of an issue than the PC itself, and although perfectly good results can be had using any built-in sound system, it's definitely better to get at least a Sound Blaster Live or Audigy, or – even better – a pro solution such as M-Audio's Audiophile 2496, which offers incredibly low latency and stunning sound quality, and can be bought for around £150.'

'In terms of software, the budding dance music producer is spoilt for choice. Propellerhead Software's Reason offers a beautifully conceived and totally self-contained virtual studio for around £220, featuring synths, samplers and effects and sounding simply awesome. Cakewalk's Project5 (£249) takes a similarly dance-oriented approach but has the added benefit of being compatible with DirectX and VST plug-ins. And for what qualifies as unarguably the cheapest way into computer-based music production, Computer Music magazine features the CM Studio free on the cover CD of every issue. This comprises a powerful sequencer (Computer Muzys), a virtual analogue synth, a drum synth, a drum sample player, a sampler and a rack of effects.'

CHAPTER 10
Music Theory

'All you need is a feel for the music. There are people that have been to college to study music and they can't make a simple rhythm track, let alone a hit record...'

Farley 'Jackmaster' Funk

BASIC MUSIC THEORY

Having discussed the technology behind the creation, processing and recording of dance music, before we move onto creating different genres of music, it's helpful to have a basic understanding of some musical theory. While some dance musicians have released records without any prior knowledge of music theory whatsoever, at least some familiarity can help you to understand why some different genres of dance may use a different time signature.

It's important to note right from the start that this is a basic introduction to musical theory aimed at those who have little or no musical knowledge. Delving into the finer points of music theory is beyond the scope of this book, so we're only going to look at areas that are relevant to the dance musician. This information will be developed upon as in the discussions of different musical genres in later chapters, so if you feel lost with what follows it will all come together by the end of the book.

We need to begin by first examining the building blocks of music, starting with the musical scale.

THE MAJOR SCALE

The major scale consists of a specific pattern of pitches that are named after the first seven letters of the alphabet. These always follow a specific sequence and always begin and end on the same letter to create an octave. For example, the C major scale always begins and ends with a C and the distance between the two

Table 10.1	The C Major Scale						
Vocal	'Do'	'Re'	'Mi'	'Fa'	'So'	'La'	'Ti'
Key	C	D	E	F	G	A	B
Degree	1	2	3	4	5	6	7
Name	Tonic	Supertonic	Mediant	Subdominant	Dominant	Submediant	Subtonic

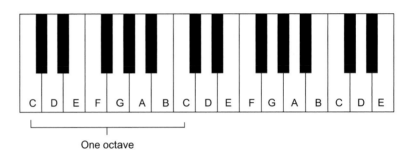

FIGURE 10.1
Layout of the keyboard

One octave

Cs is always one octave. For those who've endured Julie Andrews warbling in the *Sound of Music* or attended singing lessons, we can relate this to *Do – Re – Mi – Fa – So – La – Ti – Do* (the final *Do* being one octave higher than the previous). What's more, each pitch in the octave is also given its own name and number, the latter of which is referred to as a degree. These relationships are shown in Table 10.1.

More important, though, is the distance between each of these notes, as most of musical theory is based around understanding this relationship. To comprehend the space between notes in a scale we need to examine the layout of a typical keyboard and the note placements (Figure 10.1).

Each key on the keyboard relates to the pitch of that particular key and is equal to 1 semitone. Looking at the keyboard, we can see there are both white and black keys. These black keys are called sharps, and in written notation they are identified by the # symbol, so we can see that to the right of C is a raised black note, and we call this C#. Likewise, the black note to the right of D is called D# and so on (excluding the notes E and B as these do not have sharp notes associated with them).

Because each note of a particular pitch is equal to a semitone, in music these can be added together and classed as one tone. Because the note of C is followed by C#, the two can be added together to produce one tone (1 semitone + 1 semitone = 1 tone). Consequently, the notes C, D, F, G and A, which all have associated sharps, are known as whole tones, while the keys E and B, which don't have sharps, are referred to as semitones. From this we could view the C Major scale as:

Tone – Tone – Semitone – Tone – Tone – Tone – Semitone

This pattern of tones and semitones is the pattern that defines a major scale. If, rather than starting at C, we start at D through to the D an octave higher, this arrangement changes to what's known as Dorian mode spacing:

Tone – Semitone – Tone – Tone – Tone – Semitone – Tone

Although this may not seem significant on paper, it has an impact on the sound because the key has changed; that is, D is now our root note. The best way to understand this is to import, or programme, a melody into a sequencer and then pitch it up by a tone. Although the melody remains the same, the tonality will change because it's now playing different pitches than before.

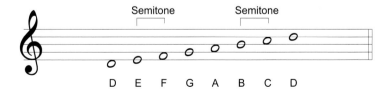

There's much more to this than simply pitching melodies up and down in a sequencer, however, because we need to take the sharps into account. For instance, with D major there are two sharps (F and C). Looking back at the keyboard layout shown in Figure 10.1, these correspond to the F and C.

Say that you've written a melody in C and you pitch it up 5 semitones to F. This essentially means that F becomes the root note and the pattern of tones changes to the Lydian mode spacing:

Tone – Tone – Tone – Semitone – Tone – Tone – Semitone

Again, this looks fine on paper, but there's a problem with simply pitching it up to F because if the original riff in the key of C is simply pitched up to F, any note that was *originally* an F would become an A# and any note that was *originally* an E would become an A. In other words, F has become A sharp. In this event, the black keys are no longer referred to as sharps; instead they're referred to as flats, which are identified by a '*b*', as in B*b*.

Because the black keys on the keyboard can be either sharps (*#*) or flats (*b*), depending on the key of the song or melody, they are sometimes referred to as 'enharmonic equivalents'. Of course, depending on where the scale is constructed the black keys will be sharp or flat, but they can never be a mixture of the two.

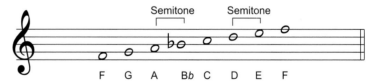

MINOR SCALES

Along with the major scale, there are three minor scales consisting of the 'harmonic' minor, 'natural' minor and the 'melodic' minor. These are based around exactly the same principles as the major scale, with the only difference being the order of the tones and semitones. As a result, each of these minor scales has a unique pattern associated with it.

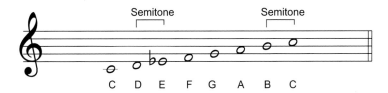

The harmonic minor scale has semitones at the scale degrees of 2–3, 5–6 and 7–8. Thus they all follow the pattern:

> Tone – Semitone – Tone – Tone – Semitone – Tone – Semitone – Semitone

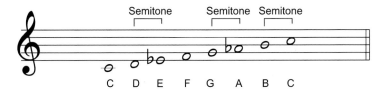

The natural minor scale has semitones at the scale degrees of 2–3 and 5–6. Thus they all follow the pattern:

> Tone – Semitone – Tone – Tone – Semitone – Tone – Tone

The melodic minor has semitones at the scale degrees of 2–3 and 7–8 when ascending, but reverts back to the natural minor when descending. Thus they will all follow the pattern of:

> Tone – Semitone – Tone – Tone – Tone – Tone – Semitone

Generally speaking, the harmonic minor scale is used to create basic minor chord structures that are used to harmonize with riffs or melodies that are written using the melodic and natural minor scales. Every major key will have a related minor and every minor will have a related major, but this doesn't mean that the closest relationship to the key of C major, for example, would be C minor. Instead, the major and minor keys that have the most notes in common with one another are most closely related.

For instance, the closest related minor to C major is A minor because it has the most notes in common. As a guideline, you can define a minor from a major by building it from the sixth degree of the scale; a major can be constructed from the

Table 10.2	Modes	
Key	**Name**	**Mode**
C	Tonic	Major (Ionian)
D	Supertonic	Dorian
E	Mediant	Phrygian
F	Subdominant	Lydian
G	Dominant	Mixolydian
A	Submediant	Minor (Aeolian)
B	Subtonic	Locrian

minor scale by building from the third degree of the minor scale. The structures of major and minor scales are often referred to as modes. Rather than describing a melody or song by its key, it's usual to use its modal term (Table 10.2).

Modes are incredibly important to grasp because they will often determine the emotion the music conveys. This isn't due to some scientific reasoning but because it's our instinctive reaction to subconsciously reference everything we do, hear or see with past events. Indeed, it's impossible for us to listen to a piece of music without subconsciously referencing it against every other piece we've ever heard.

This is why we may feel immediately attracted to or feel at 'home' with some records but not with others. For instance, most dance music is written in the major Ionian mode. If it were written in the minor Aeolian mode the mood would seem much more serious.

CHORDS

There's much more to music than understanding the various scales and modes since no matter what key you play if only one note was played at a time it would be rather boring to listen to. Indeed, it's the interaction between numerous notes played simultaneously that provides music with its real interest. More than one note played simultaneously is referred to as a harmony, and the difference in the pitch between the two played notes is known as the 'interval'. Intervals can be anything from just one semitone to above an octave apart. It should go without saying that the sonic quality of a sound varies and multiplies with each additional note that's played.

Each different interval is given its own name. For example, harmonies that utilize three or more notes are known as chords. The size of these intervals influence the 'feel' of the chord being played and some intervals will create a more pleasing sound than others. Intervals that produce a pleasing sound are called consonant chords, while those less pleasing are called dissonant chords.

Table 10.3	Interval Relationships	
Interval	**Semitones Apart**	**Description**
Unison	Same note	Strongly consonant
Minor second	One	Strongly dissonant
Major second	Two	Mildly dissonant
Minor third	Three	Strongly consonant
Major third	Four	Strongly consonant
Perfect fourth	Five	Consonant or dissonant
Tritone	Six	Mildly dissonant
Perfect fifth	Seven	Strongly consonant
Minor sixth	Eight	Mildly consonant
Major sixth	Nine	Consonant
Minor seventh	Ten	Mildly dissonant
Major seventh	Eleven	Dissonant

The combination of the two together can be used to create tension in music, so each is of equal importance when you are writing a track. Indeed, creating a track using nothing but consonant chords results in a track that sounds quite insipid, while creating a track with only dissonant chords will make the music sound unnatural. The two must be mixed together sympathetically to create the right amount of tension and release. To help clarify this, Table 10.3 shows the relationship between consonant and dissonant chords, and the interval names over an octave.

Intervals of more than 12 semitones that span the octave have a somewhat different effect.

Nevertheless, to better understand how chords are constructed we need to jump back to the beginning of the chapter and look at the layout of C major.

Using this chart we can look at the simplest chord structure – the triad. As the name suggests, this consists of three notes and is constructed from the first, third and fifth degree notes from the major scale that forms the root note of the chord.

You can see from Figure 10.2 that the root note of the C major scale is C, which corresponds to the first degree. To create the C major triad, the third and fifth degree notes must be added, giving us the chord C–E–G (or first, third and fifth

Pitch	C	D	E	F	G	A	B
Degree	1	2	3	4	5	6	7

FIGURE 10.2
Grid of the C major scale

degree). This major triad is the most basic and common form of chord and is often referred to as 'CMaj'.

As before, this chord template can be moved to any note of the scale; thus there is a whole scale of chords available. For example, a chord from the root key of G would consist of the notes G–B–D. Similarly a triad in the root key of F would result in F–A–C. This also means that in a triad with a root key of D, E, A or B, the third tone will always be a sharp.

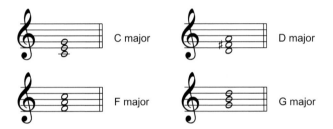

In addition, there are three variations to the major triad chord: the 'minor', 'diminished' and 'augmented' triads. Each of these works on a similar principle as the major triad, with the difference that for the minor and diminished triad, the third tone is lowered by one semitone (Figure 10.3).

FIGURE 10.3
Variations to the major triad

For instance, taking the major triad C–E–G from the root key of C and lowering the third tone gives us the notes in the minor triad C–Eb–G. For a diminished triad we do the same again, but this time both the third and fifth tone are lowered, resulting in C–Eb–Gb. Finally, to create an augmented triad we increase the fifth note by a semitone, producing C–E–G#. These and other basic chord groups are shown in Table 10.4.

So, to recap:

- To create a major triad, take the first-, third- and fifth-degree notes of the scale.
- To create a minor triad, lower the third degree by a semitone.

Chord/Root	Major	Minor	Diminished	Augmented
Table 10.4	Basic Chord Groups			
C	C–E–G	C–Eb–G	C–Eb–Gb	C–E–G#
D	D–F#–A	D–F–A	D–F–Ab	D–F#–A#
E	E–G#–B	E–G–B	E–G–Bb	E–G#–C
F	F–A–C	F–Ab–C	F–Ab–B	F–A–C#
G	G–B–D	G–Bb–D	G–Bb–Db	G–B–D#
A	A–C#–E	A–C–E	A–C–Eb	A–C#–F
B	B–D#–F#	B–D–F#	B–D–F	B–D#–G
C# (enharmonic Db)	C#–E# (F)–G#	C#–E–G#	C#–E–G	C#–E# (F)–A
Db	Db–F–Ab	Db–E–Ab	Db–E–G	Db–F–A
Eb	Eb–G–Bb	Eb–Gb–Bb	Eb–Gb–A	Eb–G–B
F# (enharmonic Gb)	F# A# C#	F#–A–C#	F#–A–C	F#–A#–D
Gb	Gb–Bb–Db	Gb–A–Db	Gb–A–C	Gb–Bb–D
Ab	Ab–C–Eb	Ab–B–Eb	Ab–B–D	Ab–C–E
Bb	Bb–D–F	Bb–Db–F	Bb–Db–E	Bb–D–F#

- To create a diminished triad, lower the third and fifth degrees by a semitone.
- To create an augmented triad, raise the fifth degree by a semitone.

These examples in Table 10.4 show single chords. In a musical context, the transition between these chords is what gives music its unique character and sense of movement. To introduce this sense of movement, the centre note of the chord (the third degree) is inverted or moved as the chords progress. This means that while both the first and fifth degrees of the chord remain 7 semitones apart, the third often moves around to create intervals. For instance, moving the third degree up a semitone creates an interval known as a major third, whereas if it were positioned or moved to 3 semitones above the root note, it would become a minor third. Inversions are created in much the same way and provide a great way for the chords to move from one to the next.

Inversions are created when one of the notes is moved by a melodically suitable degree. How this is achieved is largely down to experimentation. As an example, if we are writing a tune using the chord of C, we can transpose the first degree (C) up an octave so that the E becomes the root note, to create what's called a 'first inversion'. Next, we can transpose the E up an octave, leaving the G as the root. This creates a second inversion.

C maj   First   Second

This kind of progression is especially suitable for pads or strings that need to smoothly change as the track progresses without capturing too much unwanted attention. That said, any movement of any degree within the chord might be used to create chord progressions, which, as mentioned, is down to experimentation.

Simple triad chords, such as those shown in Table 10.4, are only the beginning. The next step up is to add another note, creating a four-note chord. The most commonly used four-note chord sequences extend the major triad by introducing a seventh degree note, which results in a 1–3–5–7 degree pattern (*C–E–G–B in C major*).

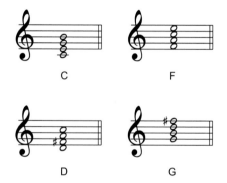

C   F

D   G

It is, however, uncommon for four-note chords to be employed in dance music because major and minor sevenths are more synonymous with Jazz music. That said, you should experiment with different chord progressions as the basis for musical harmony. For those who still feel a little unsure about the construction of chords, listed below are some popular chords:

C–Eb–G–A
C–Eb–G–D
C–Eb–G–Bb–F
C–Eb–G–Bb–D
C–Eb–G–Bb–D–F–A
C–Eb–G–B–D
C–Eb–F#–Bb–D
C–Eb–F#
C–Eb–F#–A–B
C–E–G#–B

C–E–G–B
C–E–G#
C–F–G–Bb–C#–A
C–F–G
C–F–G–Bb–D
C–E–G–B–D
C–E–G–B–D–F#
C–E–F#–Bb
C–E–G#–Bb
C–E–G–Bb–C
C–E–F#–Bb–C#
C–E–G#–Bb–C#
C–E–G–Bb–D–F#
C–E–G–Bb–Eb–F#
C–E–G–Bb–C#–A
C–E–G–A
C–E–G–D
C–E–G–Bb–D
C–Eb–G–A–D
C–Eb–G–Bb
C–Eb–G–Bb–A
C–Eb–G–Bb–D–F
C–Eb–G–B
C–Eb–F#–Bb
C–Eb–F#–Bb–D–F
C–Eb–F#–A
C–E–F#–B
C–E–G–B–F#
C–E–G–B–A
C–F–G–Bb–C#
C–F–F#–B
C–F–G–Bb
C–F–G–Bb–D–A
C–E–G–B–D–A
C–E–G–B–D–F#–A
C–E–F#–Bb–D
C–E–G#–Bb–D
C–E–G–Bb–Eb
C–E–Ab–Bb–Eb
C–E–G–Bb–F#
C–E–G–Bb–C#–F#
C–E–F#–Bb–D–A
C–E–G–Bb–D–F#–A
C–E–G–D–A
C–E–G–Bb
C–E–G–Bb–D–A

Once the basic chords are laid down, you can begin to construct a bass line around them. This is not necessarily how all music is formed, but it is common practice to derive a chord structure from the bass or melody or vice versa, and it's useful to prioritize the instruments in the track in this way. Generally, whichever instrument is laid down first depends on the genre of music.

While there are no absolute rules to how music is written – or at least aren't any as far as I know – it's best to avoid getting too carried away programming complex melodies when they don't form an essential part of the music. Although music works on the principle of contrast, having too many complicated melodies and rhythms playing together will not necessarily produce a great track; you'll probably end up with a collection of instruments that all are fighting to be noticed.

For example, a typical *hands in the air* trance with big, melodic, synthetic leads doesn't require a complex bass melody because the lead is so intricate. The bass in these tracks is kept very simple. Because of this, it makes sense to work on the melodic lead first, to avoid putting too much rhythmical energy into the bass. If you work on the bass first you may make it too melodic and harmonizing the lead to sit will be much more difficult. In our example, it may detract from the trance lead rather than enhance it.

If, however, you construct the most prominent part of the track first – in this case the melodic synthesizer motif – it's likely that you'll be more conservative when it comes to the bass. If the chord structure is fashioned first, this problem does not rear its ugly head because chord structures are relatively simple and this simplicity will give you the basic 'key' from which you can derive both the melody and bass to suit the genre of music you are writing.

FAMILIARITY

As already discussed, the subconscious reference we apply to everything we hear is inevitable, so it pays to know the chord progressions that we are all familiar with. Chord progressions that have never been used before tend to alienate listeners. Arguably most clubbers want familiar sounds that they've heard a thousand times before and are not interested in originality – an observation that will undoubtedly stir up some debate. Nevertheless, it's true to say that each genre – techno, trance, drum 'n' bass, chill out, hip-hop, and house – is based around same theme. Indeed, whether we choose to accept it or not, this is how they are categorized into sub genres. This is not to say that you should attempt to copy other artists but it does raise a number of questions about originality.

The subject of originality is a multifaceted and thorny issue and often a cause of heated debate among musicians. Although the finer points of musical analysis are beyond the scope of this book, we can reach some useful conclusions by looking at it briefly. L. Bernstein and A. Marx are two people who have made large discoveries in this area and are often viewed as the founders of musical analysis. They believed that if any musical piece is taken apart there will be aspects of it that are similar to most other records. Whether this is because musicians are often influenced by somebody else's style is a subject for debate, but while the similarities are not immediately obvious to the casual listener, to the musician these similarities can be used to form the basis of musical compositions.

We can very roughly summarize their discoveries by examining the musical scale. As we've seen, there are 12 major and 12 minor keys, each of which consists of eight notes. What's more, there are three basic major and three basic minor chords in each key. Thus, sitting at a keyboard and attempting to come up with an original structure is unlikely. When you do eventually come up with a progression that you believe sounds 'right', it sounds this way because you've subconsciously recognized the structure from another record.

This is why chord structures can often trigger an emotional response in you. In fact, in many cases you'll find that most dance hits have used a series of very familiar underlying chord progressions. While the decoration around these progressions may be elaborate and different in many respects, the underlying harmonic structure is very similar and seldom moves away from a basic three-chord progression. Whether you wish to follow this train of thought is entirely up to you, but there's nothing wrong with forming the bass and melody around a popular chord structure then removing the chords later on. In fact, popular chord structures are often a good starting point if inspiration takes a holiday and you're left staring at a blank sequencer window.

MUSICAL RELATIONSHIP

It's vital that the chords, bass and melody of any music style work together and complement one another. Of these three elements, the most important relationship is between the bass and the melody, so we'll examine this first. Generally, the bass and melody can work together in three possible ways: parallel, oblique or contrary motion.

Parallel motion is when the bass line follows the same direction as the melody. This means that when the melody rises in pitch the bass rises too, but it does not follow it at an exact interval. If both the melody and bass rise by a third degree, the resulting sound will be too synchronized and 'programmed' and will lose all of its feeling. Rather, if the melody rises by a third and the bass by a fifth or alternate thereof, the two instruments will sound like two voices rather than one. Also, the bass should borrow some aspects from the progression of the melody but should not play the same riff.

When the melody and bass work together in oblique motion, either the melody or the bass moves up or down in pitch while the other instrument remains where it is. When the music is driven by the melody, such as trance, the melody moves while the bass remains constant. In genres such as house, where the bass is often quite funky with plenty of movement, the melody remains constant while the bass dances around it.

The relationship known as contrary motion provides the most dynamic movement and occurs when the bass line moves in the opposite direction to the melody: if the melody rises in pitch, the bass falls. Usually if the melody rises by a third degree, the bass falls by a third.

This leads onto the complex subject of 'harmonization' – how the chord structure interacts with both the bass and melody. Despite the fact that dance tracks don't always use a chord progression or in some cases even a melody, the theory of how chords interlace with other instruments is beneficial on a number of levels. For instance, if you have a bass and melody but the recording still sounds drab, you could add harmonizing chords to see whether they help. Adding these chords will also give you an idea of where to pitch the vocals.

From a theoretical point of view, the easiest way to construct a chord sequence is to take the key of the bass and/or melody and build the chords around it. For instance, if the key of the bass were in E, then the chords' root could be formed around E.

However, it is unwise to duplicate *every* change of pitch in the bass or melody as the root for your chord sequence. While the chord sequence will undoubtedly work, the mix of consonant and dissonant sounds will be lost and the resultant sound will be either manic or sterile. If the chord structure is to instil emotion, the way in which the bass and melody interact with the harmony must be more complex. Attention to this will quite often make the difference between great work track and an average one.

Often, the best approach for developing a chord progression is one based around any of the notes used in the bass. For instance, if the bass is in E, then you shouldn't be afraid of writing the chord in C major because this contains an E anyway. Similarly, an A minor chord would work just as well because it is the minor to the C major (come on, keep up at the back…).

For the chords to work, you need to ensure that they are closely related to the key of the song, so it doesn't necessarily follow that because a chord utilizes E in its structure that it will work in harmony with the key of the song. For a harmony to really work, it must be inconspicuous, so it's important that you use the right chord. A prominent harmony will detract from the rest of the track so they're best kept simple and without frequent or quick changes. The best harmony is one that is only noticeable when removed.

While the relationship between the bass and chord structures is fundamental to creating music that gels properly, not every bass note or chord should occur dead on the beat, every beat, or follow the other's progression exactly. This kind of sterile perfection is tedious to listen to and quickly bores potential listeners. It's a trap that can be easy to fall into and difficult to get out of, particularly with software sequencers. To avoid this you must experiment by moving notes around so that they occur moments before or after each other. This deliberately inaccurate timing, often referred to as the 'human element', forms a crucial part of most musical styles.

Indeed, it's such a natural occurrence that our brains tune into this phenomenon, refusing anything that repeats itself perfectly without any variation. Although you may not recognize exactly what the problem is, your brain instinctively knows that a machine has created the performance. It could be argued that the human element is missing from many forms of dance music, but your brain assumes that the human element cannot be recreated electronically. Offsetting various notes deliberately or utilizing various quantize functions in the sequencer writes the slight inaccuracies we expect into the music, and it is these techniques that define the groove in the most well-known dance records. Before we look at groove, however, we need to first understand tempo and time signatures.

TEMPO AND TIME SIGNATURES

In all popular music the tempo is measured by counting the number of beats that occur per minute (BPM). BPM measurements are also sometimes referred to as 'metronome mark' and musicians along with some producers use these as well as other musical terms to describe the general tempo of any given genre of music. For instance the hip-hop genre is characterized by a slow, laid-back beat, but individual tracks will use different tempos that can vary from 80 BPM through to 110 BPM. Thus, it's usual that different musical styles are described using general musical terms. Using musical terms, then, hip-hop can be described as 'andante' or 'moderato', meaning at a walking pace or at a moderate tempo. Similarly, the house or trance genres may be described as 'allegro', meaning quickly, while drum 'n' bass may be described as 'prestissimo', meaning very fast.

Several of the most common musical terms are listed below.

- Largo – Slowly and broadly.
- Larghetto – A little less slow than largo.
- Adagio – Slowly.
- Andante – At a walking pace.
- Moderato – At a moderate tempo.
- Allegretto – Not quite allegro.
- Allegro – Quickly.
- Presto – Fast.
- Prestissimo – Very fast.

It is important to bear in mind that the actual number of beats per minute in a piece of music marked presto, for example, will also depend on the music itself. A track that is constructed of half notes can be played a lot more quickly (in terms of BPM) than one that consists almost entirely of 16th notes, but it can still be described using the same word. To make sense of this we need to examine time signatures and different note lengths.

FIGURE 10.4
Typical musical staff

Time signatures determine the rhythmic feel and 'flow' of the music, so an understanding of them is essential. To understand the principle of time signatures, you first have to learn how to count. Of course, we all learnt how to do this in primary school, but it's not quite the same in music because of the way music is transcribed onto a musical staff. So let's begin by looking at a typical musical staff (Figure 10.4).

Figure 10.4 shows the note of the C major scale on the musical staff. Written in this way, each note's pitch can be seen in relation to other notes according to their vertical position, but more importantly, the symbol used to represent the notes. In this case, the notes shown are quarter notes or crochets, which to the musician is an important indication of their duration.

Whole Note – Semibreve
Half Note – Minim
Quarter Note – Crotchet
Eighth Note – Quaver
Sixteenth Note – Semi Quaver
Thirty-second Note – Demisemiquaver

o **Whole note**

Is equivalent in timing to

 2 half notes

Is equivalent in timing to

4 quarter notes

Is equivalent in timing to

8 eighth notes

Is equivalent in timing to

16 sixteenth notes

Is equivalent in timing to

32 32nd notes

Symbol	Name	Common Name	Value in Relation to a Whole Note	Equivalent in Timing to...
•	Whole note	Semibreve	1	♩♩
♩	Half note	Minim	1/2	♩♩♩♩
♩	Quarter note	Crotchet	1/4	♪♪♪♪♪♪♪♪
♪	Eighth note	Quaver	1/8	♪♪♪♪♪♪♪♪♪♪♪♪♪♪♪♪
♪	Sixteenth note	Semiquaver	1/16	♪♪♪♪♪♪♪♪♪♪♪♪♪♪♪♪ ♪♪♪♪♪♪♪♪♪♪♪♪♪♪♪♪
♪	Thirty-second note	Demisemiquaver	1/32	–

If notes were simply dropped onto a musical stave, as they were in Figure 10.3, the musician would play them one after the other, but there would be no rhythm to the piece. Each note would just be played in one long sequence. To avoid this, music is broken down into a series of bars and each bar has an associated rhythm. Generally, this rhythm remains the same throughout every bar of music, but in some instances the rhythm can change during the course of a song.

Nevertheless, all music must have rhythm, even if it isn't written down, because all music is based around a series of repetitive rhythmic units (i.e. bars, sometimes known as measures) – it's what gets your feet moving on the dance floor.

To grasp this, we need to be able to count like a musician. This can be accomplished by examining the ticking of an ordinary clock (provided that it isn't a digital, of course!). Every tick of the clock obviously corresponds to a number from 1 to 60 s, but rather than count up to 60 we can just count up to four and then start all over again from one. For instance:

> One...two...three...four...one...two...three...four...one...two...three...four... and so forth

Now suppose we were to put an accent (i.e. say it louder) on every number one of our count we would have:

> ONE...two...three...four...ONE...two...three...four...ONE...two...three...four... and so forth

Next, rather than count up to four, count up to three instead while still placing an accent on the number one.

> ONE…two…three…ONE…two…three…ONE…two…three…ONE…two…three…etc.

You'll notice there is a distinct change in the rhythm. Counting up to four produces a rather static rhythm (typical of most dance music), while counting up to three produces a rhythm with a little more swing to it (typical of some chill out, trip-hop and hip-hop tracks). This is the basic concept behind time signatures; it allows the musician to determine what kind of rhythm the music employs.

Time signatures are placed at the beginning of a musical piece to inform the musician of the general rhythm – how many and what kind of notes there are per bar. They are written with one number on top of another and look like a fraction, such as 1/4 or 1/2 used in mathematics.

In music, the number at the top of the time signature indicates the number of 'beats' per bar (do not confuse this with beats per minute!), and the bottom number indicates which note sign represents the beat.

To explain this, let's consider the time signature of 3/4 (pronounced 'three-four'). The top number determines that there are three beats to the bar, while the bottom number tells us the length of these beats. To determine the size of these beats we can divide a whole note (1) by this bottom number (4) and as we saw before

> 1 whole note = 2 half notes = 4 quarter notes = 8 eighth notes = 16 sixteenth notes

Therefore if we divide a whole note by four, we get:

> whole note/4 = 4 quarter notes (i.e. 4 crotchets)

This gives us the beat size of one crotchet. From this we can determine that each bar of music can accept no more than the sum of three crotchets. Of course, this doesn't limit us to using just three crotchets, it simply means that no matter what notes we decide to use we cannot go above the total sum of three crotchets.

This technique can be applied to other time signatures. If we work in the time signature of 5/2 (five-two), we can tell from the top note that there are five beats to the bar. As before, we work out the size of these beats by dividing a whole note by the bottom number in the time signature, in this case two. This gives us five minim's (half notes) to the bar.

If applied to the time signature 6/8 (six-eight) 6 there are, eighth notes (quavers) in the bar. These, different time signatures, while appearing quite abstract and meaningless on paper produce very different rhythmic results when played. Although most dance music use 4/4 or 3/4, writing in different time signatures can occasionally aid in the creation of grooves.

CREATING GROOVE

While it is possible to dissect and describe the principles behind how each genre of music is generally programmed, defining what actually makes the groove of one record better than another isn't so straightforward. Indeed, creating a killer dance floor groove is something of a Holy Grail that all dance music producers are continually searching for. Although finding it has often been credited to a mix of skill, creativity and serendipity, there are some general rules of thumb that often apply.

Injecting groove into a performance is something that any good musician does naturally, but it can be roughly measured by two things: timing differences and variations in the dynamics. A drummer, for instance, will constantly differ how hard the kit is hit, resulting in a variation in volume throughout, while also controlling the timing to within microseconds. This variation in volume (referred to as dynamic variation) injects realism and emotion into a performance while the slight timing differences add to the groove of the piece.

By adjusting the timing and dynamics of each instrument, we can inject groove into a recording. If a kick and hi-hat play on the beat or a division thereof, moving the snare's timing forward or backward will make a huge difference to the feel. Similarly, programming parts to play slightly behind the beat creates a bigger, more laid-back feel, while positioning them to play in front of the beat creates a more intense, almost nervous feel. These are the principles behind which swing and groove quantize in sequencers operate, both of which can be used to affect rhythm, phrasing and embellishments.

Using swing quantize, the sequencing grid is moved away from regular slots to alternating longer and shorter slots. This can be used to add an extra feel to music, or, if applied heavily, can change a 4/4 track into a 3/4 track. Groove quantize is a more recent development that allows you to import third-party groove templates and then apply them to the current MIDI file. This differs from all other forms of quantize because it redefines the grid lines over a series of bars rather than just one bar, thereby recreating the feel of a real musical performance. In many instances, groove templates also affect note lengths and the overall dynamics of the current file, creating a more realistic performance. In more adept sequencers, groove templates extracted from audio files can be onto a MIDI file to recreate the feel of a particular performance.

To get the best results from swing and groove quantizing features, they should be used creatively. For instance, complex drum loops can be created from two or three drum loops that each use different quantize settings. This approach works equally well when creating complex bass lines (if the track is bass driven), lead lines and motifs. It is important to note, however, that quantize shouldn't be relied on to instantly add a human feel, so creating a rigid bass pattern in the hope that quantize will introduce some feel later isn't a good idea. Instead, you should try to programme the bass with plenty of feel from the start: by physically moving notes by a few ticks and then experimenting further with the quantize options to see if it adds anything extra.

In addition to groove, another important aspect of dance music is the drive, which gives the effect of pushing the beat forward, as if the song is attempting to run faster than the tempo. This feeling of drive comes from the position of the bass and drums in relation to the rest of the record. For instance, moving the drums and bass forward in time by just a couple of ticks gives the impression that they are pushing the song forward, in effect, producing a mix that appears as if it wants to go faster than it actually is. This feeling can be further accentuated if any melodic elements, such as pianos, are programmed to sit astride the beat rather than dead on it and different velocities are used to accentuate parts of the rhythm.

This brings us onto the subject of dynamics and syncopation, both of which play a vital role in capturing the requisite rhythm of dance. All musical forms are based upon patterns of strong and weak working together. In classical music, it is the constant variation in dynamics that produces emotion and rhythm from these contrasts. In dance music, these dynamic variations are often accentuated by the rhythmic elements. By convention, the first beat of the bar is the strongest. Armies on the march provide a good example of this, with the 2/4 rhythm, Left–Right–Left–Right–Left–Right. In this rhythm, there are two beats (Left and Right) and the first beat is the strongest with the second beat being a little weaker. With the typical dance music 4/4 kick drum (often referred to as 'four to the floor'), the first beat in the bar is usually the strongest (i.e. the greatest velocity) with subsequent hits varying in dynamics (Table 10.5).

Syncopation also plays an important role in the rhythm of a track. This is where the stress of the music falls off the beat rather than on it. All dance music makes use of this, and it's often a vital element that helps tie the whole track together. While the four to the floor rhythm pounds out across the dance floor, a series of much more elaborate rhythms interweave between the beats of the main kick drum. For instance if we count in the same manner as we did for the time signatures:

ONE…two…three…four…ONE…two…three…four…ONE…two…three…four… and so forth

Table 10.5	Syncopation	
No. of 'Kicks' in the Bar	Beat Pattern	Pattern Over 4 Bars
1	Strong	S–S–S–S
2	Strong–weak	SW–SW–SW–SW
3	Strong–medium–weak	SMW–SMW–SMW–SMW
4	Strong–weak–medium–weak	SWMW–SWMW–SWMW–SWMW

Each number corresponds to the kick drum, but if we were to add 'and' into the counting we get:

> ONE *and* two *and* three *and* four...ONE *and* two *and* three *and* four...

Each *and* occurs on the off the beat, or, in musical terms, on the quaver of the beat. A good example of this is found in trance music, where the bass notes occur between the kicks rather than sat on the beat.

This emphasis can be added using velocity commands, which simulate or re-create the effect of hitting some notes harder than others, thereby creating different levels of brightness in the sound. These are commonly used with bass notes. Using velocity commands you can keep the first note of the bar at the highest velocity and then lower the velocity for each progressive note in the bar and at the next bar, repeat this pattern again. It's important to also consider here that the length of any bass notes can also have an effect on the perceived tempo of the track. Shorter, stabbed notes, such as sixteenths, will make a record appear slightly more frantic than using notes that are longer.

While it is impossible to define a groove precisely, the best way to know whether you've hit upon a good vibe is to try dancing to it. If you can, the groove is working. If not, you'll need to look at the bass and drums again before moving any further. Keep in mind that the groove forms the backbone of all dance records; therefore it's imperative that the groove works during this early stage. Don't hope that adding additional melodic lines later will help it along: if it sounds great at this point then further melodic elements will only help improve. If you're stuck for inspiration or are struggling to create a groove, try downloading MIDI files of other great songs with a similar feel and incorporate and manipulate the ideas to create your own groove. This isn't stealing its research and *every* musician on the planet started this way!

The only real 'secret' behind creating that killer groove is to never settle for second best, and many artists will freely admit to spending weeks making sure that the track has a good vibe. This includes giving it a quick mix-down so that it sounds exactly as you would envisage it when the track is completed. If at this point, the vibe is still missing, go back and change the elements again until it sounds exactly right. Take into consideration that the mixing desk, as well as being a useful creative tool, only polishes the results of a mix and hence sounds going into a mix must be precise in the first place.

CREATING MELODIES/MOTIFS

Although it's true to say that melodies and motifs often take second place to the general groove of the record, they still have an important role in dance music. Notably, very few dance tracks employ long, elegant, melodic lines (some trance being an exception) since these can deter from the rhythmic elements of the music. Consequently, most dance music use short motifs, as these are simple yet powerful enough to drive the message home. Generally speaking a good motif is a small, steady, melodious idea that provides listeners with an immediate reference point in the track. For instance, the opening strings in the Verve's *Bitter Sweet Symphony* could be considered a motif, as can the first few piano chords on Britney Spears *Baby, Hit Me One More Time*. As soon as you hear them you can identify the track, and if programmed correctly they can be extremely powerful tools.

Knowing what makes a good motif is central to the ability to write one, so what are the principles behind their creation? Well, motifs follow similar rules to the way that we ask and then answer a question. In a motif, the first musical phrase 'asks' the question and this is then 'answered' by the second phrase. This is known as a 'binary phrase' and some of the best examples can be found in nursery rhymes such as, 'Baa baa black sheep' asking the question, answered by, 'Yes sir, yes sir, three bags full'.

This is a simple and well-known motif, and it's important to observe how the two lines balance each other out. Musically the first line consists of CCGGABCAG while the reply consists of FFEEDDC. Also, notice how the first phrase rises but the second phrase falls, while maintaining the similarities between the two phrases. This type of balance underlies the most memorable motifs and is key to their creation. Whether a four to the floor remix of Baa baa black sheep would go down well in a club is another question altogether, but as a motif it contains the three key elements:

- Simplicity
- Rhythmic repetition
- Some variation

The use of repetition in music is not exclusive to dance music. Most classical music is based entirely around the repetition of small flourishes and motifs, so much so, that many dance musicians listen to classical works and derive their ideas from its form. In fact, the principle behind writing a memorable record is to bombard the listener with the same phrases so that they become embedded in the listener's memory, although this must be accomplished in such a way that the music doesn't sound too repetitive. Indeed, one of the biggest mistakes made by musicians just starting out is to instil too many new ideas without taking any notice of what has happened in the previous bars.

Repeating the phrases AB–AB–AB–AB can result in too much repetition, so another way to create a motif is to create a third phrase, known as 'ternary' repetition. This is where you create a third answer and alternate between the three in different ways, such as ABC–ABAC.

It's also commonplace for musicians to introduce a fourth motif, resulting in two 'questions' and two 'answers'. This arrangement is called a 'trio minuet' formation and can take on various forms, such as ABA (minuet) CDC (trio) ABA–CDC and so forth. Although not all dance musicians work this way, these are generally accepted methods and two simple rules.

> If the beginning of the motif rises, the second part will fall.
>
> After moving in one direction, there is a step back in the opposite direction.

Having said that, it's not uncommon for the inverse to be a true relationship that can be used to musical effect. When any motif falls in scale it gives the impression of pulling our emotion downwards, yet if it increases in scale it is more uplifting. This is because there are higher frequencies when a note rises, as is evident when the low-pass filter cut-off of a synthesizer is opened up to allow more high-frequency content through. You can hear this in countless dance records whereby the music becomes duller, as though it's being played in another room and then gradually the frequencies increase, creating a sense of expansion as the sound builds. If you listen to most dance records, this form of building, using filter cut-off movement, becomes apparent. When the track is building up to a crescendo the filter opens, yet while the track drops away the filter closes. Similarly, the rising key change introduced at the end of many popular songs, builds the sound, giving the impression that the song is at full impact, driving the message home.

Like everything else in music, understanding the theory behind good motifs and actually creating one that captivates attention is an entirely different matter. As with creating grooves for the dance floor, great motifs are something of a Holy Grail and coming up with them is going to be a mix of creativity and luck. But as always there are some ideas that you can try out.

- One is to record MIDI information from a synthesizer's arpeggiator and then edit this in a sequencer to produce some basic ideas.

- Another technique that can produce usable results is what's often called, the 'two-fingered' approach. This involves tapping out a riff using two keys an octave apart on the keyboard. By alternating between these two keys, keeping the lower key the same and continually moving higher or lower in scale on the second key, you can develop the beginnings of interesting rhythmic movements. If the rhythm is recorded as MIDI, it can be edited further by lengthening notes to overlap each other or by adding further notes in between the two octaves to construct motifs.
- The third and most commonly used method is to continually loop the bass and drums and experiment on the keyboard until you come up with something that compliments the rhythm section. Once the first bar(s) of the phrase – the 'question' – have been fashioned, they can be pasted into the subsequent bar(s) and edited to create the 'answer'.

Above all, it's best to avoid getting carried away with any of these techniques. It's vital that they are kept simple. Many dance tracks use a single pitch of notes played every eighth, sixteenth, or quarter, or are constructed to interweave with the bass to produce a complex groove. Indeed, simple motifs have a much more dramatic effect on the music than complex arpeggios, so it's important that they capture the listener's interest.

Because the sound used for any motif should stand out, it's important to consider the frequencies that are to be, or are used in the mix thus far. In many dance tracks where three or four simple motifs play simultaneously, the combined frequencies can take up a proportionate amount of the available space within the mix. To prevent clogging up the mix, motifs are usually worked with as MIDI rather than transformed to audio, as this can be used with filter cut-off to reduce the frequencies. This method also allows you to utilize a low-pass filter cut-off and control the sound's sonic content while the other motifs are playing. When working with the optimum frequencies of a motif, you must carefully choose its rhythmical pitch.

Any motif that uses notes above C4 will contain many higher frequencies and few lower ones, so using a low-pass filter that is almost closed will reduce the frequencies to nothing. Careful use of the filters will create the final result and often you will need to compromise by cutting the frequencies of some motifs while opening them up for others. Manipulating the frequencies in this way creates the impression that the sounds are interweaving, which creates more movement and the 'energy' that is characteristic of most club tracks.

CREATING PROGRESSION

Although dance music is heavily based on repetition, it retains interest building and releasing the groove. This is achieved by adding and removing sonic elements throughout and by gratuitous use of filters. To fully understand how a groove is built and released we need to look at sections of the whole track, counting what has gone before, how the next event is prepared, what comes after it and how it ends. For example, when you hear the typical dance snare crescendo building, you automatically know that the main part of the groove will follow. The longer this is *reasonably* postponed the bigger the impact is expected to be. The best way to learn about this is to listen to and dissect popular dance tracks and then create a song map charting where the builds and drops occur, essentially charting the emotion that it creates.

There are various methods of song mapping and each is as valid as the next, but a large number of psychologists believe that we are naturally disposed to think in pictorial form. It can often prove beneficial to draw the general idea of the arrangement onto paper. How this is depicted is entirely up to the individual and can consist of simple lines to more complex drawings even involving colours to depict each instrument and the parts of the original recording. Because club mixes can be quite long, often averaging around 6–8 min, these types of maps can be a blessing. Paul Van Dyk and BT, for instance, are well known for creating a song map before beginning to build any type of track. A song map not only helps you to envisage the final product but it also helps you plot out the track in terms of crucial builds and drops. The basic principle behind using this strategy is that we can visually see where the different emotional states occur, helping you to envisage where the track will bring listeners up on a 'high' 1 min and then drop them down again the next (Figure 10.5).

It's evident from the song map shown in Figure 10.5 that the dance music differs from the usual musical structure of verse, chorus, verse, chorus, bridge and outro. Instead it derives its own set of rules based around mathematical repetition taken to extremes, usually broken into sections of 4 bars. This formal structure may seem overtly technical for what is otherwise a creative art, but try writing a 16-bar loop and introducing an instrument at bar 5, 13 or 15 and it will no doubt sound erroneous.

We can also determine that the typical arrangement consists of an intro, first body, drop, second body, second drop, reprise, main body and finally an outro.

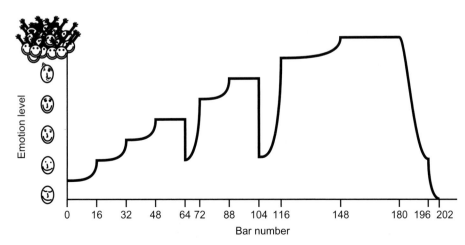

FIGURE 10.5
Example of a song map

Each of these changes relates to the emotional states that, ideally, you need to generate for the listeners. Using a map like that shown in Figure 10.5, we can see how this might be transferred and used in an imaginary club track.

The Intro

BARS 1–16

This introduces the track and consists of nothing more than a 4/4 drum loop. This is kept simple at the beginning of the track and continues unchanged for the first 16 bars. This standard introduction gives the DJ playing the music a 4/4 beat with no other instrumentation so that he can mix it in with the previous track, keeping the beat going on from one record to the next to keep everyone up on the floor. 'Dropping a beat' will easily disorientate the clubbers who are dancing; we all need a basis on which to keep time.

BARS 16–32

At this point, a new instrument is introduced. In order not to give the game away too soon, a new percussive element comes in. This could consist of another closed or open hi-hat pattern, snares or claps. This completes the drum loop that will play throughout the rest of the mix.

First Body

BARS 32–48

After the initial 32 bars, the main groove is introduced. Now both the bass and drums are playing together, laying the foundations of the groove. This plays unaltered to allow the clubbers to become 'comfortable' with the general groove of the record.

BAR 48–64

At bar 48, a motif is introduced. This interweaves with the bass and drums using slow filter cut-off movements. Applying a low-pass filter to the motif and slowly opening it introduces higher frequencies giving the effect that the song is building up towards something special. Listen to any popular record and you'll notice this effect. As the song progresses more instruments are introduced, usually at the chorus stage, which 'fills out' the mixes frequency range bringing the song towards a climax. It is these filter movements that are typically used to create tension and movement in all forms of dance music and have formed the basis of the most well-respected dance records.

The Drop

BARS 64–72

At this point, all the percussive elements, bar the kick drum, have been dropped from the mix while the motif continuously plays along. The filter is closed to help give the impression that the song is dropping emotionally. However, by the middle 4th bar, the drum rhythm picks up again, starting with the open hi-hats followed by the snares, signifying that the track is about to come back. As the rhythm picks up, motif's filter opens up again and is followed by a short snare roll, where the second body of the track is re-introduced.

The Second Body

BARS 72–88

At the final beat of the preceding bar's drop section, a crash cymbal starts a new passage of music and a new motif is introduced. The filter on both motifs is now contorted, resulting in the motifs' frequencies moving in and out of one another, which creates a flowing movement that helps the track groove along.

BARS 88–104

Again, a crash cymbal, snare roll, skip or sound effect is employed to denote the next 16 bars of the second body and some chorded strings are introduced. Low-pass filter movements are employed on these strings to prevent them from appearing too static. As before, these filter movements open giving the impression that the song is building up again, while in the final 4 bars of this section a snare roll signifies that the song is building further.

The Drop

BARS 104–108

A sudden crash cymbal at the end of the preceding bar echoes across the mix while the song drops to just the strings and the motif that was introduced in the second body. Again the filter cut-off closes to denote a break to the 2nd bar of this drop, whereby they begin to open again giving the impression that the song is once again building.

The Reprise

BARS 108–116

At the end of the drop a snare roll is introduced, slowly gaining in volume as the cut-off frequencies of both the motif and chords are opened further. This snare roll lasts the entire length of the 8 bars with the snares becoming progressively closer together (the snares being placed closer together as they grow closer to the main body). The principle aim in this section is to build emotion to a maximum, so that when the main body is reached the track is at its apex. Ideally, this is a culmination of all the instruments that have previously been introduced throughout the mix and a new motif.

The Main Body

BARS 116–148

The main body is used to drive the message of the track home and is signified by all of the previous instruments playing together, creating a crescendo of emotion. The drums, bass, previous motifs and a new motif are introduced filling up the entire frequency spectrum of the mix. Again filter cut-offs play a role here to prevent boredom, and each instrument interacts with one another by closing the filter of one motif while opening it on the other.

BAR 148–180

At this point, additional snares are introduced to add a skip to the rhythm, and a few sweeping effects are introduced that pan across the mix.

The Outro

BAR 180–196

One of the motifs is dropped from the mix after a cymbal crash and the filters on the other motifs and chords are beginning to close, bringing the track back down in emotion. Finally all the motifs and chords are removed, leaving just the drums and bass playing.

BAR 196–212

The bass is dropped from the mix, leaving the full drum rhythm playing solo for the final 16 bars, allowing the DJ to mix in the next record. The track draws to a close.

Of course, this imaginary track is quite simple; in reality building a track is much more complex. Nevertheless, this is a good example of typical dance club structure. As with all music this is open to an individual's interpretation, but bear in mind that dance music is an exact science and the audience likes to know what to expect. Building a snare crescendo and then dropping it back to how it was before this crescendo would not be greeted with smiling faces.

Also in many cases the arrangement will be different depending on the genre of music being written. While this type of arrangement may work for trance, it won't work too well for hip-hop and trip-hop, these require a more gentle approach.

PART 2
Dance Genres

CHAPTER 11

House

'By 1981 they declared that Disco was dead and there were no more up-tempo dance records. That's when I realised I had to start changing things to keep feeding my dance floor. ...'

... Frankie Knuckles

Although some house aficionados will refuse to admit it, the development of house music has much of its success accredited to the rise and fall of disco. As a result, to appreciate the history of house music, we need to look further back than the 1980s and the development of the TR909 and TR808 drum machines; we also need to examine the growth of disco during the 1970s. This is because disco still forms a fundamental part of some of today's house music and in many instances older disco records have been scrupulously sampled to produce the latest house tracks.

Pinning down an exact point in time where disco first appeared is difficult since a majority of the elements that make disco appeared in earlier records. Nonetheless, arguably it is said to have first originated in the early 1970s and was derived from the funk music that was popular with black audiences at that time. Some big name producers such as Nile Rodgers, Quincy Jones, Tom Moulton, Giorgio Moroder and Vincent Montana began to move away from recording the self-composed music and started to hire session musicians and produce hits for artists whose only purpose was to supply vocals and become a marketable commodity.

Donna Summer became one of the first *disco*-manufactured success stories with the release of *Love to Love You Baby* in 1975 and is believed by many to be the first disco record to hit and be accepted by the mainstream public. This 'new' form of music was still in its infancy, however, and it took the release of the motion picture *Saturday Night Fever* in 1977 before it eventually became a widespread phenomenon. Indeed, by the late 1970s, over 200 000 people were attending discotheques in the UK alone and disco records contributed to over 60% of the UK charts.

As with most genres of music that become popular, many artists and record labels jumped on the wave of this new happening vibe and it was soon deluged with countless disco versions of original songs and other pointless and poorly produced disco records as the genre became commercially bastardized. As a result, disco fell victim to its own success in the late 1970s and early 1980s with the campaign of '*disco sucks*' growing ever more popular. In fact, in one extreme incident Steve Vahl, a rock DJ who had been against disco from the start, encouraged people to bring their disco collections to a baseball game on the 12th July 1979 for a ritual burning. After the game, a huge bonfire was lit and the fans were asked to throw all their disco vinyl onto the fire.

By 1981, disco was dead but not without first changing the entire face of club culture, changing the balance of power between smaller and major labels and preparing the way for a new wave of music. Out of these ashes rose the phoenix that is house, but it had been a large underground movement before this and contrary to the misconceptions that are spread around, it had actually been in very early stages of evolution before disco hit the mainstream.

Although to many Frankie Knuckles is seen as the 'godfather' of house, it's true foundations lie well before and can be traced back to as early as 1970. At this time, Francis Grosso, a resident DJ at a converted church known as the Sanctuary was the first ever DJ to mix two early disco records together to produce a continual groove to keep the party goers on the dance floor. What's more, he is also believed to be the first DJ to mix one record over the top of another, a technique that was to form the very basis of dance music culture.

Drawing inspiration from this new form of mixing, DJ Nicky Siano set up a New York club known as The Gallery and hired Frankie Knuckles and Larry Levan to prepare the club for the night by spiking the drinks with lysergic acid diethylamide (LSD/acid/trips). In return he taught both all about the basics of this new form of mixing records and soon after they moved on to become resident DJs in other clubs. Levan began residency at The Continental Baths while Knuckles began at Better Days, to soon rejoin Levan at The Continental Baths 6 months down the line. The two worked together until 1977 when Levan left the club to start his own and was asked to DJ at a new club named the Warehouse in Chicago. Since Levan was now running his own club, he refused but recommended Knuckles who accepted the offer and promptly moved to Chicago.

Since this new club had no music policy, Knuckles was free to experiment and show off the techniques he had been taught by Nicky Siano. Word quickly spread about this new form of disco and The Warehouse quickly became the place to be for the predominantly gay crowd. Since no 'house' records actually existed at this time, the term house did not refer to any particular music but simply referred to the Warehouse and the style of continual mixing it had adopted. In fact, at this time the word house was used to speak about music, attitude and clothing. If a track was house, it was from a cool club and something that you would never hear on a commercial radio station, whereas if you *were* house it meant you attended all the cool clubs, wore the 'right' clothing and listened to 'cool' music.

By late 1982 and early 1983, the popularity of the Warehouse began to fall rapidly as the owners began to double the admission price; as it became more commercial, Knuckles decided to leave and start his own club known as the Powerhouse. His devoted followers went with him, but in retaliation the Warehouse was renamed the Music Box and the owners hired a new DJ named Ron Hardy. Although Hardy wasn't a doctor, he dabbled in numerous pharmaceuticals and in turn was addicted to most of them but was nevertheless a very talented DJ. While Knuckles kept a fairly clean sound, Hardy pounded out an eclectic mix of beats and grooves mixing euro disco, funk and soul to produce an endless onslaught to keep the crowd up on the floor. Even to this day, Ron Hardy is viewed by many as the greatest ever DJ.

Simultaneously, WBMX, a local radio station also broadcast late-night mixes made by the Hot Mix Five. The team consisted of Ralphi Rossario, Kenny 'Jammin' Jason, Steve 'Silk' Hurley, Mickey 'Mixin' Oliver and Farley 'Jackmaster' Funk. These DJs played a non-stop mixture of British new romantic music ranging from Depeche Mode to Yazoo and Gary Numan, along with the latest music from Kraftwerk, Yello and George Clinton. In fact, so popular was the UK new romantic's scene that a third of the American charts consisted of UK music.

However, it wasn't just the music that the people tuned in for it was the mixing styles of the five DJs. Using techniques that have never been heard of before, they would simultaneously play two of the same records to produce phasing effects, perform scratches and back spins and generally produce a perfect mix from a number of different records. Due to the show's popularity it was soon moved to a daytime slot and kids would skip school just to listen to the latest mixes. In fact, it was so popular that Chicago's only dance music store, Imports Etc, began to put a notice board up on the window, documenting all the records that had been played the previous day to prevent them from being overwhelmed with enquiries.

Meanwhile, Frankie Knuckles was suffering from a lack of new material. The 'disco sucks' campaign had destroyed the industry and all the labels were no longer producing disco. As a result, he had to turn to playing imports from Italy (the only country left that was still producing disco) alongside more dub-influenced music. More importantly, for the history of house, though, he also turned to long-time friend Erasmo Rivieria, who was currently studying sound engineering, to help him create reworks of the earlier disco records in an attempt to keep his set alive. Using reel-to-reel tape recorders, the duo would record and cut up records, extending the intros and breakbeats and layering new sounds on top of them to create more complex mixes. This was soon pushed further as he began to experiment by placing entirely new rhythms and bass lines underneath familiar tracks. While this undeniably began to form the basis of house music, no one had yet released a true house record, and in the end it was Jesse Saunders' release of *On and On* in 1984 that landmarked the first true house music record.

Although some aficionados may argue that artist Byron Walton (aka Jamie Principle) produced the first house record with just a portastudio and a keyboard, the track entitled 'Your Love' was only handed to Knuckles for him to play as part of his set. Jesse Saunders, however, released the track commercially under his self-financed label 'Jes Say' and distributed the track through Chicago's Imports Etc.

The records were pressed, courtesy Musical Products, Chicago's only pressing plant owned and run by Larry Sherman. Taking an interest in this scene, he investigated its influence over the crowds and soon decided to start the first-ever house record label 'Trax'. Simultaneously, however, another label 'DJ International' was started by Rocky Jones and the following years involved a battle between the two to release the best house music. Many of these consisted of what are regarded as the most influential house records of all time including *Music is the Key, Move Your Body, Time to Jack, Get Funky, Jack Your Body, Runaway Girl, Promised Land, Washing Machine, House Nation* and *Acid Trax*.

By 1987 house was in full swing, while still borrowing heavily from 1970s disco, the introduction of the Roland TB303 bass synthesizer along with the TR909, TR808 and the Juno 106 had given house a harder edge as it became disco made by 'amateur' producers. The basses and rhythms were no longer live but recreated and sequenced on machines resulting in a host of 303-driven tracks starting to appear.

One of these budding early producers was Larry Heard, who after producing a track entitled *Washing Machine* released what was to become one of the most poignant records in the history of house. Under the moniker of Mr Fingers he released *Can U Feel It*, the first-ever house record that didn't borrow its style from earlier disco. Instead, it was influenced by soul, jazz and the techno that was simultaneously evolving from Chicago. This introduced a whole idea to the house music scene as artists began to look elsewhere for influences.

One of these was Todd Terry, a New Yorker and hip-hop DJ, who began to apply the sampling principles of rap into house music. Using samples of previous records, he introduced a much more powerful percussive style to the genre and released *3 Massive Dance Floor House Anthems*, which pushed house music in a whole new direction. His subsequent house releases brought him insurmountable respect from the UK underground scene and has duly been given the title of Todd 'The God' Terry.

Over the years, house music mutated, multiplied and diversified into a whole number of different subgenres, each with its own name its and production ethics. In fact, to date there are over 14 different subgenres of house consisting of progressive house, hard house, deep house, dark house, acid house, Chicago house, UK house, US house, euro house, French house, tech house, vocal house, micro house and disco house…and I've probably missed some too.

MUSICAL ANALYSIS

The divergence of house music over the subsequent years has resulted in a genre that has become hopelessly fragmented and as such cannot be easily identified as featuring any one particular attribute. Indeed, it can be funky, drawing its inspiration from disco of the 1970s; it can be relatively slow and deep, drawing inspiration from techno; it can be vocal; it can be party-like or it can simply be pumping. In fact, today the word house has become somewhat of a catch-all name for music that is dance (not pop!), yet doesn't fit into any other dance category. The good news with this is that you can pretty much write what you want and as long as it has a dance vibe, it could appear somewhere under the house label. The bad news, however, is that it makes it near impossible to analyse the genre in any exact musical sense and it is only possible to make some very rough generalizations.

Firstly, we can safely say that house music invariably uses a 4/4 time signature and is produced either allegretto or allegro. In terms of physical tempo, this can range from a somewhat slow (by today's standards) 110 BPM to a more substantial 140 BPM, but many of the latest tracks seem to stay around the 127 or more recently 137 'disco heaven' BPM. This latter beats per minute is referred to as such since this is equal to the average clubber's heart rate while dancing, but whether this actually makes the music more 'exciting' in a club has yet to be proven.

Fundamentally, house is produced in one of the three ways: everything is sampled and rearranged; only some elements are sampled and the rest is programmed; or the entire track is programmed in MIDI. The approach taken depends entirely on what style of house is being written. For example, the disco house produced by the likes of Modjo, Room 5 and Daft Punk relies heavily on sampling significant parts from previous disco hits and dropping their own vocals over the top (Daft Punk's 'Digital Love' and Modjo's 'Lady' being a prime example). If you write this style of music, then this is much easier to analyse since it's based around the disco vibe. Generally, this means that it consists of a four to the floor rhythm with a heavily syncopated bass line and the characteristic electric wah guitar. On the other hand, deep house uses much darker timbres (deep bass lines mixed with atmospheric jazzy chords) that don't particularly exhibit a happy vibe but are still danceable, while acid house relies heavily on the squawking TB303. In fact, while all house music will employ a four to the floor pattern, the instrumentation used will often determine its style. Consequently, what follows is simply a guide to producing the basic elements of all types of house, and since you know how your choice of genre already sounds, it can be adapted to suit what you hear.

HOUSE RHYTHMS

When it comes to producing house rhythms, nearly all house producers will not use software sequencers preferring the AKAI MPC or E-Mu SP1200 sampling drum machines. This is simply due to the solid timing that is requisite in

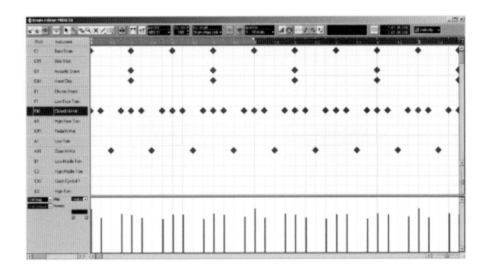

FIGURE 11.1
The cornerstone of house loops

house and can only be attained through using the internal hardware sequencers. Notably, of these the SP1200 has a maximum sampling rate of 12-bit but this isn't of any concern since it often imparts a harsher sound to the rhythms which is sometimes required. In fact, after programming a house rhythm and the subsequent timbres, it's often worth reducing the bit rate to see if it hardens up the beat for the subgenre of house you want to produce.

Generally, house relies heavily on the strict four to the floor rhythm with a kick drum laid on every beat of the bar. Typically, this is augmented with a 16th closed hi-hat pattern and an open hi-hat positioned on every 1/8th (the off-beat) for syncopation. Snares (or claps) are also often employed on the second and fourth beat underneath the kick (Figure 11.1).

This, of course, is only the start to a house loop and is based around the disco patterns from where it originated and congas, toms, bongos, tambourines and shakers are often added to create more of a feel to the rhythm but this does depend on the subgenre. Where these are placed is entirely up to your own digression, but generally they are best played live from the sampler's pads (or an attached keyboard) and left unquantized to keep the feel of the loop live. This helps in acquiring the sampled feel that's often exhibited, since more often than not the loop will have been sampled from previous records.

More importantly, though, house more than any other genre relies on plenty of drive. This can be applied in a variety of ways but the most commonly used technique is to employ a slight timing difference in the rhythm to create a push in the beat. Typically, this involves keeping the kick sat firmly on all the beats but moving the *snares* forward in time by tick or two. Then once the bass line is laid down, its timing is matched to these snares using a sequencers match quantize function. Alternatively, to provide a heavier sound, a clap is laid on beats two and four (along with the snares) but these claps are then moved forward in time by a couple of ticks. This results in the claps transient starting *just*

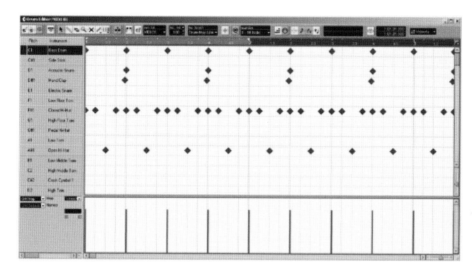

FIGURE 11.2
Thickening out the snares (note how the claps are slightly ahead of time)

before the snares which not only helps to thicken out the snares but also adds drive to the rhythm (Figure 11.2).

On this same theme, dynamics also play a vital role in acquiring the rhythmic push required in the genre. Usually, the kick drum will follow the *strong–weak–medium–weak* syncopation while the closed hi-hats follow a less-defined velocity pattern. The velocities of these will depend entirely on the feel you want to acquire, but as a general starting point the main emphasis often lies on the second and fourth 16th pulse of the bar as illustrated below:

Bar 16th Divisions															
1	2	3	4	5	6	7	8	9	10	11	12	13	14	15	16
M	S	W	S	M	S	W	S	M	S	W	S	M	S	W	S
Kick				Kick				Kick				Kick			

Of course, this is simply a convention and convention shouldn't always play a part of music, so you should feel free to experiment by placing the accents at different divisions on the beat. By doing so, the rhythm of the piece can change quite severely. If you take this approach, however, keep in mind that the open hat sits on the third (offbeat) of the bar, so generally the closed hi-hat should remain weak here to prevent frequency clashes between the two.

For house kicks, the Roland TR909 is the most frequently used drum machine but the Simmons SDS-5, Roland CR and the E-Mu Drumulator are also used to produce the timbres. To my knowledge, however, only the TR909 has ever been emulated in software form; so unless you're willing to synthesize your own kick, sample another house record, use a sample CD or source and pay for one of these archaic machines, you'll have little option but to use a TR909.

The house kick is commonly kept quite tight by employing a short decay combined with an immediate attack stage. This is to help keep the kick remain dynamic but also prevents it from blurring and destroying the all-important interplay between kick and bass. Typically, a 90 Hz sine wave produces the best starting point for a house kick with a positive pitch EG set a fast attack and a quick decay to modulate the oscillator. Although using a pitch envelope produces the best results, if the synth doesn't offer one then a self-oscillating filter will produce the requisite sound and the decay can be controlled with the filter's envelope.

In many cases, this kick will also be augmented with a square waveform to produce a heavier transient that can then be later crushed with a compressor. For this, the square wave requires a fast attack and decay stage so that it simply produces a 'click', once this is layered over the original sine you can experiment by pitching it up or down to produce the required timbre. With this basic timbre laid down, it's then worth experimenting with the decay parameters of the square and sine. Generally, a concave decay will produce the typical 'housey' character, but for a deeper or darker house vibe a convex decay may be more appropriate.

Production-wise it's worthwhile experimenting with small *controlled* amounts of distortion to give the kick a little more character in the loop, but this must be applied cautiously. If it's applied too heavily, much of the bottom-end weight of the kick can be lost, resulting in a loop with little or no energy. Also, house kicks rely very heavily on compression to produce the archetypal hardened sound. This should be applied now rather than when the rest of the loop is in place since in the context of a typical house loop, this would mean that the hi-hats, congas and toms are pumped with the compressor. This is generally avoided at this stage as the loop will be compressed along with the bass at a later stage. Instead, it's better to run just the 4/4 kick through the compressor and experiment with the settings to produce a 'hard' sounding timbre.

Although the settings to use will depend on the sound you wish to achieve, a good starting point is to use a low ratio combined with a high threshold and a fast attack with a medium release. The attack will crush the initial transient of the square producing a heavier 'thump', while experimenting with the release can further change the character of the kick. If this is set quite long, the compressor will not have recovered by the time the next kick begins and this can often add a pleasing distortion to the sound. Once the kick has been squashed, as a final stage it's often worthwhile using the PSP Vintage Warmer to add more warmth to the overall timbre.

Unlike the kicks, in the majority of cases the house snare is not sourced from the TR909 but the E-Mu Drumulator due to its warmer, rounded sound. Similar to the kick, however, the decay on the snare is kept particularly short to keep with the short, sharp dynamics of the loop. This characteristic sound can be produced through using either a triangle or square wave mixed with some pink noise or saturation (distortion of the oscillator), both modulated with an amp envelope using a fast attack and decay. Typically, pink noise is preferred over white as it produces a heavier snare, but if this proves to be too heavy for the loop, then white noise may be a better option. Much of this noise will need removing with a high-pass filter to produce the archetypal house snare but in some cases a band-pass filter may be more preferable since this allows you to modify the low and high content of the snare more accurately.

If at all possible, it's worthwhile using a different envelope on the main oscillator and the noise since this will allow you to keep the body of the snare (i.e. the main oscillator) quite short and sharp but allow you to lengthen the decay of the noise waveform. By doing so, the snare can be modified so that it rings out after each hit, producing a timbre that can be modified to suit all genres of house. Additionally (and provided that the synth allows it), it also permits you to individually adjust the decay's shape of the noise waveform. Generally speaking, the best results are from employing a concave decay but as always it's best experimenting. Some house snares will also benefit from some judicious amounts of pitch bend applied to the timbre, but rather than applying this positively, negative application will invariably produce results that are more typical to house music as it often produces a sucking/smacking sound.

If you use this latter technique, it's sometimes prudent to remove the transient of the snare in a wave editor and replace it with one that uses positive pitch modulation. When the two are recombined, it results in a timbre that has a good solid strike but decays upwards in pitch at the end of the hit. If this isn't possible, then a viable alternative is to sweep the pitch from low to high with a sawtooth or sine LFO (provided that the saw starts low and moves high) set to a fast rate. After this, small amounts of compression so that only the decay is squashed (i.e. slow attack) will help to bring it up in volume so that it doesn't disappear into the rest of the mix.

Notably, it's only the timbre for the kick and snare that truly define the house sound and while in the majority of cases the remaining percussive and auxiliary instruments such as the open and closed hi-hats, congas, cowbells, claps, tambourines, shakers and toms are sourced from the TR808 or 909, almost any MIDI module will carry sounds that can be used. Plus, many modules today carry the ubiquitous TR808 and 909 percussion sound-sets as part of the collection anyway. Normally, none of these should be compressed as it can reduce the high-frequency content, but having said that this is simply a convention and you should always be willing to 'break the rules', so to speak.

With the drum loop complete, it's worthwhile cycling it over a number of bars and experimenting with the synthesis parameters available to each instrument to create a rhythm that gels together. This includes lengthening or shortening the decay parameters, pitching the instruments up or down and applying *subtle* effects such as reverb. The basic principle here is to make the loop sound as good as possible before moving onto the subsequent processing. If it doesn't seem to gel properly together or seems wrong, then a common problem is that too many elements are playing at once. Keep in mind that although house prides itself on funky rhythms, it doesn't necessarily have to be a complex loop to achieve this and sometimes simply removing one element can suddenly make the loop work.

Once the loop is 'working' together, a common practice is to sample the loop into audio and then run it through a compressor and noise gate. Whether the gate or compressor comes first is a matter of individual taste but generally speaking the rhythm is compressed and then gated. As always, there are no definitive settings to use when compressing as it depends entirely upon the sound you're after. Having said that, a good starting point is to set the ratio quite high with a low threshold and use a fast attack with a medium release. Finally, set the make-up gain so that the loop is at the same volume when the compressor is bypassed and begin experimenting with the release settings. Shorter settings will force the compressor to pump and as a result make the kicks more prominent in the loop while longer settings will be less evident. There is no universal answer for the 'correct' settings, so you'll have to try out numerous configurations until you reach the sound you need. Ideally, the best compressors to use for this should be either valve or opto, as the second-order harmonic distortion they introduce plays a large part in capturing the sound, but if this is not possible, following the compressor with the PSP vintage warmer should help things along nicely.

Following compression, it quite usual to feed the results into a noise gate and use this to modify the loop further. One of the key elements of house loops is that the sounds remain short and sharp, as this makes the entire loop feel more dynamic since long sounds tend to mask the gaps between each hit which lessens the impact as a whole. This also helps it to accentuate the often funky bass rhythms that lie underneath as these tend to use quite long release times to help 'funk' them up. As with compression, the settings to use on a noise gate are up to your artistic interpretation, but as a start, set the threshold so that it's just lower than the kicks and snare, use a fast attack with a medium hold time and then experiment with the release setting. The faster this is set, the more the sounds will be cut, so you'll need to find a happy medium where they do not cut off too soon but do not decay for too long either. If vocals are to be employed in the mix, it's prudent to set the threshold lower so that the open hats are also gated slightly. In house these are often at a similar frequency to the vocals, and if they're left too long they can mix with the vocals which results in both drums and vocals loosing their dynamic edge. Finally, remember that house music is a popular genre to remix, so it's sensible to keep a note of all compression or effects settings along with the original MIDI files. Most studios will create these 'track' sheets and it's wise to do so too since many remixers will expect a track sheet if they're to remix the music.

Of course, it would be naive to say that all house beats are created through MIDI and a proportionate amount of rhythms are acquired from sampling previous house records. Indeed, these will have probably been sampled from other records, which will have been sampled from previous house records, which will have been sampled from other records, which will probably have been…well, you get the picture. For obvious legal reasons I can't condone this type of approach – it's illegal – but that's not to say that dance artists don't do it and so in the interests of theory it would only be fair to cover some of the techniques they use.

Almost every house record pressed to vinyl starts with just a drum loop, so sourcing one isn't particularly difficult, but the real skill comes from what you do with it after it's in the sampler. Fundamentally, it's unwise to just sample a loop and use it as it is. Even though there's a very high chance that what you're sampling is already a sample of a sample of a sample of a sample (which means that the original artists don't have a leg to stand on for copyright infringement), you might be unlucky and they may have actually written it in MIDI. What's more, it isn't exactly artistically challenging; so after sampling a loop it's always worthwhile placing it into a sequencer and moving the parts around a little to create your own variation.

Of course, this is the most simplistic approach and a number of artists will take this a step further by first employing a transient designer to alter the transients of the loop before placing it into a sample slicing programme. Using this technique, the loop does not have to be sourced from a house track as any drum break can be manipulated to create the snappy sounds used in house.

Along with working with individual loops, many house producers also stack up a series of drum loops to create heavier rhythms. Todd Terry and Armand Van Helden are well known for dropping more powerful sounding kicks over pre-sampled loops to add more of a heavy influence. If you take this approach, however, you need to exercise care that the kick starts in just the right place, otherwise you'll experience a phasing effect. Possibly the best way to circumvent this is to cut the bar into four segments where the kick occurs and then trigger it alongside the new kick at the beginning of each segment to produce a continual loop that stays in time. This also permits you to swap the quarter bars around to create variations.

BASS

After the drums, the bass is commonly laid down since many genres of house rely a great deal on the groove created from the interaction between the two. As this genre of music has its roots firmly embedded in disco, most house basses tend to follow the atypical disco funk pattern, which can range from the rather simplistic walking bass to more complex patterns produced by real bassists improvising with the drum rhythms (Figure 11.3).

The above bass line is only a very simple example, but house basses do not have to be particularly complex and in many instances you'll find that even an incredibly unsophisticated bass can have additional groove injected into it by simply changing the duration and timing of *some* notes (Figure 11.4).

It's important to note that the complexity of the bass often depends on the lead instruments that are to lie over the top. Although some house music relies on a particularly funky bass groove, the overlaying instruments are kept relatively simple by comparison, whereas if the overlying elements are more complex then the bass line should remain relatively simple. This is incredibly important to comprehend since if both the bass and overlying instruments are playing

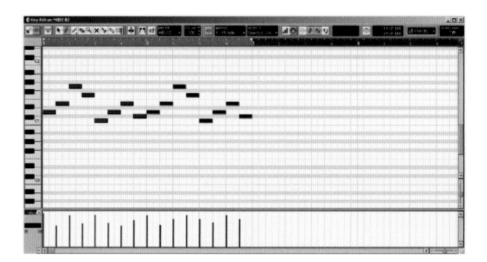

FIGURE 11.3
Disco's infamous
walking bass line

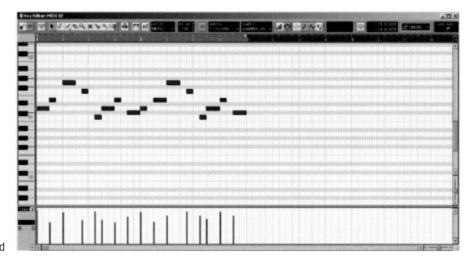

FIGURE 11.4
By changing the
duration and timing
of some notes, new
grooves can be created

complex grooves it can easily undermine the record, resulting in a miscellany of melodies that are difficult to decipher. Thus, when producing house you need to decide whether the centrepiece of the music is created by the bass *or* the overlying elements. For instance, the piano in Laylo and Bushwacka's 'Love Story' produces the centrepiece of the music, while the bass acts as an underpinning and so is kept relatively simple. Conversely, tracks such as Kid Crème's 'Down and Under', Stylophonic's 'Soul Reply' and Supermen Lovers' 'Starlight' use a more funky bass which is complimented with simple melodies playing over the top (Figure 11.5).

Some house basses, particularly those used in acid or deep house will follow a simpler pattern similar to techno. Refer to the chapter on techno for more information.

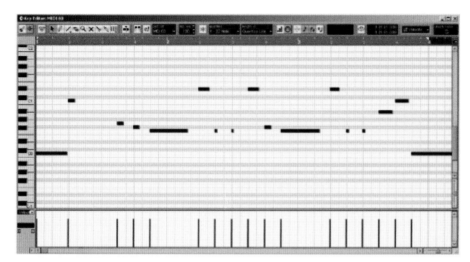

FIGURE 11.5
The bass rhythm from 'Starlight' (notice how the bass occurs before the beat for additional groove)

As with the rhythms, the timbre used for the basses in house can vary wildly from one track to the next. On the one hand it can be from a real bass guitar, while on the other it could consist of nothing more than a low-frequency sine wave pulsing away in the background to anything in between. Like much of house music, this makes it incredibly difficult to pin down any particular timbre, so what follows are a few synthesis tips to create the fundamental structure of most house basses. After you have this, experimentation is the real key and it's worthwhile trying out new waveforms, envelope settings and modulation options to create the bass you need.

The foundation of most synthetic house basses can be constructed with a sine wave and sawtooth or pulse oscillator. The main waveform (the sine) provides a deep body to the sound while the secondary oscillator can either provide raspy (saw) or woody (square) overtones. If the sound is to be quite cutting and evident in the mix then a saw is the best choice, while if you want to be more rounded and simply lay down a groove for a lead to sit over then a square is the better option. Listening carefully to the sound they produce together, begin to detune the secondary oscillator against the first until the timbre exhibits the 'fatness' and harmonic content you want. Typically, this can be attained by detuning +5 or +7 cents, but as always let your ears decide what's best.

Nearly all bass timbres will start immediately on key press, so set the amps attack to zero and then follow this with a fast decay, no sustain and a short release. This produces a timbre that starts immediately and produces a small pluck before entering the release stage. If the sound is too short for the melody being played, increase the sustain and experiment with the release until it flows together to produce the rhythm you require. If the bass sounds like it's 'running away', try moving it a few ticks forward or back and play with the amps attack stage. To add some movement and character to the sound, set the filter cut-off to low pass and set both the filter cut-off and resonance to midway, and then adjust the envelope to a fast attack and decay with a short release and no sustain. Normally, this envelope is applied positively but experiment with negative settings as this may produce the character you need. If possible, it's also worthwhile trying out convex and concave settings on the decay slope of the filter envelope as this will severely affect the character of the sound, changing it from one similar to a Moog through to a more digital nature and onto a timbre similar to the TB303. In most house tracks, the filter follows the pitch (opens as the pitch increases), so it's prudent to use filter positive key follow too.

This creates the basic timbre, but it's practical to begin experimenting with LFOs, effects and layering. Typical uses of an LFO in this instance would be to lightly modulate the pitch of the primary or secondary oscillator, the filter cut-off or the pulse width if a pulse waveform is used. On the effects front, distortion is particularly effective and an often-used effect in house as is phasing/ flanging, but if you choose this latter option, you will have to exercise care. In all dance mixes the bass should always sit in the centre of the stereo spectrum so that not only do both speakers share the energy but also any panning applied during mixing is evident (this is covered in the chapter on ambient/ chill out). Consequently, if heavy flanging or phasing is applied then the bass will be smeared across the stereo image and the mix may lack any bottom-end cohesion. Of course, if the bass is too thin or doesn't have enough body then effects will not rescue it, so it may be sensible to layer it with another timbre or in some cases five or six others first. Above all, bear in mind that we have no expectations of how a synthesized bass should sound, so you shouldn't be afraid of stacking up as many sounds as you need to build the required sound. EQ can always be used to remove harmonics from a bass that is too heavy, but it cannot be used to introduce harmonics that are not already present.

Alongside the synthetic basses, many house tracks will employ a real bass. More often than not, these are sampled from other records or are occasionally taken from sample CDs, rather than programmed in MIDI. However, provided that you have (a) plenty of patience, (b) a willingness to learn MIDI programming and (c) a good tone or bass module (such as Spectrasonics Trilogy), it is possible to programme a realistic bass that could fool most listeners. This not only prevents any problems with clearing copyright but it also allows you to model the bass to your needs.

The key to programming any real instrument is to take note of how they're played and then emulate this action with MIDI and a series of CC commands. In this case, most bass guitars use the first four strings of a normal guitar E–A–D–G, which are tuned an octave lower, resulting in the E being close to three octaves below middle C. Also they are monophonic, not polyphonic, so the only time notes will actually overlap is when the resonance of the previous string is still dying away as the next note is plucked. This effect can be emulated by leaving the preceding note playing for a few ticks while the next note in the sequence has started. The strings can either be plucked or struck and the two techniques produce different results. If the string is plucked, the sound is much brighter and has a longer resonance than if it were simply struck. To copy this, the velocity will need to be mapped to the filter cut-off of the bass module so that higher values open the filter more. Not all notes will be struck at the same velocity, though, and if the bassist is playing a fast rhythm, the consecutive notes will commonly have less velocity since he has to move his hand and pluck the next string quickly. Naturally, this is only a guideline and you should edit each velocity value until it produces a realistic feel.

Depending on the 'bassists', they may also use a technique known as 'hammer on' whereby they play a string and then hit a different pitch on the fret. This results in the pitch changing without actually being accompanied with another pluck of the string. To emulate this, you'll need to make use of pitch bend, so this will first need setting to a maximum bend limit of 2 semitones, since guitars don't 'bend' any further than this. Begin by programming 2 notes, for instance, an E0 followed by an A0, and leave the E0 playing underneath the successive A0 for around a hundred ticks. At the very beginning of the bass track, drop in a pitch bend message to ensure that it's set midway (i.e. no pitch bend), and just before where the second note occurs, drop in another pitch bend message to bend the tone up to A0. If this is programmed correctly, on play back you'll notice that as the E0 ends the pitch will bend upwards to A0, simulating the effect. Although this could be left as it is, it's sensible to drop in a CC11 message (expression) directly after the pitch bend as this will reduce the overall volume of the second note so that it doesn't sound like it has been plucked. In addition to this, it's also worthwhile employing some fret noise and finger slides. Most good tone modules will include fret noise that can be dropped in between the notes to emulate the bassist's fingers sliding along the fret board.

Whether you decide to use a real bass or a synthesized one, after it's written, it's an idea to compress the previous drum loop against the bass. The principle here is to pump the bass to produce the classic bottom-end groove of most house records – they tend to pump like crazy. This is accomplished by feeding both bass and drum loop (preferably with the hi-hats muted) into a compressor and then set the threshold so that each kick registers approximately −6 dB on the gain reduction metre. Use a ratio of around 9:1 with a fast attack and set the gain make-up so that it's at the same volume level when the compressor is bypassed. Finally, set the release parameter to 200 ms and then experiment by increasing and reducing this latter parameter. The shorter the release becomes, the more the kick will begin to pump the bass, becoming progressively heavier the more that it's shortened. Unfortunately, the only guidelines for how short this should be set are to use your ears and judgement but try not to get too excited. The design is to help the drums and bass gel together into a cohesive whole and produce a rhythm that punches along energetically. On the same note, it should not be compressed so heavily that you loose the excursion of the kick – you have to reach a happy medium!

MELODIES AND MOTIFS

With the bottom-end groove working, the lead instruments can be dropped over the top. As touched upon when discussing basses, the lead instruments' melody will depend entirely on the type of house being produced. Funky bass lines will require less active melodies to sit over the top while less active basses will require more melodic elements. Additionally, since these leads are often sampled from other records, it's naïve to suggest that they are programmed with MIDI and therefore it's impossible to offer any guidelines apart from to listen to the latest house records to see where the current trend is.

This also means that there are no definitive sounds that characterize the genre and as such practically everything should be seen as a fair game. In fact, a proportionate amount of house producers feel the same as many have sampled heavily from previous records (particularly disco) to produce the sound. With that said, there are some timbres that have always been popular in house including the Hoover, plucked leads and pianos from the DX series of synthesizers. These have all already been discussed in detail in the chapter on sound design, so here we'll look at the general synthesis ideals.

Fundamentally, synthetic house leads will more often than not employ sawtooth, triangle and/or noise waveforms to produce a harmonically rich sound that will cut through the mix and can be filtered if required. Depending on how many are employed in the timbre, these can be detuned from one another to produce more complex interesting sounds. If the timbre requires more of a body to the sound, then adding a sine or pulse wave will help to widen the sound and give it more presence. To keep the dynamic edge of the music, the amplifiers attack is predominantly set to zero so that it starts upon key press but the decay, sustain and release settings will depend entirely on what type of sound you require. Generally speaking, it's unwise to use a long release setting since this may blur the lead notes together and the music could lose its dynamic edge, but it's worth experimenting with the decay and sustain while the melody is playing to the synth to see the effect it has on the rhythm. As lead sounds need to remain interesting to the ear, it's prudent to employ LFOs or a filter envelope to augment the sound as it plays. A good starting point for the filter EG is to set the filter cut-off to low pass and set both the filter cut-off and resonance to midway, and then adjust the envelope to a fast attack and decay with a short release and no sustain. Once this is set, experiment by applying it to the filter by positive and negative amounts. On top of this, LFOs set to modulate the pitch, filter cut-off, resonance and/or pulse width can also be used to add interest. Once the basic timbre is down, the key is, as always, to experiment.

For chords, the most common instrument used in-house is the Solina String Machine which has made an appearance on hundreds of house records including those by Daft Punk, Air, Joy Division, Josh Wink, STYX, New Order, Tangerine Dream, Roger Sanchez, Supermen Lovers and nearly every disco track ever released. Although this synth is now out of production, a similar timbre can be created in any analogue-style synth. Begin by using three sawtooth waveforms and detune two from one of the oscillators by ±3. Apply a small amount of vibrato using an LFO to these two detuned saws, and then use an amplifier envelope with a medium attack and release and a full sustain (there is no decay since sustain is set to full and it would have nothing to drop down to). Now, experiment with the sawtooth that wasn't detuned by pitching it up as far as possible without it becoming an individual sound (i.e. less than 20 Hz), and if possible, use two filters – one set as a low pass to remove the low-end frequencies from the two detuned saws and the other as a high pass to remove some of the high-frequency content from the recently pitched saw.

Once this basic sound is created, it may be prudent to make it appear as though it's been sampled and subsequently affected. To do this, it's worthwhile copying the timbre to three sequencer tracks and removing all the bottom-end frequencies of the first track, all of the midrange from the second and everything but the high frequencies on the third with a good EQ unit. Following this, insert a flanger or phaser across the audio track that only has the high frequencies and then mixing the three tracks together under it produces a timbre that sounds appropriate for the track. Depending on the music so far, this may also involve applying the flanger (or phaser) to the mid-range or the low-range track or perhaps even all of them. In this case, the basic principle is to have the effects sweep the frequencies present in each track by differing amounts to construct a big textural timbre. Once complete, export these three files into a single audio file and then place it into a sampler and create a chord structure to suit the music.

Even if this sound is not used, effects can play an important role in creating interesting and evolving pads for the genre. Wide chorus effects, rotary speaker simulations, flangers, phasers and reverb can all help to add a sense of movement to help fill out any holes in the mix. A noise gate can also be used creatively to model the sound. For instance, by setting a low threshold and using an immediate attack and release, the pad will start and stop abruptly producing a 'sampled' effect that cannot be replicated through synthesis parameters. What's more, if you set the hold time to zero and adjust the threshold so that it lies just above the volume of the timbre, the gate will 'chatter' which can be used to interesting effect. Overall, compression is generally not required on pads since they'll already be compressed at the source, but if distortion or heavy filtering or EQ is used then it may be prudent to follow them with a compressor to prevent them from clipping the desk.

With all the parts programmed, we can begin to look at the arrangement. As touched upon in the musical analysis, house is an incredibly diverse genre, so there are no definite arrangement methods. Nevertheless, as much of house relies on the sampling ethics introduced by Todd Terry, a usual approach is to arrange the instruments in different configurations to produce a series of different loops. For instance, a loop could consist of just the drums, another of the drums and bass mixed together, another consisting of the drum, bass and leads and another made from just the leads. Each of these are then sampled (or exported from an audio sequencer) and set to a specific key on a controller keyboard connected to the sampler. By then, hitting these keys at random you can produce a basic arrangement plus if one-shot mode is disabled on the sampler it's possible to create the stuttered mix effect used on some house tracks.

That said, as with all genres this is subject to change, so the best way to obtain a good representation of how to arrange the track is to listen to what are considered the classics and then copy them. This isn't stealing; it's research and nearly every musician on the planet follows this same route. This can be accomplished using a method known as '*chicken scratching*'. Armed with a piece of paper and a pen, listen back to the track, and at every bar, place a single scratch mark on the paper. When a new instrument is introduced or there is a change in the rhythmic element, place a star below the scratch. Once you've finished listening back to the song, you can refer to the paper to determine how many bars are used in the track and where new instruments have been introduced. You can then follow this same arrangement and, if required, change it slightly so that it follows a progression to suit.

> The data CD contains a full mix of a typical house track with narration on how it was constructed.

RECOMMENDED LISTENING

Ultimately, the purpose of this chapter has been to give some insight into the production of house and, as ever, there is no one definitive way to produce the genre. Indeed, the best way to learn new techniques and production ethics is to actively listen to the current market leaders and be creative with your processing, effects and synthesis. In fact, in many cases it's this production that makes one track better than the other, and if you want to write house you need to be on the scene and listen to the most influential tracks. If you *listen* closely to these, they will invariably reveal significant information about how they were made. Keep in mind that the timbres are unimportant since this genre does not necessarily rely on a particular sound; it's more to do with *feel* and *groove*. With this in mind, what follows is a short list of some of the artists that, at the time of writing, are considered the most influential in this area:

- Daft Punk
- Room 5
- The Supermen Lovers
- Modjo
- Stardust
- Kid Crème
- Stylophonic
- Some of Moby's work
- Laylo And Bushwacka
- Roger Sanchez

And of course, some disco artists who *could* be used as a source of inspiration are:

- Alec Costandinos
- Bohannon

- Bootsy Collins
- Cerrone
- Chic
- Deodato
- Donna Summer
- Eddie Kendricks
- Evelyn King
- France Joli
- George McRae
- Gino Soccio
- Giorgio Moroder
- Gregg Diamond
- Gwen McRae
- Harold Melvin & the Blue Notes
- Heatwave
- LaBelle
- Larry Levan
- Lime
- Love Unlimited Orchestra
- Meco Monardo
- Patrick Adams
- The Salsoul Orchestra
- Shalamar
- Sister Sledge
- Slave
- Sylvester
- Sylvia
- Tavares
- Thelma Houston
- Tom Moulton
- Van McCoy
- Vicki Sue Robinson
- Vincent Montana Jr.

CHAPTER 12

Trance

'The "soul" of the machines has always been a part of our music. Trance belongs to repetition and everybody is looking for trance in life... in sex, in the emotional, in pleasure... So, the machines produce an absolutely perfect trance...'

Ralf Hütter (Kraftwerk)

Trance is possibly one of the most ambiguous genres of dance music because it appears in so many different forms and few can actually agree on exactly what makes the music *trance*. However, it's fairly safe to say that it can be roughly generalized as the only form of dance music that's constructed around glamorous melodies which are either incredibly vigorous, laid-back or pretty much anywhere in between. In fact, it's this 'feel' of the music that often determines whether it's *Progressive, Goa, Psychedelic, Acid* or *Euphoric*. Even then, some will place euphoric trance with Progressive and others believe that Acid trance is just another name for Goa. To make matters worse, some forms of trance are given numerous titles by both DJs and clubbers. For example, euphoric trance may be subdivided into *Commercial* or *Underground*, each of which is different yet again. This obviously presents a dilemma when trying to encapsulate it for the purpose of writing an example track since it becomes a Hobson's choice.

Nevertheless, much of how the music is written can be determined by looking at its history and one thing that is certain is that trance, in whichever form, has its roots embedded in Germany. During the 1990s a joint project between DJ Dag and Jam El Mar resulted in Dance2Trance and their first track labelled "We came in Peace" is considered by many to be the first ever 'club' trance music. Although by today's standards it was particularly raw, consisting solely of repetitive patterns (as Techno is today), it laid the basic foundations for the genre with the sole purpose of putting clubbers into a trance-like state.

The ideas behind this were nothing new; tribal shamans had been doing the same thing for many years, using natural hallucinogenic herbs and rhythms pounded out on log drums to induce the tribe's people into trance-like states. The only difference with Dance2Trance was that the pharmaceuticals were man-made and the pounding rhythms were produced by machines rather than skin stretched across a log. Indeed, to many trance aficionados (if there is such a thing) the practice of placing clubbers into a trance state is believed to have formed the basis of Goa and Psychedelic trance.

Although both these genres are still produced and played in clubs to this day, the increased popularity of 3,4-methylenedioxy-*N*-methylamphetamine (MDMA or 'E') among clubbers inevitably resulted in new forms of trance being developed. Since this drug stimulates serotonin levels in the brain it's difficult, if not impossible, to place clubbers into states of trance with tribal rhythms; instead, the melodies became more and more exotic slowly taking precedence over every other element in the mix. The supposed purpose was to no longer induce a trance-like state but emulate or stimulate the highs and lows of MDMA. The basic principle is still the same today in that the chemicals, arguably, still played a role in inducing a state, albeit a euphoric one, and the name euphoric trance was given to these tracks.

This form of music, employing long breakdowns and huge melodic reprises, became the popularized style that still dominates many commercial clubs and the music charts today. As a result, when the term trance is used to describe this type of music most view it as the typical anthemic "hands in the air" music that's filled with big breakdowns and huge melodic leads. As this remains one of the biggest and most popular genres of dance music to date, this form of trance will be the focus of the chapter.

MUSICAL ANALYSIS

Possibly the best way to begin writing any dance music is to find the most popular tracks around at the moment and break them down to their basic elements. Once this is accomplished we can examine the similarities between each track and determine exactly what it is that differentiates it from other musical styles. In fact, all music that can be placed into a genre-specific category will share similarities in terms of the arrangement, groove and sonic elements.

In the case of euphoric trance, it is viewed as being an anthemic form of music, which essentially means that it has an up tempo, uplifting feel that's very accessible to most clubbers. As a result, it can best be illustrated as consisting of a relatively melodic synth and/or vocal hook line laid over a comparatively unsophisticated drum pattern and bass line. The drums usually feature long snare rolls to signify the build-up to the reprise and breakdowns, alongside small motifs and/or chord progressions that work around the main melody.

Consequently, all euphoric trance music will utilize a four-to-the-floor time signature to help clubbers keep time while 'stimulated' and be produced either allegretto or allegro. In terms of physical tempo, this can range from a rather contained 125 BPM and move towards the brain mashing upper limits of

150 BPM, but most tend to stay around 137–145 BPM mark. This, of course, isn't a demanding rule but it is important not to go too fast as it needs to remain anthemic and the majority of the public can't dance to an incredibly fast beat.

THE RHYTHM SECTION

The best place to start when programming any dance music is with the rhythm section as this will define the overall pace of the record. With most trance this is invariably kept quite simple and will almost always rely on a four-to-the-floor pattern. This means a kick drum is placed on every beat of the bar, along with snares or claps laid on the second and fourth beat to add some expression to these two beats. To complement this basic pattern closed hi-hats are commonly placed on every 16th division, or variation of 16ths, while to introduce some syncopation open hi-hats are often employed and placed on every 1/8th division of the bar. Notably, any closed hi-hats that occur in this position should be removed to prevent any frequency clashes between the closed and the open hi-hat (Figure 12.1).

This, of course, is only a general guideline to the drum patterns used and is open to your interpretation. For instance, occasionally a set of triplet closed hats may sit at the end of every few bars to add some extra variation, or alternatively, the snare/clap may be placed a 1/16th before the kick occurs on the last beat of the fourth bar. Additionally, the kicks are sometimes used to add variation and a double kick at the end of the fourth bar can be used to signify a change, for example, when introducing a new instrument into the mix (Figure 12.2).

Although these techniques produce what is essentially an incredibly basic loop, this is exactly what much of trance relies on. Indeed, bongos, congas and most other percussive instruments are not commonly employed since a more complex drum loop will not only reduce the room left for the lead melody but may also take the attention away from it. On this same theme, despite

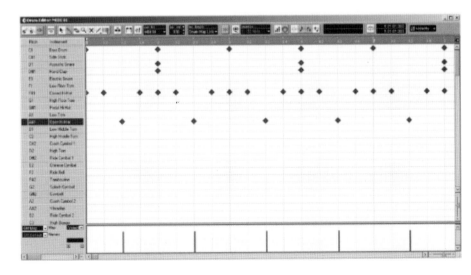

FIGURE 12.1
Typical trance drum loop

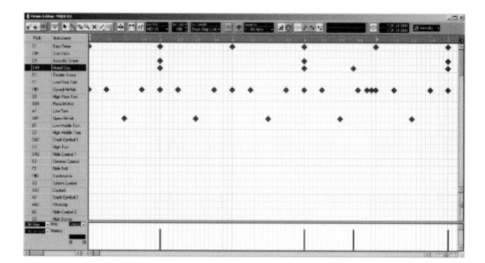

FIGURE 12.2
A variation on the trance loop

mentioning 'dynamics' in a previous chapter, both the kick and the snare in trance will often remain at the highest velocity rather than following the *strong – weak – medium – weak* syncopation. Again this is simply because the drums should pull very little attention away from the main melodic lead. The closed hi-hats, however, often employ different velocities throughout the bar to add some interest to the patterns. As a general rule of thumb, the main emphasis is commonly on the first and fourth (remember that the open hat sits on the third!) 16th division of the bar as illustrated below:

Bar 16th Divisions

1	2	3	4	5	6	7	8	9	10	11	12	13	14	15	16
S	M	W	S	S	M	W	S	S	M	W	S	S	M	W	S
Kick				Kick				Kick				Kick			

Of course, this is simply convention and convention shouldn't always play a part in your music, so you should feel free to experiment by placing the accents at different divisions on the beat. By doing so the rhythm of the piece can change quite severely, so it is worth experimenting.

RHYTHM TIMBRES

Simply playing this rhythm through a GM module you can expect the results to sound absolutely nothing like the pounding beats you hear in the clubs, so it's vital to use the right instruments to produce the sounds. Predictably, all of these are usually derived from the Roland TR909 but the Waldorf Attack, Novation Drum Station or a drum sampler and a series of 909 samples will perform the job just as well. Alternatively, they can be programmed in any capable synthesizer.

The trance kicks tend to be kept quite tight rather than 'boomy' since this keeps the entire rhythm section sounding fairly strong to produce the four-to-the-floor solidity. To reproduce this try using a 90 Hz sine wave and modulate it with a positive pitch envelope with no attack and a medium release. If you have access to the pitch envelopes shape then generally a concave release is more typical than convex but that's not to say convex doesn't produce the results. Finally, experiment with the decay setting as the faster the decay is set, the tighter the rhythm will become but there is a limit as to how far you should go. If the decay is set too short then the kick will become a 'blip', so you should look towards making a kick that has enough body to pull out of the mix but not so much that it begins to boom. Generally, this, as with most instruments, is best left as MIDI triggering the drum synth rather than bounced down to audio as soon as it sounds appropriate, as it allows you to further contort the kick when the rest of the instrumentation is in place.

As discussed in earlier chapters, compression can also play a large role in getting the kick to sound 'right' and should be applied to the 4/4 kick loop, rather than when the rest of the loop elements are in place. Although many engineers apply compression after the loop is complete (and feel free to experiment), keep in mind that the principle here is to 'pump' the drums, and if the loop only has high-frequency elements sat off-beat – the compressor will pump them. This can work musically with the more expensive compressors but most budget units will destroy the high frequencies of the hats resulting in a duller sounding loop. What's more, as the kick and snare may occur on the same beat a compressor with the required fast attack setting will capture the higher frequencies of the snare, dulling them too.

This can be avoided by applying compression just to the 4/4 kick loop. Although there are no underlying elements for the compressor to pump, the attack is set as fast as possible so that the transient is compressed and some of the high-frequency content is removed to produce a more substantial 'thud'. The lower the threshold and higher the ratio, the more this will be evident, but by increasing the release parameter so that when the compressor is nearing the end of its release the next kick starts, a good compressor will distort in a pleasing manner. Generally speaking, most valve-based compressors will pump musically, but the most commonly used compressors for this are the *Joe Meek SC 2.2, UREI LA 3* or the *UREI 1176 LN* due to the amount of second-order harmonic distortion they introduce. If you don't have access to a good valve compressor, then after compressing it with a standard unit it's certainly worth throwing the loop through the *PSP Vintage Warmer* to recreate the requisite sound. In fact, even if it is put through a valve compressor it's often worth placing it through the vintage warmer anyway.

Once the kick is sounding 'right' the snares can be added and further modified to suit. Notably, these are not always used in the production of trance and claps are occasionally used in their place – it depends entirely on what your creative instincts tell you. Nevertheless, whichever is used in the track they will, more often than not, be sourced from the TR909 or a sample CD (or more commonly another record). They can, of course, also be created in a synthesizer but along with the triangle wave it's worth employing a pink noise waveform rather than white as it produces a thicker timbre. The pitch envelope is generally set to positive modulation with a fast attack and a long decay but it's also worthwhile creating a slurred effect by using negative modulation. This latter snare can be used as part of a build-up to new instruments.

If a sample has been used and the decay cannot be made any longer, sending them to a reverb unit set to a short pre-delay and tail will increase the decay. This is also worth experimenting with even if the decay is long enough as it can help 'thicken' the timbre. If this latter approach is taken, though, it's prudent to employ a noise gate to remove some of the reverb's tail off. After this, much of the low-frequency content will need removing to provide the basic snare timbre, so it's sensible to employ a high-pass filter to remove some of the low end and then experiment with the concave and convex decay parameter (of the amp envelope). Convex decays tend to be more popular in most trance tracks, but concave may produce the results to suit what you have in mind.

As previously mentioned, compressors are often used on the snares to help create the timbre, but rather than compress the transient it's prudent to set the attack so that it just misses and captures the decay stage instead. This will prevent the high frequencies of the snare from being compromised, but by increasing the make-up gain, the decay stage is increased in volume, helping to produce the often used and distinctive *thwack* style timbre. Similar to the kick, the lower the threshold and higher the ratio, the more this will be evident so experiment.

Finally, the closed and open hi-hat patterns will need some modifications to suit the kick and snare. Yet again, these tend to be sourced from the TR909, a sample CD or another record. However, it is sometimes prudent to programme your own as these will play a large part in the rhythm of the record.

Ring modulation is possibly the easiest method for creating hi-hats and is easily accomplished by ring modulating a high-pitched triangle wave with a lower pitched triangle. The result is a high-frequency noise type waveform that can then be modified with an amplifier envelope set to a zero attack, sustain and release with a short-to-medium decay. If there isn't enough noise present, it can be augmented with a white noise waveform using the same envelope. Once this basic timbre is constructed, shortening the decay creates a closed hi-hat while lengthening it will produce an open hat. Similarly, it's also worth experimenting by changing the decay slope to convex or concave to produce fatter or thinner sounding hats.

As both these timbres depend on the high-frequency content to sit at the top of the mix, compression should be avoided but adjusting the amplifier decay of synthesized hi-hats can help to change the perceived speed of the track. If the hi-hats are sampled then this can be accomplished by importing them into a 'drum sampler' such as *Native Instruments Battery*, or alternatively you can use a transient designer such as *SPL's Transient Designer* or the software alternative available in the *Waves Transform Package*. Using these to keep the decay quite short they will decay rapidly which will make the rhythm appear faster. Conversely, by increasing the decay, you can make the rhythm appear slower. Most trance music tends to keep the closed hi-hats quite short, though, using a short decay but the open hats often benefit from lengthening the decay. This can often help the higher frequencies of the loop gel together more appropriately for the genre.

With the loop complete, it's worthwhile looping it over a number of bars and experimenting with the synthesis parameters or a transient designer to each instrument in the loop to create a loop that gels together. This includes lengthening or shortening the decay parameters, pitching the instruments up or down and applying effects such as reverb (*subtly!*). Although at this stage it's inadvisable to use compression to warm up the loop, noise gates can be particularly useful for creating loops with a different rhythmic flow. By feeding the entire loop into the gate and reducing the threshold so that the transients of all the instruments pull though, shortening the release parameter on the gate will cut some instruments short.

BASS RHYTHM

With the drum groove laid down, most artists then move onto the bass rhythm. This, in many trance records, commonly consists of simple 8th notes with very little movement in pitch or in the timbre itself. The reason for this is to leave more 'room' for other elements such as the inter-melodic lines and the chorded reprise after the middle eight of the track. For this technique to work the bass most usually sits *off* the beat rather than on it since they're kept relatively short and play a very simple pattern to help keep the focus on the melodic lead. If they occurred *on* the beat the groove would sound similar to a march because there would be no low-frequency energy (LFE) in between the kicks. As a result, the rhythm would have a series of 'holes' in between each kick/bass hit.

Kick	....	Kick	....	Kick	....	Kick	....
Bass	....	Bass	....	Bass	....	Bass	....

To avoid this, the bass commonly sits an eighth *after* the kick resulting in:

Kick		Kick		Kick		Kick	
	...Bass		...Bass		...Bass		...Bass

What's more, by adopting this technique there is a more substantial groove to the music since the two LFE signals are in effect working as a very basic Da Capo (binary phrase) throughout the track. That is, the first low-frequency signal (the kick) is asking a question which is answered by the second low frequency (the bass).

Of course, this, like everything, is open to artistic interpretation, and while it is generally recommended that the bass occurs after the kick, it isn't necessary to place it an eighth after. Indeed, by experimenting through moving it subtly around the off-beat, the groove can become more intense or relaxed. Also, it isn't compulsory to use notes that are an eighth in length. While these are commonly used to allow the timbre's characteristics to be evident, smaller notes can be used to great effect. Similarly, if the track is quite simple even with the main melody, increasing the length of the notes to a quarter allows more of the basses character to be evident which can be especially useful if the bass features lengthy timbre augmentation with LFOs or envelopes.

Generally the bass, as with all the melodic elements of trance, works on a looped eight-bar basis – that is the melodies, chords and so forth all loop around every eight bars (usually with a cymbal crash or short snare roll at the end of the eighth bar to denote the end of the musical passage). This helps to keep the repetition that forms the basis of the music. Thus, the bass can be programmed over two bars as a simple series of one pitch notes that can then be pasted to create the other six bars. Working this way, you can select and pitch each consecutive two *bars* of notes up and down to create some movement in the rhythm.

The technique of pitching an entire bar rather than any individual notes within bars is very common since pitch-shifting constituent notes creates a much more noticeable bass rhythm. This not only dissuades the focus from the main melody but also lays down a series of musical ground rules as to how the melody can be written.

Keep in mind that if the bass features plenty of movement in a bar, for the track to remain musical, the melody should harmonize with this progression. However, by keeping the bass at one pitch throughout a bar and only moving an entire bar up or down the only ground rules to the melody are to harmonize with the movements of the entire bar allowing for more experimentation on the lead. Of course, this again is merely the conventional approach used by most trance artists and the example track, and it's certainly open to artistic licence. For example, some tracks will keep the bass off-beat but not maintain even spaces between each hit, thus creating a rhythmic drive to the bass. Tracks of this nature, however, will usually employ less melodic leads (Figure 12.3).

BASS TIMBRE

For the bass timbre, analogue synthesizers (or DSP equivalents) are invariably chosen over digital since the bass is exposed from the kick and the 'distortion' introduced by analogue produces a warmer tone to complement the kick.

Trance basses are sometimes sourced from sample CDs but many artists have commented that they prefer to create their own in a synth. Generally speaking, any analogue synth will produce a good analogue bass but many trance musicians often rely on the *Novation Bass Station* (VSTi), the *Novation SuperNova*, *Novation V-Station* (VSTi) or the *Access Virus* to produce the timbres. Using any of these, possibly the easiest solution for this timbre is to dial up a preset bass in a synth and then contort the results to suit the tune. If, however, you wish

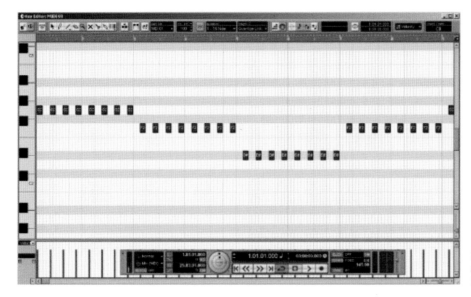

FIGURE 12.3
A typical trance bass
rhythm

to program your own a good starting point is to employ a pulse and sine (or triangle) oscillator and detune them + or − 5 to produce the basic tone. In general, a heavy transient pluck is not entirely necessary since it isn't competing with the kick. Thus, the amp and filters attack and decay parameters can be quite long and, in many instances, the decays actually act as the release parameters due to the relatively short duration of the notes.

When using the decay as a release parameter, the amplifier's release can then be used as a 'fine tune' parameter to help the bass sit comfortably in with the drum loop. Additionally, by shortening the attack and decay parameter of the filter envelope, set to modulate at a positive depth, more of a pluck can be introduced into the sound so that the tone also complements the drum rhythm. In fact, it's essential to accomplish this rhythmic and tonal interaction between the drum loop and the bass by playing around with both the amp and the filters A/D EG before moving on. This is the underpinning of the entire track, and if it's weak, any instrumentation dropped on top will not rescue it.

By and large, most trance music doesn't employ any velocity changes on the bass rhythm as this is usually associated to control the filter cut-off, so the harder it's struck, the more the filter opens and the more of a pluck it exhibits. This type of timbral variation is often avoided in order that the bass doesn't demand too much attention and rather simply 'pulses' along to lay down a solid foundation for the lead and counter melodies. It is, of course, always prudent to experiment, and it's also worth editing the synth so that velocity controls other parameters.

On the subject of experimentation, delay, distortion and compression are often used on basses. While most effects should be avoided since they tend to spread the sound across the image (which destroys the stereo perspective) small amounts of controlled distortion can help to pull the bass out of the mix or give it a much stronger presence. Similarly, a conservatively applied delay effect can be used to create more complex sounding rhythms that will not place any musical 'restrictions' on the lead. Any effects should be applied cautiously, though, as the purpose of the bass is not to draw attention to itself but simply underpin the more important elements.

Also, while it's true to say that the sounds from any synthesizer are already heavily compressed at the source, the principle behind a compressor here is to control the results from any preceding effects and to be used as an 'effect' itself. Using the exactly same methods as compressing the kick, lengthening or shortening the release parameter can introduce a different character to the bass. More importantly, though, with both the drums and bass laid down, if they're both fed into a compressor (with the hi-hats muted), it can be set to activate on each kick which results in it pumping the bass. Generally, the best compressors to use for this should be either valve or opto due to the second-order harmonic distortion they introduce, as this helps the mix to pump more musically. Again this means using the *Joe Meek SC 2.2*, *UREI LA 3* or the *UREI 1176 LN*, but if you don't have access to these, the Waves *C1*, *C4* or *Renaissance* compressors will do the trick or the *PSP Vintage Warmer* if you need more warmth.

Naturally, the amount of compression applied will depend upon the timbres used, but as a general guideline, start by setting the ratio to 9:1, along with an attack of 5 ms and a medium release of 200 or so milliseconds. Set the threshold control to 0 dB and then slowly decrease it until every kick registers on the gain reduction meter by at least 3 dB. To avoid the volume anomaly (i.e. louder invariably sounds better!), set the make-up gain so that the loop is at the same volume as when the compressor is bypassed and then start experimenting with the release settings. By shortening the release the kicks will begin to pump the bass, which becomes progressively heavier the more that the release is shortened. Unfortunately, the only guidelines for how short this should be set are to use your ears and judgment but try not to get too excited. The idea here is to help the drums and bass gel together into a cohesive whole and produce a rhythm that punches along energetically. On the same note, it should not be compressed so heavily that you lose the excursion of the kick altogether!

Once these two elements are working together, it's prudent to export the groove as the four separate two-bar loops of audio and drop them into a sampler. This allows you to trigger the groove from numerous points along the arrangement and also permits you to experiment with different progressions by simply hitting the appropriate key on the sampler. It's also sensible to export the drum track alone and keep a note of all compression or effects settings along with the original MIDI files to come back to later. Most studios will create these 'track' sheets, and it's wise to do so since many remixers will expect a track sheet if they're to remix the music. What's more, having the groove as a cohesive whole and sat in a sampler can often speed up the production process since you already have a groove you're happy with and therefore are less likely to be tempted to make more 'tweaks' when they aren't required. In other words, the rest of the instruments have to be programmed to fit the groove, rather than constantly tweaking all instruments to make them all fit together. Not only this prevents you from spending months constantly making unnecessary tweaks, but also most tracks will start with just the groove!

TRANCE MELODIES

With the basic groove laid down a good approach is to programme the melody before any further elements are programmed since this will often dictate the direction of any other instrumental counter melodies.

Creating a melodic lead line for trance is possibly the most difficult part, as a good lead is derived from not only the melody but also the timbre and both of these have to be 'accurate,' so close scrutiny of the current scene and market leaders is absolutely essential in acquiring the right feel for the dance floor. Unfortunately, in terms of MIDI programming, trance leads follow very few 'rules,' so how to programme one is entirely up to your own creative instincts. That said, as ever there are some general guidelines that can be applied.

Firstly, in many cases the lead is constructed using a 'chorded' structure so that the notes alternate between two notes. This creates the results of jumping from the 'key' of the song to a higher note before returning to the main key again (Figure 12.4).

Also, as the above diagram shows, this lead continues over eight bars before being repeated again and each consecutive two bars of music will tend to rise in pitch rather than fall. Indeed, this building up the scale plays a key role in creating trance leads for the dance floor. By progressively moving up the scale

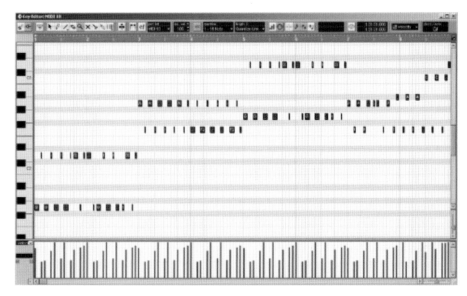

FIGURE 12.4
A trance melody

throughout the bars and then setting the few final notes of the last bar higher in scale than any of the preceding notes an uplifting feel results. Conversely, by setting the final notes lower than that of the start of a bar the feeling is reversed, resulting in a track that pulls the listener down.

To keep the relative speed and energy of trance, it's better to keep the notes short and stabby. This means using a mix of 32nd, 16th and/or 8th notes and then using the decay and release envelope of the amplifier to lengthen and contort the timbre if required. This gives much more control over the groove, allowing you to immediately adjust the 'feel' of the music by simply adjusting the amplifier's release envelope rather than having to 'kick' MIDI notes around.

With these guidelines in mind, a lead can be constructed in any manner, but if inspiration has left you, there are some techniques that can sometimes produce good results. The first and possibly easiest method is to begin by fashioning a chord structure from the previously programmed bass line. This can follow the progression exactly, but generally it's worthwhile experimenting with a mix of parallel, oblique or contrary motions. For instance, in the example track the bass played over eight bars and consisted of:

Bar 1	Bar 2	Bar 3	Bar 4	Bar 5	Bar 6	Bar 7	Bar 8
G	G	F	F	D#	D#	F	F

Thus, the key of the lead could follow this progression exactly in a parallel motion which would produce musically 'acceptable' results, even though it would be somewhat a little boring to the ear. However, by following the progression in a parallel motion for only the first, second and fourth bar and using a contrary motion on the third, the interaction between the lead and the bass exhibits a much better result:

Bar No	1	2	3	4	5	6	7	8
Bass	G	G	F	F	D#	D#	F	F
Lead	G and D	G and D	F and A	F and A	G and D	G and D	F and A	F and A

Of course, it isn't imperative that the third bar becomes contrary and any or all bars could be contrary, or parallel, or oblique, provided that it sounds right to you.

Once this basic chord is constructed to complement the bass, using a noise gate or MIDI CC messages you can cut up each bar in a rhythmic manner. If using CC messages, though, a much better approach is to write a complex rhythmic hi-hat pattern and then use the sequencer to convert the hi-hats notes to CC11 messages rather than programme all the CC's in by hand. This can often lead to a robotic nature, whereas using a hi-hat pattern you're much more likely to produce a rhythmic flowing gate. Alternatively, you can use a side chain from the hi-hats or kick drum to pump the chords. This gated effect, once suitable, can be applied physically to the notes (i.e. cut them up to produce the effect) so that not only does each note retrigger the synth but also it permits you to offset the top notes of a chord against the lower ones to produce the alternating notes pattern that's often used for the leads.

Alternatively, another frequently used method to create a trance melody is to begin with a synthesizer's arpeggiator, as this can help to get the general idea down. With some trial and error it's possible to find both arpeggiator pattern and chord combination to produce some inspiration for a lead melody. Once recorded as MIDI, this can be further attuned in a sequencer's key editor.

Another technique that's often referred to as the 'two fingered' approach can produce useable results too. This involves tapping out a riff using just two keys an octave apart on the keyboard. By alternating between these two keys keeping the lower key the same and continually moving higher in scale on the second key while playing the audio through a delay unit can produce interesting rhythmic movements. If this is recorded as MIDI information, it can be edited further by lengthening notes so that they overlap each other slightly or by adding further notes in between the two octaves to construct a melody that builds.

MELODIC TIMBRES

Just as the melody for the lead is important, so is the timbre, and it's absolutely vital that time is taken to programme a good one – the entire track will stand or fall on its quality. As discussed in the chapter on sound design, the lead is the most prominent part of a track and hence the main instrument that sits in the mid-range/upper mid-range. Consequently, it should be rich in harmonics that occupy this area and should also exhibit some fluctuations in the character to remain interesting. This can sometimes be accomplished through modulating the pitch with an LFO on a digital synth, but the phase initialization from analogue synths, more often than not, will produce much better, richer results. Consequently, the preferred synths by many artists are the *Access Virus*, the *Novation SuperNova*, *Novation A-Station* or the *Novation V-Station* (VSTi).

We've already covered the basic principles behind creating a harmonically rich lead, so rather than go into too much detail again here, what follows is a quick overview:

- Oscillators:
  - 2 × pulse detuned by −5 and +4 cents
  - 1 × sawtooth (try pitching this up or down an octave)
  - 1 × pink noise
- Amp and filter envelope:
  - Zero attack
  - Medium decay

- Small sustain
- Short release
- Filter
  - Low-pass 24 dB
- Modulation
  - Sine wave LFO positively modulates pulse width of oscillator 1
  - Triangle wave LFO positively modulates pulse width of oscillator 2
  - Filter keyfollow to maximum
  - Filter cut-off modulated with velocity commands

This is naturally open to artistic licence, and once the basics of the lead are down, it's prudent to experiment by replacing oscillators, the LFO waveforms, depths and envelope settings to create variations. If this technique does not produce a lead that is rich enough, then it's worth employing a number of methods to make it 'bigger', such as layering, doubling, splitting, hocketing or residual synthesis as discussed in the chapter on sound design.

Finally, the timbre will also benefit heavily from applying both reverb and delay. The reverb is often applied quite heavily as a send effect but with 50 ms of pre-delay so that the transient pulls through undisturbed and the tail set quite short to prevent it from washing over the successive notes.

For the delay, this is best used as a send effect but the settings will depend on the type of sound you require. Generally speaking, the delays should be set to less than 30 ms to produce the granular delay effect to make the timbre appear big in the mix, but, as always, experimentation with longer settings may produce the results you prefer.

If vocals are employed in the track then there may be little need for any effects as the lead should sit under them. However, if you want to keep a wide harmonically rich sound and vocals, it's wise to employ a compressor on the lead timbre and feed the vocals into a side chain so that the lead drops when the vocals are present.

MOTIFS AND CHORD PROGRESSIONS

With both main melody and the all essential groove down, the final stage is to add motifs and any chord progressions. The latter progressions should require little explanation as, if used, it's simply a case of producing a chord structure that harmonizes with the lead and the bass. Motifs, on the other hand, are a little more complex and require more thought. These counter melodies are the small ad-lib riffs best referred to as the icing used to finally decorate the musical cake and play an essential part of any dance music, adding much needed variation to what otherwise would be a rather repetitive track. These are often derived from the main melodic riff as not only do they often play a role in the beginning part of the track before the main melodic lead is introduced but they are also re-introduced after the reprise.

There are various techniques employed to create motifs, but one of the quickest ways is to make a copy of the MIDI lead and 'simplify' it by removing notes to

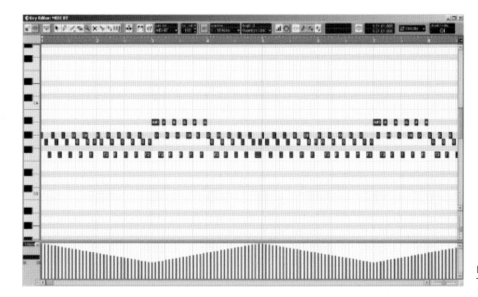

FIGURE 12.5
The motif

create a much simpler pattern. Once created, this pattern can be offset from the main lead to make the motif more noticeable, or alternatively, it can occur at the same time as the lead but the attack and release of the timbre are lengthened. In fact, this latter approach will often produce the best results since, as discussed, dance music works on contrast and so far all the instruments have had a definite attack stage. By lengthening the attack and release, the motif will take on a whole new rhythmic character which can quickly be changed by listening back to the arrangement so far and adjusting the amplifier envelope's parameters.

Above all, using any of these techniques it's sensible to avoid getting too carried away and it's vital that they are kept simple. Many dance tracks simply use a single pitch of notes, playing every eight, sixteenth or quarter, or are constructed to interweave with the bass rather than the lead. This is because not only do simple motifs have a much more dramatic effect on the music than complex arpeggios, but you don't want to detract too much from the main melodic element (Figure 12.5).

It should also go without saying that the timbre used for any motif should be different from the lead melody, and it's at this point that you'll need to consider the frequencies that are used in the mix thus far.

As touched upon, in many trance tracks the lead is harmonically rich, which reduces the available frequencies for a motif, so it's quite common to utilize a low-pass filter cut-off to reduce the harmonic content while the lead is playing, yet set this filter to open wider while there is no lead playing. Additionally, as with all dance music, this filter is tweaked 'live' to create additional movement and interest throughout the arrangement. This movement obviously has to be restrictive while the lead is playing, otherwise the mix can quickly become swamped in frequencies and difficult to mix properly, but during the beginning of the song, the filter can be opened wider to allow more frequencies through to fill up any gaps in the mix.

More importantly, the rhythmical pitch of the motif will have to be carefully chosen. A motif that is mostly written using notes above C4 will naturally contain higher frequencies and few lower ones, so using a low-pass filter that is almost closed will reduce the frequencies to nothing. Therefore, the pitch of this motif and the timbre must occupy the low to mid-range to fit with the bass and the lead melody, assuming of course, that the lead melody contains higher frequencies.

Indeed, with trance, it's careful use of the filter that creates the final results, and often you will need to compromise carefully by cutting some of the frequencies of the lead to leave room for the motif and vice versa. Having said that, if the bass is quite simple, an often used technique is to programme a motif that contains an equal amount of low frequencies as mid, and then use a filter to cut higher frequencies to mix the motif in with the frequencies of the bass, helping to enhance the low end groove, before proceeding to remove this low end interaction by cutting the lower frequencies and leaving the higher ones in as the track progresses. This creates the impression that all the sounds are interweaving with one another helping to create more movement and the typical 'energy' that appears in most club tracks.

One final, yet vital, element of trance before we come to the arrangement is the programming of rolls. Although these may not initially appear to be particularly significant, dance music relies on tension and drama, which best implemented with well-programmed snare rolls. The quick snare rolls that interrupt a groove along with the huge 4- or 8-bar snare roll that leads up to the main reprise create a build-up of tension, followed by the ecstatic release when the track returns so it's vital that these are well designed. Typically, the snares are the same as those used in the drum loop and are best programmed in step time as this allows you to programme them precisely while also allowing you to edit each note in terms of time, size and velocity. The most common snare rolls used in this genre are programmed by dropping in MIDI notes every 1/16th followed by 32nd and then drawing in velocity that continually rises. As a result the snares become progressively louder as the roll reaches its apex (Figure 12.6).

Even though many trance tracks rely on this strictly programmed 16th/32nd snare pattern these rolls form such an integral part of the emotion of the music that it's worth experimenting further to see what emotion it can convey. For instance, rather than programming a strictly quantized pattern of notes, mixing 16th and 32nd notes together with velocity that does not always climb upwards can produce interesting variations (Figure 12.7).

Along with the velocity, pitch bend can also help to add some extra anticipation by pitching the snares up as they reach the apex, or alternatively, a low-pass

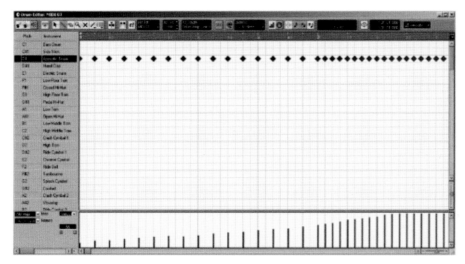

FIGURE 12.6
A typical trance snare roll

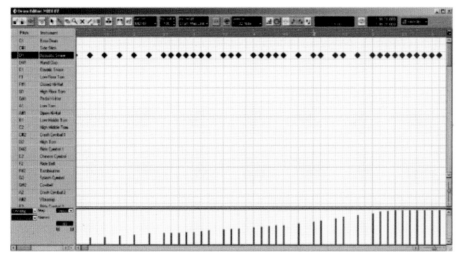

FIGURE 12.7
A more intense snare roll

filter across the snares which opens as the roll progresses can also help creating a building sensation (as used in the example track).

Snare rolls should not just be restricted to the main reprise either, as shorter rolls can be used to denote a new passage of music or the introduction of a new instrument. This commonly consists of placing two snares near the end of the bar a 1/16th apart. The timbre of these two snares is usually different to indicate a form of signing off the previous bar and introducing a new one. This is best accomplished by pitching the second snare a semitone up from the first while also placing the kick drum to play at the same time, in effect creating a small kick/snare roll (Figure 12.8).

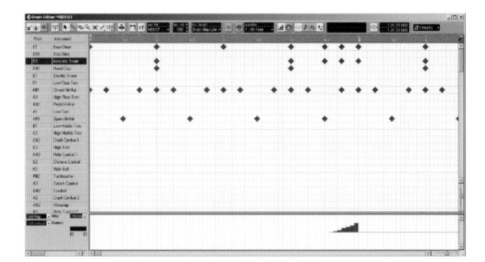

FIGURE 12.8
Using a skip in the rhythm to introduce new instruments

Ultimately, these are only suggestions and you should be willing to spend as much time on the snare rolls as every other instrument. The real key is to try out variations at building and applying effects to the snares to create as much tension as possible within the music.

THE ARRANGEMENT

With all the basic parts laid down you can begin to look at the arrangement. Unlike most other genres of dance music, this doesn't commonly rely on repeating phrases but generally follows a more musical structure. Possibly the best way to accomplish this is to mute all the tracks bar the drums, begin playback, close your eyes and *listen*. If you've listened to trance before (and if not it is *strongly* recommended that you do!) you'll instinctively know when a new instrument should be introduced. However, if you're still stuck for where to go, listen to other trance tracks and physically count the bars, making note of where new instruments are introduced and then implement this same structure in your own music. While this may seem like stealing, it isn't. Many trance tracks follow the very same arrangement principles (and remember that arrangements cannot be copyrighted!).

Indeed, although it would seem unreasonable to say that all euro-trance tracks follow a set pattern and rather that you should use your own integrity to produce the arrangement, trance, like most forms of dance music, is an exact science and it is important that you don't stray too far from what is considered the standard. Staying with this principle, it tends to follow a particular blueprint that normally consist of a combination of bass, drums and inter-melodic lines that culminate into the main uplifting climax. More often than not, this final climax is a culmination of the first part of the track mixed with the main melodic reprise. Occasionally, this main reprise is different thematically from the beginning section but the timbre used will have been present beforehand to retain some continuity throughout.

With this in mind it's generally best to first plan the track using a song map. Trance, like all dance music, deals mostly in emotional waves, consisting of building and dropping the arrangement to generate an emotional state in the audience. Mathematics play a part in this, created by collecting and introducing sounds every 4, 8, 16, or 32 bars, depending on the individual elements used within the track and the required length. This may sound overly mechanical but it's a natural progression that we have all come to expect from music. As an example, try introducing a new element into an arrangement at the seventh bar rather than the eighth and it'll sound off beam.

Generally speaking most trance tracks will begin with just a basic drum loop playing over the first 16 bars to allow the DJ to mix the record in. At the end of the sixteenth bar, a new instrument is often introduced which could be signified by a cymbal crash, or short snare roll. This instrument is often another percussive element to complement the drum loop or the bass itself to generate the groove. This is often left playing as is for the next sixteen or thirty-two bars to allow the clubbers to become more comfortable with the groove of the record. After these bars, the first motif tends to be introduced with another crash or snare roll and this motif continues for the next 16 or 32 bars. Notably if this is to play for a more prolonged period of time, it's prudent to employ filter movements to prevent the track from becoming tedious.

After this, the first drop of the record commonly appears. This often consists of dropping the percussive elements bar the kick drum and motif and may continue for 4 or sometimes even 8 bars. At the end of the final bar of this drop, a crash or short snare roll is used again to signify the groove of the record returning along with another new element. In many instances, this new element is the lead motif which has been sliced up and simplified. This is to give the audience a 'taster' of what is to come later in the record. How long this plays for depends on the track itself but generally it can range from 16 to 32 bars with a cymbal crash or small snare roll/double kick placed at the end of every fourth bar to keep the interest. Occasionally, the lead riff may also be filtered with a low-pass filter which is gradually opened at the final 4 bars of this section, often complemented with a 4-bar snare roll and perhaps a pad or string section laid underneath.

A crash cymbal placed at the end of the final bar, treated to reverb and delay so that it echoes across the track, is often implemented as the rest of the mix drops leaving just the filtered down lead or the previously introduced pads. These pads/leads continue for 4 to 8bars whereby a 4- or 8-bar snare roll is built behind the instruments creating an emotional rise towards the reprise of the track. This breakdown and build-up forms an essential part of trance and has to be executed carefully, as the whole idea is to link one fast section to another without losing the speed or feel of the track.

At the end of the snare roll, the same crash that was used to signify the start of the breakdown is commonly used to echo across the introduction of the full track and all of the previous instruments are playing together to create a crescendo of emotion. At this point, additional snares are often introduced to add a skip to the rhythm and a few sweeping effects or an additional motif is introduced. This often plays over 32 bars but sometimes 64 with a cymbal crash placed at the end of every fourth bar. Finally a 2-bar snare roll often signifies the removal of instruments such as the lead and melody as the track is slowly broken down again to end with 16 bars of the drum loop.

Notably, trance relies heavily on drive and, if at this stage, it seems to meander along, it's prudent to inject some additional drive into the patterns. Typically this can be accomplished by changing the position of some elements in relation to the rest of the rhythm. For instance, keeping the kicks sat firmly on the first and third beats but moving the kicks (and snares) that occur on the second and fourth beats forward by a tick or two (so that they occur earlier in time) will add more drive to the piece. Alternatively, moving the bass a tick or two forward in the arrangement can add a feeling of drive.

> The data CD contains a full mix of a typical Trance with narration on how it was constructed.

RECOMMENDED LISTENING

Ultimately, the purpose of this chapter has been to give some insight into the production of trance and there is no one definitive way to produce the genre. Indeed, the best way to learn new techniques and production ethics is to actively listen to the current market leaders and be creative with processing, effects and synthesis. Phaser, flangers, chorus, delay, reverb, distortion, compression and noise gates are the most common processor and effects used, so experiment by placing these in different orders to create the 'sound' you want. While the arrangement and general premise of every trance track is very similar, it's this production that differentiates it between all the others. With this in mind, what follows is a short list of artists that, at the time of writing, are considered the most influential in this area:

- Ferry Corsten
- Redd Square
- Binary Finery
- Sasha
- DJ Sammy

'When I first started to make Speed Garage, I didn't term it as Speed Garage. I'd been into drum and bass for years. The scenario was, I'm not gonna try and make drum and bass, I'm gonna take it and put it with house and see what happens. That's all it is, that's the birth of Speed Garage...'

Armand Van Helden

UK garage has its roots firmly set within the development of jungle music; therefore, before we analyse this genre, we need to examine the roots and development of jungle.

Jungle was a complex infusion of breakbeat, reggae, dub, hardcore and artcore, but in the majority, it is the complex rhythms which defined the genre. These can be traced back to the 1970s and the evolution of breakbeat. Kool Herc, a hip-hop DJ, began experimenting on turntables by playing only the exposed drum loops (breaks) and continually alternating between two records, spinning back one while the other played and vice versa. This created a continual loop of purely drum rhythms, allowing the breakdancers to show off their skills. DJ's such as Grand Wizard Theodore and Afrika Bambaata began to copy this style, adding their own twists by playing two copies of the same record but delaying one against the other, resulting in more complex, asynchronous rhythms.

It was early 1988 and the combined evolution of the sampler and the rave scene really sparked the breakbeat revolution. Acid house artists began to sample the breaks in records, cutting and chopping the beats together to produce more complex breaks that were impossible for any real drummer to play naturally. As these breaks became more and more complex, a new genre evolved known as hardcore. Shifting away from the standard 4/4 loops of typical acid house, it featured lengthy complex breaks and harsh energetic sounds that were just too 'hardcore' for the other ravers.

Although initially scorned by the media and record companies as drug-induced rubbish that wouldn't last more than a few months, by 1992 the entire rave scene was being absorbed in the commercial media machines. Riding on this 'new wave for the kids', record companies no longer viewed it as rubbish but a cash cow and proceeded to dilute the market with a continuous flow of watered-down rave music. Rave became commercialized and this was taking its toll on the nightclubs.

In response to the commercialization, in 1992 two resident DJs – Fabio and Grooverider – pushed the hardcore sound to a new level by increasing the speed of the records from the usual 120 to 145 BPM. The influences of house and techno were dropped and quickly replaced with ragga and dancehall, resulting in mixes with fast complex beats and a deep bass. Although jungle didn't exist as a genre just yet, these faster rhythms mixed with deep basses inspired artists to push the boundaries further and up the tempo to a more staggering 160–180 BPM.

To some, the term 'jungle' was attributed to racism but the name was derived from the 1920s. It was used on flyers to describe music produced by Duke Ellington. This featured exotic fast drum rhythms and when Rebel MC sampled an old dancehall track with the lyrics 'Alla the Junglists', 'jungle' became synonymous with music that had a fast beat and a deep, throbbing bass. This was further augmented by pioneers of the genre such as Moose and Danny Jungle.

Jungle enjoyed a good 3–4 years of popularity, but as the beats and bass line became darker and more minimal, and drum 'n' bass took over, the music lost a majority of its appeal to the more mainstream UK audiences. Drum 'n' bass and jungle were described as far too aggressive and clubbers wanted a happier vibe to dance to.

This is where the history of garage becomes disjointed to many. Garage had already existed before the jungle revolution but not in the US term of garage. This was a term derived from Larry Levan's *Paradise Garage*, which incorporated a broad range of musical styles. The earlier UK garage was a laid-back form of house music with mellow beats and vocals. Often referred to as 'Sunday music' because it was music that wasn't popular enough to be played to the mainstream on Friday and Saturday nights, it was an evolution of jungle that bought garage to the masses.

To many, it was Armand Van Helden's remix of *Sugar Is Sweeter* by CJ Bolland in 1996 that started the garage phenomenon. By mixing house beats with the slow dub/warping bass lines of jungle and time-stretched vocals, the genre was soon labelled 'speed garage'. This was further augmented with his next release of *Spin Spin Sugar* by the Sneaker Pimps. Further artists such as RIP, Dreem Teem, Booker and 187 Lockdown began to add their own unique styles to the music.

Todd Edwards, a producer from New Jersey, began to experiment further with the garage sound. He started to strip down the verse and chorus and instead picked out vocal phrases, choosing to play them like instruments and experimenting with the latest sampling technology of the time to make them stand out above the crowds. One such technique was pitch-shifting individual syllables, a style that has become characteristic of the whole UK garage scene.

By 1997, the genre of speed garage began to diversify with the evolution of 2-step garage. This broke down the 4/4 rhythms of speed garage by removing the second and fourth bass kick from each bar, placing an open high-hat on the upbeat and introducing syncopated bass lines, altogether giving the rhythms a funkier feel. MJ Cole, The Artful Dodger, Ed Cas, Shanks and Bigfoot and the Dreem Teem are all accredited with evolving this new genre, although for many it was *Baby Can I Get Your Number* by Silo that pioneered and paved the way for 2-step garage.

In 2000, 2-step garage hit the mainstream charts with releases from Craig David and Daniel Beddingfield. However, it wasn't long before the record companies took their hold on this new genre and the charts were bombarded with track after track of 2-step 'pop' garage for the masses. As those passionate about the music – not the money – tried to distance themselves, breakstep garage was born. This exhibited heavier break beats than before and deeper bass lines. This was further evolved into dub-step, a much more minimalist approach to the rhythms but still featuring heavy dub bass lines.

MUSICAL ANALYSIS

The most proficient way to approach writing garage is to find the most popular tracks around at the moment and break them down to their basic elements. Once this is accomplished we can begin to examine the similarities between each track and determine exactly what it is that differentiates it from other musical styles. In fact, all music that can be placed into a genre-specific category will share similarities in terms of the arrangement, groove and sonic elements.

However, in the case of garage, the divergence of the genre over the subsequent years has resulted in a genre that has become heavily fragmented and as such cannot be easily identified as featuring any one particular attribute. It can feature house rhythms with slow pounding bass lines, it can incorporate more funky rhythms or it can be extremely minimalist. Indeed, since garage takes its influence from just about every other form of music it is near impossible to analyse the genre in any exact musical sense. It is only possible to make some very rough generalizations unless we select a specific genre to work with.

While garage is influenced by most other forms of music, its strongest roots lie within its history: jungle, hip-hop, dancehall and R 'n' B. Generally the music will use a 4/4 time signature, but sometimes with the more complex 2-step a 3/4 time signature is used. In terms of tempo, this can range from 94 to 145 BPM. For the purpose of this chapter, we'll examine what has been termed speed garage (more recently reinvented and renamed 'niche'), a genre featuring heavily warped bass lines with straight 4/4 rhythms, a motif and time-stretched vocals. Although we will be focusing on this genre of music, if you want to produce

2-step, these techniques can be mixed with drum 'n' bass and hip-hop/rap, which are discussed in detail in other chapters.

DRUM RHYTHMS

Much of niche garage relies on a 4/4 rhythm with a slight skip in the beats, usually generated by the snares.

Generally, a kick drum is laid on every beat of the bar and is augmented with a 16th closed hi-hat pattern and an open hi-hat positioned on every 1/8th (the off beat) for syncopation. Snares (or claps) are often employed on the second and fourth beat underneath the kick but these are often skipped. This can be accomplished by laying a snare just before or after the second and fourth kick and also dropping a snare in between beats 3 and 4. As always, experimentation is the key so it is worth further experimentation by moving the snares around the beats (not on the beat) until the rhythm displays a skipping beat (Figure 13.1).

This should only be viewed as the start. Both tambourines and shakers are often added to create more of a feel to the rhythm but this does depend on the subgenre. Where these are placed is entirely up to your own discretion, but generally they are best played live and left unquantized to keep the feel of the loop live.

More importantly, though, like house music, garage relies on plenty of drive. This can be applied in a variety of ways. The most commonly used technique is to employ a slight timing difference in the rhythm to create a push in the beat. Typically, this involves keeping the kick sat firmly on all the beats but moving the *snares* forward in time by a tick or two. Then once the bass line is laid

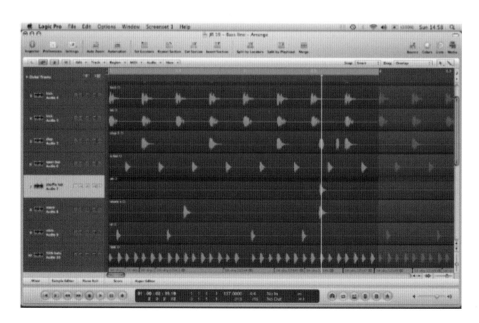

FIGURE 13.1
The cornerstone of
niche loops

down, its timing is matched to these snares using a sequencers match quantize function.

For garage kicks, samples are often used. These can be sampled from specific garage sample CDs or, more often, other records. It is important to note that the kick and snare should not be heavy enough to detract from the bass.

Generally, the kick is commonly kept quite tight by employing a short decay combined with an immediate attack stage. This is to help keep the kick dynamic but it also prevents it from blurring and destroying the all important interplay between kick and bass. Typically, a 120 Hz sine wave produces the best starting point for a garage kick with a positive pitch EG set, a fast attack and a quick decay to modulate the oscillator.

Production-wise the kicks rely very heavily on compression to produce the archetypal hardened sound. Although the settings to use will depend on the sound you wish to achieve, a good starting point is to use a low ratio combined with a high threshold and a fast attack with a medium release. The attack will crush the initial transient of the square producing a heavier 'thump', while experimenting with the release can further change the character of the kick. If this is set quite long, the compressor will not have recovered by the time the next kick begins and this can often add a pleasing distortion to the sound. Once the kick has been squashed, as a final stage it's often worthwhile using the PSP Vintage Warmer to add more warmth to the overall timbre.

The snares should be kept particularly snappy and pitched quite high, reminiscent of drum 'n' bass. To accomplish this, keep the decay on the snare particularly short to keep with the short, sharp dynamics of the loop. This characteristic sound can be produced by using either a triangle or a square wave mixed with some pink noise or saturation (distortion of the oscillator) both modulated with an amp envelope using a fast attack and decay. White noise should be used to keep the snare quite bright, but much of this noise will need removing with a band-pass filter to allow you to remove the high and low content of the timbre to produce the effect you need.

If at all possible, it's worthwhile using a different envelope on the main oscillator and the noise since this will allow you to keep the body of the snare (i.e. the main oscillator) quite short and sharp but allow you to lengthen the decay of the noise waveform. By doing this you can modify the snare so that it rings out after each hit. It also permits you to individually adjust the decay's shape of the noise waveform. Generally speaking, the best results are from employing a concave decay but as always it's worth experimenting. The snares may also benefit from some pitch bend applied to the timbre.

If you use this latter technique, it's prudent to remove the transient of the snare in a wave editor and replace it with one that uses positive pitch modulation. When the two are recombined, the timbre will have a good solid strike but decay upwards in pitch at the end of the hit. If this isn't possible then a viable alternative is to sweep the pitch from low to high with a sawtooth or sine LFO (provided that the saw starts low and moves high) set to a fast rate. After this, small amounts of compression so that only the decay is squashed (i.e. slow attack) will help to bring it up in volume so that it doesn't disappear into the rest of the mix. If you're still finding it difficult to produce a snappy snare, a transient designer can be used to alter the transients of the snare hits.

Notably, it's only the timbre for the kick and snare that truly define the niche sound. While in the majority of cases the remaining percussive and auxiliary instruments such as the open and closed hi-hats, tambourines and shakers are sourced from sample CDs or the TR808 or 909, almost any MIDI module will carry sounds that can be used. Plus, many modules today carry the ubiquitous TR808 and 909 percussion sound sets as part of the collection anyway. Normally, none of these should be compressed as thus can reduce the high-frequency content. But having said that, this is simply convention and you should always be willing to 'break the rules', so to speak.

Of course, it would be naive to say that all beats are created through MIDI and a proportionate amount of rhythms are acquired from sampling records with a drum break. Indeed, these will have probably been sampled from other records, which will have been sampled from previous records, which will have been sampled from other records, so you should feel free to experiment with your sampler and sequencer.

Along with working with individual loops, some producers have been known to stack up a series of drum loops to create heavier rhythms. Armand Van Helden is well known for dropping more powerful-sounding kicks over pre-sampled loops to add more of a heavy influence.

If you mix a number of loops together to create a rhythm, EQ is an important tool. You can use it to reduce the frequency content of the kicks to prevent attaining a beat that is too heavy. Indeed, it's preferable to just have one kick playing, so the kick should be removed from all the other loops. Once a few basic loops have been merged, effects and aggressive EQ cuts or boosts are employed on each individual element of a sliced loop to create interesting interacting timbres.

THE NICHE BASS

As previously mentioned, garage, particularly niche, relies heavily on the bass rhythm and timbre so it is vital that these are correct for the genre. Typically the rhythm of the bass is kept fairly simple and will often consist of nothing more than a few 1/4 or 1/2 notes per bar that will tend to move very subtly in pitch by no more than 3 or 5 semitones (Figure 13.2).

In the above example, the bass moves from C to A, but many of the notes are kept quite long allowing full pitch bend of the notes. Note also how in the above example some of the notes overlap the ones following those overlaps. The portamento control on a synthesizer, creates a timbre that will rise during its period and it's this type of movement that is fundamental to the bass.

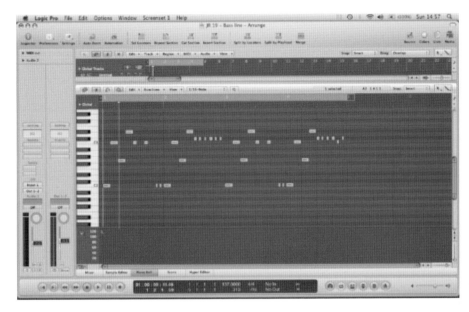

FIGURE 13.2
A typical niche bass line

For the bass timbre, just about any synthesizer is fair game. Native Instruments Massive has been used to its full extent in this genre, but any analogue synthesizer (or DSP equivalent) can be used so long as it contains enough frequencies to be swept with a filter. Indeed, the filter envelope is the key to producing the warping bass timbre. Using two square waves, detune one from the other by + or − 5 and set the amp envelope to a fast attack and short decay. Set the filter envelope to a medium attack and full modulation. Finally, set the LFO to modulate the cut-off frequency and experiment with all the settings until you accomplish the sound you require.

Another typical timbre can be attained by combining two oscillators, one set to a sine wave to add depth to the timbre while the other set to a sawtooth to introduce harmonics that can be swept with a filter. These are commonly detuned from one another; the amount varies depending on the type of timbre required. Hence, it's worth experimenting by first setting them apart by 3 cents and increasing this gradually until the sound becomes as thick as you need for the track.

Set the amp EG's attack to zero along with sustain but set the release and decay initially to midway. The decay setting provides the 'pluck' while the release can be modified to suit the motif being played from the sequencer. The filter cut-off is set to a low pass, as you want to remove the higher harmonics from the signal (as opposed to removing the lower frequencies first). This along with the resonance are adjusted so that they both sit approximately halfway between fully exposed and fully closed. Ideally, the filter should be controlled with a filter envelope using the same settings as the amp EG, but to increase the 'pluck' of the sound it's beneficial to adjust the attack and decay so that they're slightly longer than the amplifier's settings. Finally, positive filter key follow should be employed so that the filter will track the pitch of the notes being played, which helps to add more movement.

Finally, you can pitch shift the timbre by modulating both oscillators with a pitch envelope set to a fast attack and medium decay. If possible, the pitch-bend range should be limited to 2 semitones to prevent it from going too wild. If you decide to modulate the filter, then it's best to use an LFO with a saw-tooth that ramps upwards so that the filter opens, rather than decays, as the note plays. The depth of the LFO can be set to maximum so that it's applied fully to the waveform, and the rate should be set so that it sweeps the note quickly. What's more, if the notes are being played in succession it's prudent to set the LFO to retrigger on key press. Otherwise it will only sweep properly on the first note and any successive notes will be treated differently, depending on where the LFO is in its current cycle.

This will not suit all niche music, though. Some artists ignore synths and construct their own basses from culminating samples together or pitching a sample down the keyrange because just about anything sounds great when it's pitched down a couple of octaves into the bass register. The key is to experiment in creating a low tone that has the energy to add some bottom-end weight to the music but that at the same time does not demand too much attention.

Both distortion and compression are often used on niche basses. While most effects should be avoided since they tend to spread the sound across the image (which destroys the stereo perspective) small amounts of controlled distortion can help to pull the bass out of the mix and often give it the much needed rawness. Similarly, compression after the distortion can be used to even out the volume, keep the effect under control and bring the overall levels up. As a general guideline, start by setting the ratio to 4:1, along with an attack of 5 ms and a medium release of 150 ms or so. Set the threshold control to 0 dB and then slowly decrease it until the bass registers on the gain reduction meter by at couple of decibels. Finally, set the make-up gain so that the loop is at the same volume as when the compressor is bypassed and then start experimenting with the attack and release settings until it produces the level and 'sound' you require.

MOTIFS

The motif in garage depends entirely on the type of garage being produced. Pretty much anything should be considered fair game. Very generally speaking, though, the lead is often the exact opposite of the bass. Whereas the bass is very low and boomy, the lead riff is often incredibly bright and pitched high up the range. 187 Lockdown's Gunman used a sample of a watch from an old western film to great effect. Indeed, many leads are often sampled from other records or films and it would be naive to suggest that they are always programmed with MIDI. Consequently, it's impossible to offer any guidelines apart from 'to listen to the latest records to see where the current trend currently lies'.

This does, however, also mean that there are no definitive motif sounds that can characterize the genre (bar the heavy bass line). That said, there are some timbres that have always been popular, including bright pianos, angel bells and fairy type timbres.

Triangle and noise waveforms can produce a harmonically rich, bright sound that will cut through the mix. Depending on how many are employed in the timbre, these can be detuned from one another to produce more complex, interesting sounds. If the timbre requires more of a body to the sound then adding a sine or pulse wave will help to widen the sound and give it more presence. To keep the dynamic edge of the music, the amplifier's attack is predominantly set to zero so that it starts upon key press, but the decay, sustain and release settings will depend entirely on what type of sound you require.

Generally speaking, you should avoid long release settings since this may blur the lead notes together and much of this music relies on sharp attacked timbres. Since the bass encapsulates the music, the lead should not detract from it; therefore the motif should remain quite steady with little modulation, but that's not to say you shouldn't experiment. If you want to add movement to a timbre, you can employ LFOs or a filter envelope to augment the sound as it plays.

If you need a starting point for a high motif, start with two oscillators; – one set to a sine wave and the other set to a triangle wave, – with one of the oscillators detuned so that it's at a multiple of the second oscillator. The amp envelope is then set to a fast attack, short decay and release and a medium sustain with the filter key-tracking switched on. To produce the initial transient for the note, a third sine wave pitched high up on the keyboard and modulated by a one-shot LFO (i.e. the LFO acts as an envelope – fast attack, short decay, no sustain or release) will produce the desired timbre.

Another technique is to use a pulse and a triangle oscillator. As the sound starts on keypress, the amp uses a zero attack with a full sustain and medium release (note that there is no decay since the sustain parameter is at maximum). A low-pass filter is used to shape the timbre with the cut-off set to zero and the resonance increased to about halfway. A filter envelope is not employed since the sound should remain unmodulated. If you require a 'click' at the beginning of the note you can turn the filter envelope to maximum, but the attack, release and sustain parameters should remain at zero with a very, very short decay stage.

CHORDS

Generally, chords are not used within speed garage. Chord progressions are usually employed to fill up any empty spaces in the mix, but in speed garage, the space within the mix is as important as the bass. If you fill up the frequency spectrum, the music doesn't have as much impact as it would if there were holes in the frequency range of the mix.

Nonetheless, if you wish to employ chords in the music, they should act as a harmony to the bass and lead. This means that they should fit in between the rhythmic interplay of the instruments without actually drawing too much attention. To accomplish this, they need to be very closely related to the key of the track and not use a progression that is particularly dissonant. Generally, this can be accomplished by forming a chord progression from any notes used in the bass and experimenting with a progression that works. For instance, if the bass is in E, then C major would produce a good harmony because this contains an E (C–E–G). The real solution is to experiment with different chords and progressions until you come up with something that works.

Since it is important not to completely fill the frequency range of the mix, the chord timbre should be quite thin. If necessary it can be filled out later with flangers or phasers. Pulse waves have less harmonic content than saws and tri-angles and do not have the 'weight' of a sine wave, so these are the best choice. Generally only one pulse wave is required with the amp envelope set to a medium attack, sustain and release but a fast decay. If this timbre is to sit in the upper mid-range of the music then it's best to use a 12 dB high-pass filter to remove the bottom end of the pulse. Otherwise, use a low, pass to remove the top end and then experiment with the resonance until it produces a general static tone that suits the track.

Following this, set an LFO to positive modulation with a slow rate on the pulse width of the oscillator and the filter to add some movement. Which waveform to use for the LFO will depend on the track itself, but by first sculpting the static tone to fit into the track you'll have a much better idea of which wave to use and how much it should modulate the parameters. If the timbre appears too 'statically modulated' in that it still seems uninteresting, use a different rate and waveform for the oscillator and filter so that the two beat against each other. Alternatively, if the timbre still appears too thin even after applying effects, add a second pulse detuned from the first by 3 cents with the same amp envelope but use a different LFO waveform to modulate the pulse width.

Alternatively, try using three sawtooth waveforms and detune two from one of the oscillators by + and −3. Apply a small amount of vibrato using an LFO to these two detuned saws. Then use an amplifier envelope with a medium attack and release and a full sustain (there is no decay since sustain is set to full and it would have nothing to drop down to). Experiment with the sawtooth that wasn't detuned by pitching it up as far as possible without letting it become an individual sound (i.e. less than 20 Hz). If possible use two filters – one set as a low pass to remove the low-end frequencies from the two detuned saws and one set as a high pass to remove some of the high-frequency content from the recently pitched saw.

SOUND EFFECTS

One final aspect is the addition of sound effects and sometimes vocals. Often the vocals are nothing more than an MC speaking, but commercial vocals have been used (*Spin Spin Sugar* remix). Vocals are rarely used as is, however, and often benefit from time stretching in a sampler/audio workstation. No real advice can be given here since it depends entirely on the vocals on how they should be effected, but as always, listening to the current market leaders will give you ideas on how the vocals are being effected in the current climate.

The sound effects also play a large role in the production of speed garage. Indeed, they often help to define the genre (the gunshot and siren in 187 Lockdown's *Gunman* play a large role). These effects can be sampled from movies, or sample CD's or created in any competent synthesizer. What follows, therefore, is a quick rundown on how to create some of the most used sound effects.

The siren is possibly the easiest sound effect to recreate. Set one oscillator to produce a sine wave and use an amp envelope with a fast attack, sustain and release and a medium decay. Finally, use a triangle wave or sine wave LFO to modulate the pitch of the oscillator at full depth. The faster the LFO rate is set, the faster the siren will become.

Another popular effect is the gunshot. Just about every sound effects sample CD has these, but if you don't have one you can create your own in any analogue-style synthesizer with two oscillators. One oscillator should be set to a saw wave while the other should be set to a triangle wave. Detune the triangle from the saw by +3, +5 or +7 and set a low-pass filter to a high cut-off and resonance (but not so high that the filter self oscillates). Set the amp's envelope to a fast attack, sustain and release but with a medium decay and copy these settings to the filter envelope. Make the decay a little shorter than the amp's EG. Finally, use a sawtooth LFO set to a negative amount and use this to control the pitch of the oscillators along with the filter's cut-off.

Yet another effect commonly used is the rising effect. To recreate this use a sawtooth oscillator and set both the amp and filter EG to a fast decay and release but a long attack and high sustain. Use a triangle or sine LFO set to a positive mild depth and very slow rate (about 1 Hz) to modulate the filter's cut-off. Finally, use the filter's envelope to also modulate the speed of the LFO so that as the filter opens the LFO speeds up. If the synthesizer doesn't allow you to use multiple destinations, you can increase the speed of the LFO manually and record the results into a sampler or audio sequencer.

ARRANGEMENTS

With all the parts programmed we can begin to look at the arrangement. As touched upon in the musical analysis, garage is an incredibly diverse genre so there are no definite arrangement methods.

Nevertheless, as much of house relies on the sampling ethics of drum 'n' bass, a usual approach is to arrange the instruments in different configurations to produce a series of different loops. For instance, one loop could consist of just the drums, another of the drums and bass mixed together, another of the drum, bass and leads, and another of just the leads. Each of these are then sampled (or exported from an audio sequencer) and set to a specific key on a controller keyboard connected to the sampler. By then hitting these keys at random you can produce a basic arrangement.

That said, as with all genres this is subject to change, so the best way to obtain a good representation of how to arrange the track listen to what are considered the classics and then copy them. This isn't stealing – it's research and nearly every musician on the planet follows this same route. This can be accomplished by using a method known as *chicken scratching*. Armed with a piece of paper and a pen, listen to the track and for every bar place a single scratch mark on the paper. When a new instrument is introduced or there is a change in the rhythmic element, place a star below the scratch. Once you've finished listening, you can refer to the paper to determine how many bars are used in the track and where new instruments have been introduced. You can then follow this same arrangement and, if required, change it slightly so that it follows a progression to suit.

> The data CD contains a full mix of a typical Speed Garage track with narration on how it was constructed.

RECOMMENDED LISTENING

Ultimately the purpose of this chapter has been to give some insight into the production of UK garage and bassline but as ever, there is no one definitive way to produce the genre. Indeed, the best way to learn new techniques is to actively listen to the genre and experiment with the tools you have to hand. With this in mind what follows are some of the more popular UK Garage tracks from the past, along with some more popular bassline releases:

- Mis-Teeq
- Double 99
- Monsta Boy
- Scott & Leon
- Shola Ama
- Kristine Blonde
- Wideboys
- 187 Lockdown
- Sneaker Pimps – (Spin Spin Sugar remixed by Van Helden)
- MJ Cole
- Dreem Team
- Daniel Bedingfield – (Gotta Get Thru This)
- Architechs

- Lonyo
- Tru Faith
- NRG
- Truesteppers
- H Two O & Platnum
- Kristine Blond
- Agent X – Boops
- Scandalous Unlimited & Carly Bond
- Duggan, Jamie & Tezz Kidd
- Dizzee Rascal
- Shaolin Master & Flirtations
- Jones, Dezz & Gia Mia

CHAPTER 14

Techno

'Techno is a complete mistake. It's like George Clinton and Kraftwerk stuck in an elevator with only a sequencer to keep them company...'

Derrick May

To the uninitiated, techno is used to describe any electronic dance music, and although this was initially true, over the years it has evolved to become a genre in its own right. Originally, the term techno was coined by Kraftwerk in an effort to describe how they mixed electronic instruments and technology together to produce 'pop' music. However, as the following years were riddled by numerous artists taking the idea of technology on board to write their own music, the true foundations of where techno, as we know it today, originated is difficult to pinpoint accurately.

To some, the roots of this genre can be traced back to as early as 1981 with the release of *Shari Vari* by A Number of Names, *I Feel Love* by Donna Summers (and Giorgio Moroder) and *Techno City* by Cybotron. To others, it emerged in the mid-1980s when the 'Belleville Three' collaborated together in Detroit. These three high school friends – Juan Atkins, Kevin Saunderson and Derrick May – used to swap mix tapes with one another and religiously listen to the Midnight Funk Generation on WJLB-FM. The show was hosted by DJ Charles *'Electrifying Mojo'* Johnson and consisted of a 5 h mix of electronic music from numerous artists including Kraftwerk, Tangerine Dream and George Clinton.

Inspired by this eclectic mix, they began to form their own music using cheap second-hand synthesizers such as the Roland TR909, TR808 and TB303. The music they produced was originally labelled as *House* and both May and Saunderson freely admit to gaining some of their inspiration from the Chicago clubs, particularly the Warehouse and Frankie Knuckles, and the house music they played. In fact, Derrick May's 1987 hit *Strings of Life* is still viewed by many as house music, although to Derrick himself and many other aficionados it was an early form of Detroit techno.

This form of music was characterized by its mix of dark pounding rhythms mixed with a soulful feel and a stripped down vibe. This latter stripped down feel was a direct result of the limited technology available at the time. Since the TB303, TR909 and TR808 were pretty much the only instruments obtainable to those without a huge budget, most tracks were written with these alone which were then recorded directly to two-track tape cassettes.

It wasn't until late 1988, however, that techno became a genre in its own right when Neil Rushton produced a compilation album labelled *Techno – The New Dance Sound of Detroit* for Virgin records. Following this release, techno no longer described any form of electronic music but was used to describe minimalist, almost mechanical house music. Similar to most genres of dance, this mutated further as more artists embraced the ideas and formed their music around it. By 1992 and the evolution of the new 'rave generation' techno bore almost no relationship to the funky beats and rhythms of house music as it took on more drug-influenced hypnotic tribal beats.

As technology evolved and MIDI instruments, samplers, sequencers and digital audio manipulation techniques became more accessible, techno began to grow increasingly complex. While it still bore a resemblance to the stripped down feel of Detroit techno consisting solely of rhythms and perhaps a bass, the rhythmic interplay became much more complex. More and more rhythms were laid atop one another and the entire studio became one instrument with which to experiment.

Of course, Detroit techno still exists today but it has been vastly overshadowed by the tribal beats of 'pure' techno developed by numerous artists including Thomas Krome, Redhead Zync, Henrik B, Tobias, Carl Craig, Kenny Larkin and Richie Hawtin. Each of these artists has injected their own style into the music while keeping with some of the original style set by their contemporaries.

MUSICAL ANALYSIS

Techno can be viewed as dance music in its most primitive form since it's chiefly formed around the cohesion and adaptation of numerous drum rhythms. Although synthetic sounds are also occasionally employed, they will, more often than not, remain atonal as it's the abundance of percussive elements that remains the most vital aspect of the music. In fact, in most techno tracks additional synthetic instruments are not often used in the 'musical' form to create bass lines or melodies; instead, the genre defines itself on a collection of carefully programmed and manipulated textures rather than melodic elements.

Fundamentally, this means that it's produced with the DJ in mind and, in fact, most techno is renowned for being 'DJ friendly' and is formed and written to allow him (or her) to seamlessly mix all the different compositions together to produce one whole continuous mix to last through the night. As such, techno will generally utilize a four-to-the-floor time signature but it isn't unusual to employ numerous other drum rhythms written in different time signatures which are then mixed, effected and edited to fit alongside the main 4/4 signature. Tempo-wise, it can range from 130 to 150 BPM, and although some techno has moved above this latter figure, it is in the minority rather than the majority.

Techno is also different from every other genre of music covered so far since it does not rely on programming and mixing in the 'conventional' manner (if there is such a thing). Rather, it's based around using the entire studio as one interconnected tool. While a sequencer (hardware or software) is still used as the centrepiece, it's commonly only used to trigger numerous drum rhythms contained in the connected samplers and drum machines. Each of these rhythms have been previously edited and manipulated with effects to produce new variations which are then layered with others or dropped in and out of the mix to produce the final arrangement.

These rhythms are layered on top of one another so that they all interact harmonically to produce interesting variations of the original patterns. As more of these patterns are laid together, they create an often syncopated feel as the rhythmic harmony becomes progressively more complex. The mixing desk is then used to not mix the rhythms together in a conservative manner but as a creative tool with EQ employed to enhance the interesting harmonic relationships created from this layering, or to prevent the cohesive whole from becoming too muddy or indistinct.

This method of working is analogous to subtractive synthesis, whereby you build a harmonically rich sound and then employ filters to shape the results. With techno, you construct a hectic, yet harmonically rich rhythm by layering loops, and then proceed to deconstruct them with filters and EQ until you're left with some interesting harmonic interplay and rhythm. This produces incredibly complex beats that could never be 'programmed' through MIDI or played live and that also subtly move from one interesting rhythmic interaction to another with some live (or automated) tweaking of EQ or additional effects.

The obvious place to start with any techno is to begin by shaping the loops and, generally speaking, occasionally they may be programmed in MIDI but are more often sourced from sample CDs or other records. These are then subsequently diced, sliced and manipulated with programs such as Wavesurgeon to create rhythms that are very different from the original. However you decide to create the loops, the individual rhythms begin quite simply, and the beginnings consist of nothing more than a kick drum, snare, closed and open hi-hats along with an occasional cymbal crash employed every 4 or so bars to mark the end of a musical segment (Figure 14.1).

Although this generally provides a good starting pattern, any further loops should be programmed differently and/or use different timbres so that when the rhythms are overlaid with one another a more complex sound and rhythm evolves. To accomplish this, it's often worth experimenting by programming tribal rhythms to mix with the previous loop. How these are programmed is entirely up to your own creativity, but a good starting point is to employ side

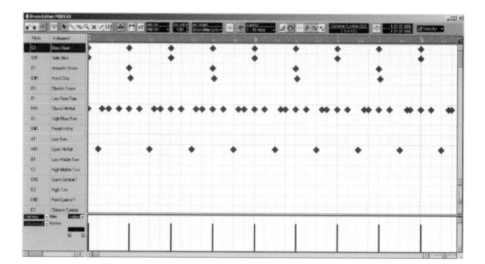

FIGURE 14.1
The beginnings of a techno loop

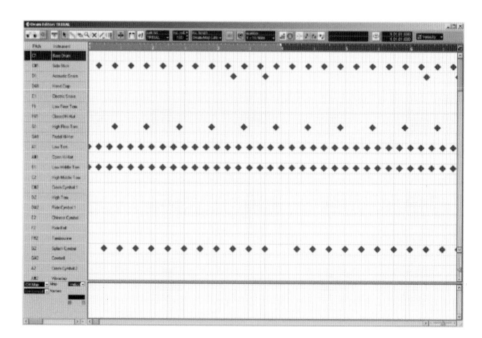

FIGURE 14.2
A tribal techno loop

sticks, splash cymbals, tom drums and more snares that can dance around the main starting loop. An example of a typical tribal loop is shown in Figure 14.2.

Note that in the above example, very few of the instruments actually land on the beat; rather they are all offset slightly. This helps them to combine more freely with the initially created loop, creating thicker textures as the sounds play slightly earlier or later than the first loop, thus preventing the loops from becoming too metronomic. Also, note how this second loop does not contain

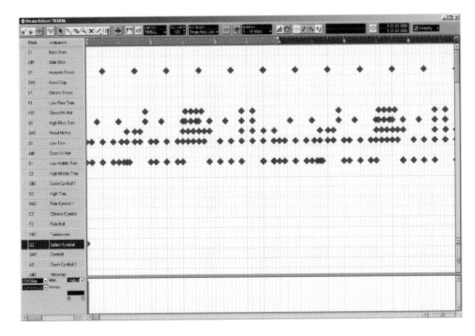

FIGURE 14.3
Another tribal techno
loop to mix with the
previous two

any kicks. This is simply because if each consecutive loop contained a series of offbeat kicks, the rhythm would lose its four-to-the-floor feel, while if the kicks were all laid on the beat of the bar, they would all amalgamate producing a mix with a very heavy bottom end. The principle is to create a series of loops, each of which focuses on a different timbre to carry the loop forward.

These two loops could be further augmented with another tribal loop, this time much more complex than the last, containing a number of closely paired instruments that all complement the previous two loops and create an even more complex drum arrangement (Figure 14.3).

Yet again, note how none of the instruments land on the beat. This prevents the rhythm from becoming too metronomic and allows these new timbres to mix with the previous two loops to thicken out those timbres already present. Additionally, it is also worthwhile experimenting by mixing loops with different time signatures together. For instance, two 4/4 loops mixed with two 3/4 loops can produce results that cannot be easily acquired any other way.

Once a number of MIDI files have been created in this manner, they can be exported/imported as audio and sliced, diced and EQ'd to modify the sounds further to create abstract timbres suitable for techno. Although for this example we've only discussed three loops, you should feel free to programme and layer as many loops together as you feel are necessary to accomplish the overall sound. This can range from using just three to layering over eight together and then progressively reducing the content of each until you are left with harmonically rich and interesting sound. The layering of these should also not just be restricted to dropping them dead atop one another, and in many cases, moving one of the two loops forward or back in time from the rest can be used to good effect.

Once a few basic loops have been created, effects and aggressive EQ cuts or boosts are employed on each individual element of a sliced loop to create interesting timbres. Although the premise of techno is to keep a 4/4 kick pumping away in mix, other percussive elements are commonly heavily affected, processed and EQ'd to produce timbres that add to its tribal nature. Although any effects can produce good results, of particular note the *Sherman Filterbank* is almost a requisite for creating strange evolving timbres. That said, the effects and processing applied are, of course, entirely open to artistic licence as the end result is to create anything that sounds good but what follows is a list of possibilities to start with:

- Use a transient designer to remove the transients of a snare or hi-hat and then increase the sustain.
- Reverse a sample, apply reverb with a long tail and then reverse the sample so that it plays the correct way again (albeit with a reversed reverb effect).
- Use a noise gate to shorten sounds in the loops.
- Apply heavy compression to squash the transients of some sounds.
- Apply heavy compression but only to the sustain (i.e. use a long attack).
- With the aid of a spectral analyser, identify the frequencies that contribute to the body of a sound and reduce their volume while increasing the volume of those surrounding them.
- Merge two timbres (such as a snare and hi-hat) together and use EQ to remodel the sound.
- Pitch shift individual notes up and by extreme amounts.
- Apply heavy chorus or flangers/phasers to singular hi-hats or snares.
- Write a hi-hat pattern, export it as audio (if required), apply heavy delay and then cut the resulting file up to produce a new pattern.
- Apply heavy delay/chorus/flanging or phaser to an entire drum loop and cut segments out to mix with the rest of the loops.
- Constantly time stretch and compress audio to add some digital clutter and then mix this with the other loops.

The main principle of applying all these processes is to generate timbres that are tonally different from every other loop so that layered together they combine to create an interesting sound. Using mixing desks EQ and low-, high- or band-pass filters, you can then amplify or filter this area and slowly evolve its progression over the length of the mix with some automation. This live 'tweaking' forms a fundamental part of the music, so if the rhythms are contained in a software audio sequencer, it's of paramount importance to use an external controller so that you can modulate the various parameters in real time. Obviously, the more parameters this controller offers, the more creative you can become, so ideally, you should look towards a controller that has numerous options. Novation's *ReMote* 25 can be particularly useful since this offers a plethora of real-time controllers consisting of knobs, faders, a touch pad and a yoke (a control stick that can be used to control both pitch and modulation simultaneously). All of these can then be recorded as CC data in real time, permitting you to edit the automation further if required.

Naturally, techno relies on compression to produce the harsh, heavy beats as much as any other dance genre but its production ethics are slightly different. Whereas in much of dance you usually wish to refrain from pumping the hi-hats or additional percussion by compressing the 4/4 kick singularly, it isn't uncommon in this genre to actually pump these percussive elements on some of the individual loops to create new sounds. Additionally, the compressor is often treated as an effect as well as a processor and it isn't considered unusual to access a valve compressor as a send 'effect'. With the setting of a low threshold and high ratio, the returned signal can be added to the uncompressed signal, while experimenting with the attack and release will produce a series of different timbres. If you do not have access to a valve compressor, then sending the signal out to the PSP vintage warmer can often add more harmonics to the signal similar to using a valve compressor.

Of course, once the beats are finally laid down, with all the frequency, EQ and pitch interaction the loop will need compressing to prevent any potential clipping and, in techno, this is applied very heavily indeed. Generally, valve-based compressors are used since these tend to pump musically. The most commonly used compressors for this are the *Joe Meek SC 2.2*, *UREI LA 3* or the *UREI 1176 LN* due to the amount of second-order harmonic distortion they introduce. If you don't have access to a good valve compressor, then after compressing it with a standard unit it's certainly worth throwing the loop through the PSP vintage warmer to recreate the requisite sound. In fact, even if it is put through a valve compressor it's often worth placing it through the vintage warmer anyway. The amount of compression to apply will, as always, depend heavily upon the timbres used, but as a general guideline, start by setting the ratio to 12:1, along with an attack of 5 ms and a medium release of 200 ms or so. Set the threshold control to 0 dB and then slowly decrease it until both the kick and second loudest elements of the loop (commonly the snare) register on the gain reduction meter by at least 5 dB. To avoid the volume anomaly (i.e. louder invariably sounds better!), set the make-up gain so that the loop is at the same volume as when the compressor is bypassed and then start experimenting with the release settings. By shortening the release, the loop will become progressively heavier the more that this is shortened. Unfortunately, the only guidelines for how short this should be set are to use your ears and judgement but try not to get too excited. Keep in mind that it should not be compressed so heavily that you lose the excursion of the kick, otherwise the loop will lack any real punch.

As previously mentioned, techno commonly consists of drums alone but some may also include a bass rhythm to help the music groove along. In these instances, the bass is kept incredibly simple so as not to detract from the fundamental groove created by the drums. In other words, the bass commonly

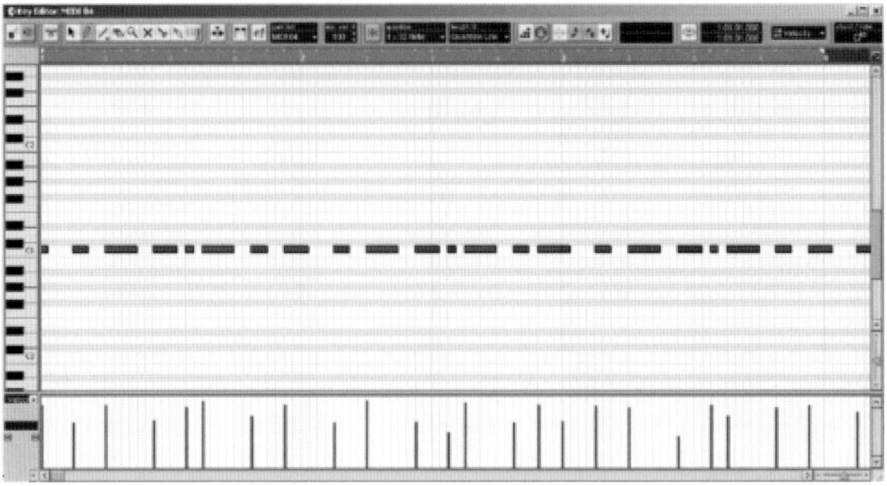

FIGURE 14.4
A typical techno bass line

consists of noting more than a series of 1/16th, 1/8th or 1/4th notes (sometimes consisting of a mix between them all) with none or very little movement in pitch (Figure 14.4).

In the above example, the bass remains atonal but movement is provided by lengthening and shortening the bass notes, while the velocity controls the filter cut-off allowing the groove to move in and out of the drum rhythms. Some tracks may also employ some pitch movement in the bass, but if this is the case then the timbre used is invariably mono and makes heavy use of portamento so that that any overlapping notes slide into one another (Figure 14.5).

Note how in the above example a secondary note overlaps the first. This creates a timbre that rises and falls during its period and it's this type of movement that is fundamental to a techno bass. Since the bass will generally only consist of one bar that is continually looped over and over, it's the harmonic and timbral movement that plays a primary role in attaining the groove. This is accomplished not only by using portamento but also by adjusting various synthesis and effects parameters as the bass plays alongside the drum track. The basic belief here is to manipulate the frequencies contained in the bass so that it augments the frequencies in the drum track. That is, it adds to the harmonic relationship already created through manipulating the drums to create a cohesive whole that pulses along. The bass should still remain a separate element to the drum track but, nevertheless, any frequency-dependent movement should be to bring further interest to the harmonic interaction with the drums than to bring attention to itself.

For the bass timbre, the TB303 is the most widely used instrument since this is what was originally used by the techno originators; however, more lately, any analogue synthesizers (or DSP equivalents) are used so long as they contain enough frequencies to be swept with a filter so that they interact with the rhythms. If you want to stay with the roots of techno, though, the TB303 is

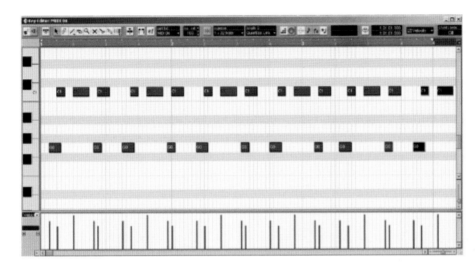

FIGURE 14.5
Using portamento on a techno bass

the bass to use but this timbre can be created in almost any subtractive synthesizer. A good starting point is to employ both a saw and a square oscillator, set the amplifiers EG to zero attack and sustain with a medium release and a fast decay. Use a low-pass filter with both the resonance and cut-off set midway and then adjust the filters envelope to a short attack, decay and sustain with no release. Finally, as ever, experiment with all these settings until you obtain the sound you require.

For those who are a little more adventurous, try combining four sawtooth waves, or a couple of sawtooths and a sine wave to add some bottom end if required. If a sine wave is used, detune this by an octave below the other oscillators and then proceed to detune each saw from one another by ± 3, ± 5 and ± 7. The amp envelope for all the waveforms is commonly set to a fast attack with a medium-to-long decay and no sustain or release. A filter envelope is not commonly used as this adds harmonic movement through the period of the sound that may conflict with the already programmed/manipulated rhythms, and the oft preferred option is to keep it quite static and employ filter movement manually to suit the constantly changing frequencies in the rhythms. That said, if a pitch envelope is available in the synth, it may be prudent to positively or negatively modulate the pitch of the oscillators to add some small amounts of movement. Once this basic timbre is laid down, it's prudent to experiment with the attack, decay and release of the amp's EG to help the bass sit comfortably in with the kick's drum loop. In fact, it's essential to accomplish this rhythmic and tonal interaction between the drum loop and the bass before moving on. Techno relies on a very sparse collection of 'instruments' and the interaction attained here will form a majority of the record.

As the harmonic movement and interaction with the bass and rhythms provide the basis for most techno, it's also prudent to experiment by applying effects to the bass timbre to make it richer sounding. While most effects should be avoided since they tend to spread the sound across the image (which destroys the stereo perspective), small amounts of controlled distortion can help to pull the bass out of the mix or give it a much stronger presence. Similarly, a conservatively applied delay effect can be used to create more complex sounding rhythms.

One final aspect of techno is the addition of sound effects and occasionally vocals. While the sound effects are generated by whatever means necessary, from sampling and contorting anything with effects and EQ, the vocals very rarely consist of anything more than a short sample. The verse and chorus structure is most definitely avoided and, in many cases, only very small phrases are used which are often gleaned from the old 'speak and spell' machines of the early 1980s. This particular machine isn't a requirement (with its increased use in techno, the second-hand prices of these units have increased considerably) and the same effect can be obtained from most vocoders so long as the carrier consists of a saw wave and the modulator is robotically spoken.

When it comes to the arrangement of techno, it's important to understand that it does not follow the typical dance structure. Instead, it relies totally on the adaptation and interrelationship between all the elements. This consists of dropping beats in and out of the mix, along with the bass, vocals and sound effects (if used) but it mostly centres on the real-time application of filters, effects and EQ. The plan is not to create a track that builds to a crescendo or climax but rather stay on one constant rhythmical level that warps from one rhythmically interesting collective to another. Indeed, its careful use of filters and effects creates the final arrangement, not the introduction of new melodic elements. The overall goal is to create the impression that all the sounds are interweaving with one another at different stages in the music. This helps to prevent monotony but also averts the building sensation often introduced by adding new melodic elements into the mix.

> The data CD contains a full mix of a typical Techno track with narration on how it was constructed.

RECOMMENDED LISTENING

Ultimately, the purpose of this chapter has been to give some insight into the production of techno and, as ever, there is no one definitive way to produce the genre. Indeed, the best way to learn new techniques and production ethics is to actively listen to the current market leaders and experiment widely with the mixing desk, effects and processors. With this in mind, what follows is a short list of artists that, at the time of writing, are considered the most influential in this area:

- Derrick May
- Juan Atkins

- Kevin Saunderson
- Eddie Fowlkes
- Richie Hawtin (Plastikman)
- Carl Craig
- Kenny Larkin
- Stacey Pullen
- Jeff Mills Mike Banks
- James Pennington
- Robert Hood
- Blake Baxter
- Alan Oldham

CHAPTER 15
Hip-Hop (Rap)

'Rap is like the polio vaccine. At first no one believed in it. Then, once they knew it worked, everyone wanted it...'

Grandmaster Flash

In recent years hip-hop has become a global phenomenon, so much so that what was once frowned upon for seemingly glorifying drugs, guns and general delinquency has now emerged as a multi-billion pound industry. However, it should be noted that despite the record industry's habit of pigeon-holing absolutely everything that features a rapper as hip-hop, this isn't the case.

Hip-hop is a culture, not a form of music, and although it does encompass rap it also embraces dancing, language and fashion. Consequently, producing rap music has very little to do with programming some MIDI patterns and rapping over the top. To better understand why this is, it's vital to know a little about the history and culture behind it all.

Hip-hop, as a culture, can be defined as consisting of four distinct elements: DJ'ing, breaking, graffiti and MC'ing (emceeing). The roots of the DJ'ing element can be traced back to 1950s Jamaica where the 'DJs' began to experiment with groove elements of records, resulting in the creation of reggae, ska and the rocksteady beat. In 1968, this became even more experimental when King Tubby created the first ever 'dub' record by dropping out all the vocals from the acetate discs he was to press (often called 'dub' plates).

At the same time many Jamaicans were immigrating to the United States, taking these new ideas with them to the ghettos of New York. One particular immigrant, Kool Herc, began to DJ at parties throughout the ghettos and used to chant rhymes over the top of the instrumental breaks of the records he played. As many of these breaks were short, in 1974 he decided to play two copies of the same record on two decks and then use a mixer to switch between them, in effect creating a longer break beat to rhyme over. Almost simultaneously, in another ghetto Afrika Bambaataa founded Zulu Nation, consisting of a group of DJs, break dancers, MCs and graffiti artists, and offered an alternative to the current street-gang culture, allowing the youth culture to express themselves in various ways.

Inspired by these new DJ'ing tactics and culture, DJ Grandmaster Flash adopted the style and contorted it into a continuous stream of break beats. This allowed MCs to rhyme over the top of the beats to warm up the crowds, permitting the DJ to concentrate on developing new techniques such as 'beat juggling', 'scratching', 'cutting' and 'breakdown'. Not being a DJ myself, I can't comment on what some of these techniques involve or who originally developed them but, arguably, it was Grandmaster Flash who introduced this new complex form of DJ'ing to the mass market with the release of *The Adventures of Grandmaster Flash on the Wheels of Steel* in 1981.

These continual break beats also gave rise to a new dance style known as breaking (AKA B-Boying), which consisted of a combination of fancy, complex footwork, spins and balancing on hands, heads or shoulders. Many of these moves were inspired by the relentlessly released Kung Fu movies in the 1970s but the inspiration could be drawn from anywhere, including their rivals during 'battles'. This is where 'crews' would compete against one another to see who was the better breaker, and in many instances a breaker had to face off against a crew to be accepted into the clan.

This form of dancing was categorized and renamed by the media as 'break dancing', but not all forms of dancing at this time were breaking. Two other styles had developed – 'locking' and 'popping' – these involved waving arms and sharp robotic movements which were not classed as part of the hip-hop scene.

Alongside this new music grew another part of hip-hop culture, graffiti. Many credit the explosion of graffiti to TAKI 183 and the publicity he received in the *New York Times* after 'tagging' numerous trains in the subway.

The term 'tag' is used since graffiti refers to any unwelcome defacing of property with spray paint or markers the true hip-hop was much more artistic. A tag is essentially the writer's signature expressed in an artful and creative way, consisting of three areas; the 'tag', the 'throw up' and the 'piece'. The tag is classed as the simplest form of graffiti and consists of a signature in just one colour, written using a marker. As time moved on, spray cans were introduced and the tag moved up a step to the throw up, which is made up of two colours, resulting in more complex styles. The final style, a piece, is the most complex form of graffiti, which people like Lee Quinones have made a living from selling to art galleries. On the streets, however, to be referred to as a piece it must consist of at least three colours and preferably be 'end-to-end' art. As most graffiti was sprayed onto subway trains, this latter term should be self explanatory. This type of graffiti is an incredibly complex form of art which, although viewed by many as destructive, takes a lot of planning, people and an artistic flair (*although for obvious reasons I cannot condone the defacing of any property*).

The last element of hip-hop is MC'ing. Although the media considers rap to be the same as MC'ing, rap is only one element of it. Indeed, MC'ing encapsulates everything from simply talking over the beats to rapping or using your voice as an instrument (human beat box). As touched upon, originally MC'ing was used to entertain the crowds by accompanying the break beats rather than taking the focus away from them. The original form was known as call-and-response whereby the MC would typically shout out 'Everybody in the house say yeah' to which the audience would respond with a resounding 'Yeah'.

Although it would be easy to say that rap developed from this basic form of MC'ing, to many it actually existed long before Kool Herc, Afrika Bambaataa or Grandmaster Flash began to rhyme over the breaks. Indeed, it's believed to have originated in Jamaica, where stories were told in rhymes otherwise known as 'toasts'.

In 1974 these were developed into the very first forms of rapping where the youth would put together boastful rhymes to sit over the top of break beats in an effort to upstage the previous rapper. The first commercial pressing of rap music was by the Fatback Band in 1979 with the title *King Tim III*, but it took the Sugar Hill Gang's *Rappers Delight* released later that same year to bring rapping to the attention of larger record labels as a viable and acceptable (in other words lucrative) form of music.

Over the following years, rap acts such as N.W.A, Ice T and Public Enemy brought rap to the forefront of music and demanded a bigger audience through their often hotly debated rhymes that were seen as glamorizing violence, prostitution and guns.

MUSICAL ANALYSIS

Although technology has moved on since the early pioneers began to produce breaks from various DJ techniques, the fundamental creation of rap still relies heavily on the sound quality generated using these techniques.

For many the turntable techniques have been replaced with more recent developments such as samplers, hardware workstations and wave editors, but the 'feel' of the early techniques is still of paramount importance to producing rap. As a result the break beats are *not* programmed through MIDI but are sampled from previous obscure records and then contorted and manipulated in sample slicing programmes such as WaveSurgeon to create new beats. This sampling is fundamental to keeping with the original roots of hip-hop by introducing the vintage 'vinyl' feel to the music that is not possible any other way. Similarly, many of the instrumental riffs and motifs are also sampled from early records. It is, however, possible to programme these through MIDI provided that you use the right instruments.

Nevertheless, the preferred option *is* to sample from original vinyl. In many cases sampling will produce all the sounds used in the genre, including basses and any 'lead' timbres. Indeed the general consensus between most professional hip-hop musicians is that if you have a good record collection, a good record deck and a sampling workstation (such as the *AKAI MPC*) you can create hip-hop.

Generally speaking, sample CDs are avoided because everyone has access to them and no matter how CDs are contorted they can often still be recognized, so obscure vinyl is the preferred choice. Of course, for legal reasons I can't condone stealing riffs, loops or entire segments from other records, but many hip-hop artists agree that so long as the samples chosen are ambiguous or manipulated enough, there is no need to clear them.

Having said that, Dr Dre (Andre Young), who's considered by many as rap's current godfather, is well known to the American courts for his countless copyright infringements – he's been sued over eight times in the last 3 years. In the most publicized case, he asked for permission to use a sample from George Lucas' THX sonic boom and when permission was refused, used it anyway. Consequently, he was sued for $10 million. London-based Minder Music also successfully sued him for $1.5 million after he sampled a bass line from one of their releases for his 2001 track 'Let's Get High'.

More recently a French Jazz musician started legal proceedings to sue Dr Dre and Eminem for $10 million because (he claims) they used his music on *Slim Shady's* 'Kill You'. As a result, Dr Dre has now apparently hired a musicologist to advise him on whether he can sample music riffs without infringing copyright. The purpose of this is not to disrespect or accuse Dr Dre or any other rap artist of serial stealing but to simply offer food for thought before you consider that samples may not need to be cleared at all. While the best approach is to create a track by whatever means you feel necessary, if the track feels right, you should always attempt to clear the samples afterwards. This is a much easier approach than asking permission first, being refused and loosing the chance to write what potentially could have been a hit.

Generally, hip-hop can be described as using a slow, laid-back beat that can vary from 80 BPM through to 110 BPM. As discussed most of the break beats used are sampled from other records to create the right vibe, while any further melodic instrumentation takes a back seat to the rapper. Consequently these instruments play very simple melodies so as not to distract too much attention from the rap. As a consequence, the drum rhythm is usually laid down first, followed by the rap vocals; finally any further melodic instruments are added to 'sit' around the rap.

To accomplish this, you will need to have a good record (sample) collection, most of which have preferably only been small releases pressed in small amounts. Not only can these be picked up for relatively small outlay at any charity/junk shop, but you're more likely to find records that no one has ever considered using before. No matter what the genre of music (unless it's choir or similar) they will all feature a break that can be sampled and manipulated into a hip-hop beat; even unlikely records such as *Don Reeve's Non-Stop Double Hammond Goes Around the World* features some good hip-hop basses if you listen closely enough. The trick is to use your instincts and develop an ear for the right sounds. This can only be accomplished from being actively involved in the hip-hop scene and gaining experience at listening out for the right sounds.

THE RHYTHMS

Primarily, there are two processes used to produce hip-hop rhythms. The first is to use sampled hits from various records and then reconstruct a rhythm live by assigning each pad of a workstation to a specific sample or, more commonly, simply sample a loop or number of loops from records and edit them using software such as WaveSurgeon. For the purpose of this chapter, we'll look at both methods, starting with sampling single hits.

Sampling individual hits is only recommended if you have access to a hardware drum sampler with pads, as creating the hip-hop vibe requires a 'live' feel. Indeed, very few professional rappers will rely on a software sequencer to produce hip-hop but instead use a hardware sampler and sequencers since the timing must be incredibly accurate. Although most of these samplers will default at 16-bit 44.1 kHz, for hip-hop it's recommended to sample the hits at 22 kHz and 12-bit. This reduces the range of the samples, giving them a grittier feel which is often essential. Once sampled, each of the hits is assigned its own pad in the sampler, allowing you to tap out a live rhythm, piece by piece.

Hip-hop rhythms are generally kept quite simple and often consist of nothing more than a kick drum, snare, and closed and open hi-hats along with an occasional cymbal crash employed every 4 or so bars to mark the end of a musical segment. The time signature can also play a fundamental role; hip-hop may use either 4/4 or 3/4 depending on the amount of swing you need the drums to exhibit. What is important, however, is that very few, or in some cases none of the hits are quantized to the grid. The idea is to create the feel of a real drummer and very few drummers play exactly to the beat. Additionally, the pattern very rarely follows the typical *kick/snare/kick/snare* configuration used in most genres of dance and often relies more on rhythmic interplay between a few kicks followed by a snare which is followed by another few kicks and a snare. On top of this, a closed hi-hat pattern often follows a standard 16th pattern ticking away in the background to keep some rhythmic drive and syncopation to the loop.

In the following example, the closed hi-hats are providing some offbeat syncopation while both the kick and the snare are occurring ahead of time to increase the feel of a laid-back groove (Figure 15.1).

In the next example, more emphasis has been placed on rhythmic interaction between the snares and the drum kick. The closed hi-hat pattern has also been reduced and does not occur on the same beat as the open hi-hat to prevent any frequency clashes between the two. More importantly, though, note how very few of the elements actually occur on the beat (Figure 15.2).

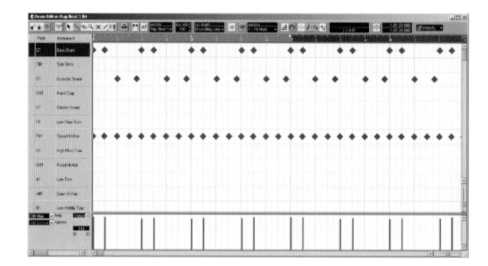

FIGURE 15.1
A typical laid-back hip-hop rhythm

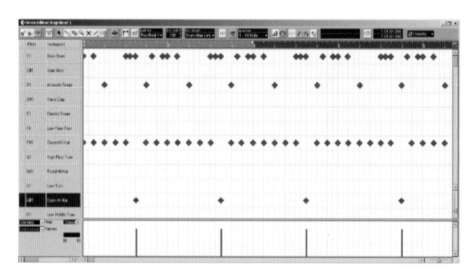

FIGURE 15.2
A more complex rhythm

The real key to producing a hip-hop rhythm in this manner is to experiment with the interplay between the kicks and snares and ensure that they are off-set from the beat to introduce the laid-back nature of the music. After this has been accomplished, the hi-hats are added to introduce some syncopation to the final groove. Also, as the above examples reveal, hip-hop tends to be based around the constant repetition of 2 bars of music, allowing for a binary phrase to be applied in the bass or chorded lead if used/required. Following this it's sensible to create a drum rhythm that continues over 2 bars before repeating again at the end.

Typically, in most music of this genre there are no variations in the rhythm, as this tends to distract from the most important element of the track, but velocity plays a large role in the creation of the rhythm to keep the realism. The most usual method to implement is the *strong–weak–medium–weak* syncopation on the kick drums. The snares, too, are often subjected to different velocities, with the first and final snares using the highest values and those in between using a mixture of different values.

As touched upon, the common approach to producing this 'live' effect is to play the sampler workstation pads live and record the subsequent MIDI data for editing later. This is one of the reasons why many hip-hop artists will use the AKAI MPC sampler workstations – the pads can respond to 16 different velocity values depending on how hard they are hit. There is no need to create the entire loop in one run. Typically the drum's kick is recorded first, then played back from the unit while the snare rhythms are tapped in over the top. This is then followed with the open hi-hat patterns, followed by the closed patterns and any additional auxiliary percussion. This approach is preferred to 'stamping' in notes in a MIDI sequencer, since the timing becomes far too accurate and the MIDI delay between sequencer and sound source can result in the groove being lost.

Of course, even with the rhythm programmed in the sampler, it's highly probable that they will need some slight quantizing so that any further elements will lock to it, but any strict quantizing should be avoided altogether. The principle is to create a rhythm that has a loose feel but at the same time not so loose that the instruments do not lock to its rhythm. Once this has the right flow, the MPC's various synthesis parameters are used to manipulate each drum timbre to produce a rhythm that flows along. This can involve anything from pitch-shifting individual snares to create more complex rhythms to tweaking the decay parameters to either lengthen or reduce them to create sharper, more distinct patterns or create slow-flowing patterns. Ultimately, experimentation is the true nature here and close listening to the most popular current hip-hop tracks will let you know whether you're on the right track.

The second process to produce a hip-hop rhythm is to sample entire breaks from records and then manipulate these so that they no longer sound like the original and are more suited towards rap. This is often the preferred approach by many artists as the principle is much closer to the music produced by its original forefathers. This must be applied cautiously, however, since many record companies do not look lightly on sampling their artists' material. Regardless, this doesn't seem to have discouraged many hip-hop artists; most will sample a loop and then manipulate it with various synthesis parameters to make it totally unrecognizable. This is accomplished through sampling the loop and importing it into an audio sequencer such as Ableton Live, Logic or Cubase.

Many programmers do not settle for simply moving parts around. In most cases they will adapt the sounds further by using the mixing desk as a sound design tool before sampling the results. Typical application of this is to EQ the snares heavily to produce hi-hats or bright kicks. Similarly, transient designers such as *SPL's Transient Designer*, the software plug-in included in the *Waves*

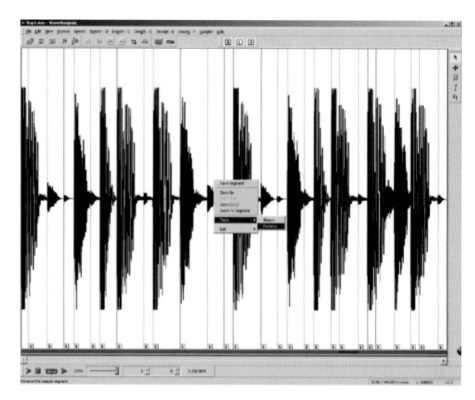

FIGURE 15.3
Cutting up and
rearranging a rhythm

Transform bundle, can be used to edit the attack and sustain of audio enve-lopes. For instance by setting the attack parameters on these quite long you can remove the initial attack of the hit to produce a different timbre, while shorten-ing the sustain parameter can make snares appear much tighter, snappier and controlled (Figure 15.3).

As an alternative to simply using and splicing one loop, it is often worth sam-pling three or four loops. Then they can each be lined up in an audio sequencer and time-stretched so that each is playing at the correct tempo. Taking this lat-ter approach, all the loops can be played simultaneously and elements from each loop can be cut up and removed to produce a rhythm composed of con-stituent parts from each loop (Figure 15.4).

This can help you create complex polyrhythms, but it's important to note that for this to work each drum loop should be quite simple, since the more com-plex each is the more the combined rhythm will sound muddy and indistinct. This can make it particularly difficult to create a good rhythm through slicing and removing samples. Also, it isn't absolutely necessary to ensure that all the sampled loops occur at the same time. In many instances by moving one of the loops forwards by a few ticks you can create a more laid-back feel, since the snares will occur off the quantize values. Indeed, this technique is applied by many hip-hop artists during sampling. Rather than sampling the beginning of

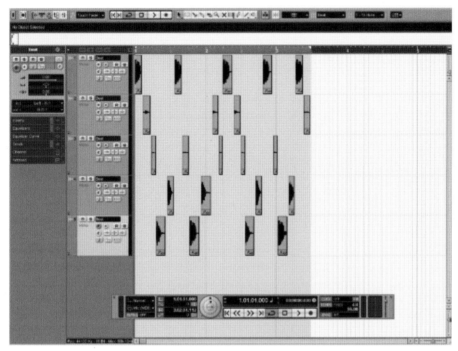

FIGURE 15.4
Lining up a series of loops and removing elements to create a basic rhythm

the drum track (i.e. from the first kick) they often sample the rhythm halfway through and then loop this to create the complete drum loop.

With this technique, the kicks are obviously positioned in a different place than in the original record, meaning that the sample may start on a snare or even a hi-hat. To overcome this, it's common practice to remove the original kicks using sequencing software and then reprogramme a new kick loop to sit over the original sample. Typically, this new kick is derived from the Roland TR808 drum synthesizer, which is sampled at 12-bit 22 kHz to produce a gritty timbre. If you don't have access to this synth, then any synthesizer can produce the requisite sounds using a 60 Hz sine wave positively modulated with a pitch envelope set to a fast attack and short decay. Although you can synthesize an additional transient to sit over this by using a square wave, generally, the hip-hop kick remains quite 'boomy' without a prominent attack so, this isn't particularly necessary but it is worth experimenting with.

Once the basic tone is present, lengthening the pitch decay will produce a timbre with a heavier boom, but this can only be lengthened so far before it begins to produce a sound that *whoops*. If this is the case, then changing the pitch decays envelope from linear to exponential (convex) will help to make it boom more. While these methods will produce a deep 808 kick, it may be worth using a self-oscillating filter and a noise gate in its place. This particular technique has already been discussed in sound design, but to recap quickly begin by increasing the resonance until it breaks into self-oscillation and produces a pure sine wave. After this, programme the kicks pattern to suit the current loop and feed the resulting audio into a noise gate with the threshold set so that only the peaks of the wave are breaching it. Finally, set the attack to zero, use the release to control the 'kicks' delay and hold the time parameter to produce a heavier sound.

Roughly speaking, hip-hop drums are not compressed nor are any effects applied, as the purpose behind the genre is to keep the sound raw. Besides, if the drums have been sampled from a record they will have already been compressed and effects applied at the original recording, mixing and mastering stages. Naturally, this is only a generalization and if the loop appears quite weak then compression may help to pull it up a little. Typically the compressors used for this are as transparent as possible so solid-state is often used, such as the Alesis 3630 or the Waves C1.

As a starting point, try a ratio of 8:1 with a fast attack, short release and a threshold set so that the kick registers on the reduction meter by 2 dB. To avoid the volume anomaly, set the make-up gain so that the loop is at the same volume as it was when the compressor is bypassed and then start experimenting with the release settings. When you shorten the release the kicks will begin to pump the rest of the drum timbres more heavily, but you need to exercise caution as to how much this pumps. Keep in mind that the hi-hats and snares are occurring in between the kicks, so that the snares and hi-hats will be pumped against the kick, which will take way from the 'raw' flavour of the timbres used! If after compression, the beats seem to exhibit a 'digital' nature then it may be worth compressing them with a valve compressor such as the UREI 1176 LN or alternatively the PSP Vintage Warmer can be used after the solid-state compression to give the sound warmth.

Above all, these are only suggestions – it's up to your own creativity to produce new variations of old loops through experimentation. One thing that is certain is that although explaining the principles of creating hip-hop loops is simple, accomplishing it proficiently is an art form. You have to listen to loops with some awareness and be willing to sacrifice some elements to make the loop work properly. Complex loops are not part of hip-hop, but it does take plenty of time and effort to produce a loop that works properly. You have to have a thorough understanding of the genre and how the loops work in the context of other instruments; this can only be accomplished by digesting both the culture and the music.

RAP VOCALS

With the drums laid down and presenting a groove, it's usual to record the rapper next, since any other instrumentation will sit around their rhymes. As these form the centrepiece of the record it's of paramount importance that you have a good rapper (anyone can rap but only a select few can actually rap well!). Equally, it is also essential that the rapper have something to say that can be associated with the genre as a whole: simply rapping about how your car broke down last week will rarely work well (unless you happen to be Eminem). This is another reason why it's vital that the rapper is actively involved in the whole culture.

Customarily, the lyrics are drawn from lyricists rhyming about their skill, or the skill of the crew they're associated with, or they are simply 'dissing' (slang for disrespecting) their rivals. This is a form of battle similar to breaking and is essentially a way of competing with other rappers on the same circuit for prominence and respect. In fact, this form of competition is deeply embedded in the history of rap and is still evident today, playing a fundamental role in hip-hop.

Alternatively, the message conveyed may have a more personal political stance, rhyming about the current state of the ghetto, the nation, guns or violence and drug usage. These are usually not to invoke shock value but rather a true story from the rapper's own experience and is the one of the key reasons why the vocalist must be actively involved in the hip-hop scene. If not, it's very doubtful that the words will have any real meaning for hip-hop aficionados and you could find yourself laughed off the circuit (Ice T?).

Whatever the subject, rapping has become an incredibly complex lyrical delivery that bases itself around sophisticated rhythms that syncopate with the drums. Although there are many different styles, ranging from almost being on the verge of singing to more poetic 'talking', the delivery and the tone are everything. Any good rapper will be able to produce complex rhythms and wordplay that complement the instrumentation while emphasizing the most important words.

Another important aspect is quick thinking and the right delivery. Most professional rappers today are able to come up with rhymes on the spot, a talent that they learned in the earlier years when they may be challenged on the street and have to come up with a rhyme quickly. Of course, this latter talent isn't an absolute necessity, and while rappers such as Run DMC can come up with a rhyme off the top of their heads in an instant, they certainly don't walk on stage and make it up on the spot. Indeed, these and other hip-hop artists often take weeks to come up with an entire song's worth of rapping since the sound and wordplay must have the right style for the music.

Usually, the rap is recorded before the instrumentation is completely laid down so that any additional instruments can also be laid down to emphasize the rap itself. Of course, this is a purely conventional approach and convention doesn't always play a part in music, so it's up to your own artistic interpretation.

The microphone typically used to record the vocals is the Shure SM58, as it produces the nasal tone typical of the genre. It should be hand-held to allow the rapper to move freely as this helps to keep the sound authentic – no rapper stands still while rhyming to a beat – and should be held approximately 10–20 cm (4–8″) away from the mouth, depending on the vocalist. You need to listen carefully to the sound produced and either increase or decrease this distance to capture the right sound. The vocal tone should be full and present – if it seems deep and boomy ask the rapper to move the microphone further away from the mouth while if it's too bright ask him or her to hold it closer.

Compression is often required while recording the vocals to prevent any clipping but this should be applied lightly. If heavy compression is used the dynamics will be too heavily restricted, resulting in a rap that lacks any emotional stature. Generally speaking, a valve compressor, such as the UREI LA 3 or LN1176, are the preferred option, but the Joe Meek SC 2.2 can also be used on vocals to add some warmth. Alternatively, a solid-state compressor followed by the PSP Vintage Warmer can often add the requisite warmth after recording.

Whichever compressor you use, a good starting point is to set a threshold of −12 dB with a 3:1 ratio and a fast attack and moderately fast release. Then, once the vocalist has started to practice, reduce the threshold so that the reduction meters are only lit on the strongest part of the performance. Although the resulting rap is best recorded directly into an audio editor or sampler, some artists will record the vocals to tape cassette before transferring this to the digital editor. This technique helps to capture a 'rough' yet warmer tone that is typical of many rap records.

Once recorded, the vocals are very rarely treated to any processing or effects as the genre tends to move towards a raw sound rather than the 'professional' polished sound of most popular music. This means that, unlike with most vocals, after recording compression is not applied unless it's absolutely necessary and effects such as reverb are applied very lightly, if at all. Additional compression may be needed if the rapper has been slightly off axis from the mic while performing; in this instance very light settings are used to keep the vocals at a similar volume, which allows them to pull through the mix. Unfortunately, there are no generic settings for this – the setting depends entirely on the vocals – but as a very rough starting point, use a threshold setting so that you have −5 dB on the reduction meter, along with a ratio of 4:1 and a fast attack and moderately fast release (approx. 200 ms). Unless you're after the very commercial form of rap, the ratio should not be set any higher than this, and even then you should avoid going over 6:1. So the vocals do not loose all of their emotion and 'raw' flavour. If they sound a little thin, then small amounts of reverb may help, but similar to compression this should be kept to a minimum to retain the feel of the music. In fact, reverb should be applied so lightly that its application is only noticeable when it's removed.

Many hip-hop tracks will also make use of double tracking the vocals to accentuate particular words or phrases. As these are always recorded in mono and sit dead centre of the mix, by double tracking the important phrases you can pan the original vocal and the double-tracked phrase left and right to fill the stereo image and bring attention to particular parts. Many rap tracks will offset one side of the vocals by a few ticks so that they occur moments later than the original, helping to thicken out the vocals and bring more attention to the phrase. These usually lie at the end of a number of bars (similar to how much of dance uses a cymbal crash) but it's up to your own artistic interpretation as to where these accentuated phrases or words should sit. The best advice that can be offered is to *listen* to the rapper and double track him (or her) wherever they place their own accents with the music.

THE BASS RHYTHM

Although most old skool hip-hop tracks consisted of nothing more than a break beat and a rapper, today's producers often drop in bass lines to add some bottom-end weight and help the track groove along. In the majority of cases these are real bass guitars that have been sourced from another record and are not normally cut up or edited in any way. In fact, quite a few of today's tracks have used bass lines (and in some cases the entire track they originally belonged too) that are instantly recognizable. Generally, it is more prudent to write your own, though, since there's no chance of being caught for copyright infringement – and it's much easier to write the bass around a rapper than have a rapper try to rhyme around the bass.

On the whole, most hip-hop basses are kept relatively simple as they merely act as the underpinning of the track and do not form a fundamentally major melodic element in themselves. This is to prevent the movements and groove of the bass from drawing attention away from the rapper's rhymes. As a result, a bass line will often consist of nothing more than a few 1/8th or 1/4th notes per bar that will tend to stay at one pitch or move very subtly in pitch by no more than 3 or 5 semitones. As the rhyme at this stage is already down, the key of the bass will most probably already be determined, and it's simply a case of repeating the first bar of the rhyme and moving down the keyboard making notes of the pitches that harmonize with the rapper. Once you have these, it's just a case of producing a groove that flows alongside and complements the rapper.

Notably, the bass groove very rarely occurs on the beat of each bar; it's quite usual for it to begin just a few ticks *before* or *after* the bar to offset it from the rest of the record. This is done to emulate the nuances of a real bassist since they very rarely play dead on the beat. What's more, offsetting the bass grove by eight or ten ticks, will make the drums appear to pull or push the record forward, helping to create a more flowing groove. This 'vibe' in between the bass and drums is an essential part of rap, but whether you should set it before or after the beat is entirely open to artistic license. After programming it's worth experimenting by moving it before and then after the beat to see which produces the best results (Figure 15.5).

For the bass timbre, real bass guitars are customarily preferred over synthetic instruments not only to keep with the original feel of the music but also because most real basses are recorded by miking up the bass cab, they tend to be particularly noisy. Although some engineers will struggle to get these sounding clean, within hip-hop this noise plays a fundamental role. While it may sound awful soloed, in the context of a mix the bass often helps to pull through and add the grit required. Recording live instruments is beyond the

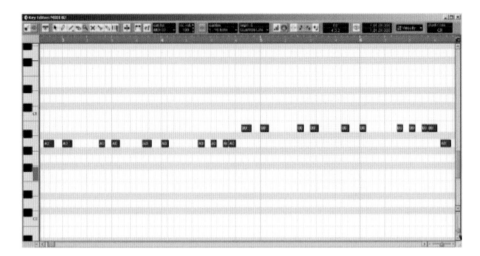

FIGURE 15.5

A typical bass rhythm

scope of this book as it requires a whole new set of production ethics that would require another book to explain. So for those who do not want to sample from another record but want the 'real' bass sound in the music, here we'll look at how to programme a 'realistic' one through MIDI.

The key to programming any real instrument is to take note of how they're played and then emulate this action with MIDI and a series of CC commands. In this case, most bass guitars use the first four strings of a normal guitar E–A–D–G, but these are tuned an octave lower, resulting in the E being close to three octaves below middle C. Also they are monophonic, not polyphonic, so the only time notes will actually overlap is when the resonance of the previous string is still dying away as the next note is plucked. This effect can be emulated by leaving the preceding note playing for a few ticks while the next note in the sequence has started.

The strings can also either be plucked or struck, and the two techniques produce different results. If the string is plucked, the sound is much brighter and has a longer resonance than if it were simply struck. To copy this, the velocity will need to be mapped to the filter cutoff of the bass module so that higher values open the filter more. Not all notes will be struck at the same velocity, though, and if the bassist is playing a fast rhythm, the consecutive notes will commonly have less velocity since he has to move his hand and pluck the next string quickly. Naturally, this is only a guideline, and you should edit each velocity value until it produces a realistic feel.

Depending on the 'bassist', they may also use a technique known as 'hammer on' whereby they play a string and then hit a different pitch on the fret. This results in the pitch changing without actually being accompanied with another pluck of the string. To emulate this, you'll need to make use of pitch bend. First set to a maximum bend limit of 2 semitones, since guitars don't 'bend' any further than this.

Begin by programming 2 notes, for instance an E0 followed by an A0, and leave the E0 playing underneath the successive A0 for around a hundred ticks. At the very beginning of the bass track, drop in a pitch bend message to ensure that it's set midway (i.e. no pitch bend), and just before where the second note occurs drop in another pitch bend message to bend the tone up to A0. If this is programmed correctly, on play back you'll notice that as the E0 ends, the pitch will bend upwards to A0 simulating the effect. Although this could be left as is, it's sensible to drop in a CC11 message (expression) directly after the pitch bend, as this will reduce the overall volume of the second note so that it doesn't sound like it has been plucked.

In addition to this, it's also worthwhile employing some fret noise and finger slides. Most good tone modules will include fret noise that can be dropped in between the notes to emulate the bassist's fingers sliding along the fret board. The pitch bending is best emulated by first programming the notes to overlap slightly and then recording movements of the pitch bend wheel live and editing them in the sequencer.

For those who break out in a sweat at the mere mention of in-depth MIDI editing, it isn't always necessary to use a real bass. Some producers do use synthetic instruments, provided that they're deep enough and have a good 'body'. Although the type of timbre obviously differs from producer to producer, the general tone can be made in any synth by using both a triangle and a pulse wave with the latter detuned from the triangle by -3 cents.

Set the amplifier and filters envelope to a fast attack, medium decay, with a short sustain and no release and use a 2-pole low-pass filter, with a low resonance setting. Finally, modulate the pulse width of the pulse with a sine, triangle or sawtooth LFO set to a slow rate and medium depth. This will produce a basic timbre typical of the genre, but it's worth experimenting with the filters decay, the LFOs rate and depth, and the shape of the decay envelopes on both the amp and the filter to mould the timbre to suit your music.

This will not suit all music, though, and some artists ignore synths and construct their own basses from culminating samples together or pitching a sample down the keyrange because just about anything sounds great when it's pitched down a couple of octaves into the bass register. The key is to experiment in creating a low tone that has the energy to add some bottom-end weight to the music but at the same time does not demand too much attention.

On the subject of experimentation, distortion and compression are often used on hip-hop basses. While most effects should be avoided since they tend to spread the sound across the image (which destroys the stereo perspective), small amounts of controlled distortion can help to pull the bass out of the mix and often give it the much-needed rawness. Similarly, compression after the distortion can be used to even out the volume, keep the effect under control and bring the overall levels up. As a general guideline, start by setting the ratio to 4:1, along with an attack of 5 ms and a medium release of 150 ms or so. Set the threshold control to 0 dB and then slowly decrease it until the bass registers on the gain reduction meter by a couple of decibels. Finally, set the make-up gain so that the loop is at the same volume as when the compressor is bypassed and then start experimenting with the attack and release settings until it produces the level and 'sound' you require. Unlike most other genres of dance music, rap is one of the few that does not use a compressor to 'pump' the low frequencies, so any compression applied should be applied lightly and you should generally try to avoid pumping the low end.

CHORDS AND MOTIFS

Similar to the bass, old skool hip-hop did not employ any chords or leads, but most tracks today will employ some sort of lead sound to sit behind the rapper. Somewhat unsurprisingly, these are usually sampled from other records too, but they can also be programmed in MIDI and then sampled and subsequently reduced in bit and sample rate to add the necessary 'grunge'.

As always there are no definitive rules as to what instruments should be used as a lead, or indeed, how they should be programmed. But, as always, there are some guidelines that will at least help keep you on the right path. Fundamentally, hip-hop tends to stay with real instruments rather than go for synthetic, and a proportionate amount of tracks will use Rhodes pianos, old 'electric' pianos, bells, orchestral strings or orchestral 'pizzicatos'. As with the bass, these are kept quite simple so as not to take the focus away from the rapper. What's more, they're not quantized and are commonly recorded live to help keep the instrumentation sounding real rather than as if were generated by a machine. As a very rough generalization, these instruments – but particularly the piano – tend to follow the da capo sequence with the first bar asking the question and the second bar providing the answer. Typically, this 'answer' plays the same melody as the question but rises in pitch by a few semitones (try semitone shifts of 3, 5 or 7) (Figure 15.6).

If more than one 'lead' instrument is used in the track, only one of these follows the da capo sequence, and any other instruments tend to remain the same throughout the bars (Figure 15.7).

Of particular note, even if the elements of the record are *not* sampled, they should nevertheless sound as though they are. This means gratuitous use of bit reduction, sample reduction along with sampling the vinyl crackle and hiss from a record. This is the preferred approach to using a vinyl plug-in as these rely on generating random noise or cycling a sample every few bars of music which often doesn't sound particularly realistic. However, by sampling vinyl hiss from a record, you can place it into a wave editor and begin cutting up, looping and creating a good 6 bars of hiss to sit in the background. If an audio sequencer is being used, this crackle can be laid on an audio track, and each 6-bar segment can be overlapped at different points throughout the arrangement to create more realistic crackle.

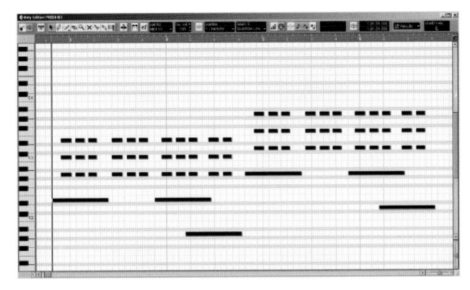

FIGURE 15.6
The piano rhythm using the da capo

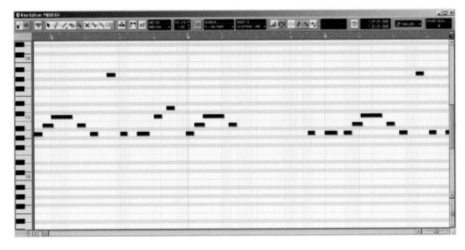

FIGURE 15.7
The pizz rhythm

ARRANGEMENTS

The arrangement of hip-hop differs from most other genres in this book in that it doesn't follow the dance builds or the verse and chorus structure. Instead, the music generally repeats the same bars over and over throughout the track and relies on the rappers to provide the movement of interest. In fact, if you were to strip the rappers away from most rap tracks, the backing would quickly become tedious to listen to since there are so few changes implemented. On occasion, some tracks will 'drop' the beat close to the end of a bar while the rapper rhymes. This not only accentuates the rapper but also creates a rush of emotion as the track returns on the next bar. As ever, if you're unsure about how to produce a typical rap arrangement, listen to other similar tracks and physically count the bars, making note of where new instruments are introduced and then implement this same structure in your own music. This isn't stealing, it's research and every musician on the planet does it. In fact you'll find that many rap tracks follow the very same arrangement principles (and remember that arrangements cannot be copyrighted!).

> The data CD contains a full mix of a typical Rap track with narration on how it was constructed.

RECOMMENDED LISTENING

Ultimately, the purpose of this chapter is to give some insight into the production of hip-hop, and there is no one definitive way to produce the genre. Indeed, the best way to learn new techniques and production ethics is to actively listen to the current market leaders and be creative with processing, effects and synthesis. Chorus, delay, reverb, distortion, compression and noise gates are the most common processor effects used within rap music, so experiment by placing these in different orders to create the 'sound' you want. While the arrangement and general premise of each track is similar, it's this production that differentiates it between all the others. With this in mind, what follows is a short list of some of the artists that, at the time of writing, are considered the most influential in this area:

- Africa Bambaataa (often seen as the original godfather of hip-hop)
- Eminem
- Dr Dre
- Public Enemy
- Run DMC
- El-P
- LL Cool J
- Cyprus Hill
- Ice T

CHAPTER 16

Trip-Hop

'Trip-hop is British hip-hop that lacks the lyrical skills of the U.S. counterparts, but British kids have got the musical side...'

James Lavelle (Mo' Wax label)

Trip-hop is, in the context of dance music, a relatively new genre that evolved out of Bristol in the early 1990s. During this time, American rap was the predominant musical style that was taking Europe by storm. Since this genre of music requires a heavy involvement in the entire hip-hop scene (hip-hop is actually a culture of which rap is only a part) British DJs and musicians contorted it further. The principle elements behind the construction of American hip-hop were fully embraced, with the emphasis remaining on slow, laid-back heavy beats mixed with the gratuitous sampling of old records, but the vocals were left out. This resulted in a genre that was termed British hip-hop by the media, but at this stage it lacked any real diversity, consisting mostly of slow 'tripped out' beats and bass lines mixed in with samples of old jazz records.

In 1991, the release of Bristol-based Massive Attack's *Blue Lines* album marked the first serious release of British hip-hop and also revealed its close connections with American hip-hop (the single *One Love* displaying a remarkably similar feel to its relative). It wasn't until 1994, however, that British hip-hop was coined as 'trip-hop' by UK's *Mixmag* magazine with the release of Massive Attack's second album entitled *Protection* alongside the appearance of Portishead and Tricky. Portishead defined a new style of trip-hop music labelled 'lo-fi' through a mixture of Beth Gibbon's brooding vocals mixed among samples of 1960s and 1970s jazz music which were, in the most, left with a predominantly 'raw' feel.

In fact this raw approach to music was deliberately encompassed by Portishead as they made an effort to record all their instruments into old analogue tape recorders rather than straight to digital media. Tricky's style was somewhat different again, branded by low-pitched singing and an overall cleaner sound but like

Portishead and Massive Attack the style often exhibited a slow, almost depressing feel. Although these three acts did not necessarily aim to create music that was particularly dark, it just so happened that the brooding attitude of the music often oozed a dismal feeling. Part of this may have been attributed to the fact that they had all worked in the same circle. Massive Attack and Tricky originally produced music together under the moniker 'The Wild Bunch,' and Portishead's founder Geoff Burrow aided Massive Attack in producing *Blue Lines*.

On the back of this relatively new genre, more trip-hop artists began to emerge, each putting their own distinctive twist on the music. Artists such as Red Snapper, Howie B, Baby Fox, Lamb, Sneaker Pimps and the Brand New Heavies mutated the genre by mixing it with break beat, ambience, house and acid jazz. The vocals became more upbeat and lively to encapsulate a wider audience, resulting in trip-hop being associated with more energetic music rather than the dark and gloomy vibe. Indeed, because trip-hop is often associated with a dark brooding atmosphere, many artists do not appreciate being placed under the trip-hop tag and will describe their music as 'Illbient', 'Ambient hip-hop', 'British hip-hop' or 'Jazz hop'.

MUSICAL ANALYSIS

Actually defining trip-hop for musical analysis is difficult because, as mentioned, most artists will flatly deny that they produce this particular style of music. In fact, only Massive Attack, Portishead and Tricky don't seem to mind being labelled as producing the genre. Nevertheless, it can be roughly summarized that trip-hop is commonly produced using an eclectic mix of acoustic and electronic instruments, combining ideas from R 'n' B, hip-hop, dub, ambient, industrial and jazz. This means that it often features instruments such as acoustic pianos, Rhodes pianos, upright basses, clavinets, horns and flutes, along with electric and acoustic guitars. Principally, these are combined to produce an often nostalgic or dark ambience which is helped along further with haunting vocals and samples taken from vintage radio and films.

On the subject of samples, in keeping with its original roots of hip-hop many of the instrumental riffs, melodies and drums are commonly sampled for old records. It's this approach that is often accredited to the creation of 'lo-fi' since these samples are not respectively 'cleaned up' and are often left dirty and gritty even to the point that the vinyl crackle is left evident in the background of the music. This approach has meant that even if the sounds are recorded from a live instrumentalist (which is progressively becoming more common), it's quite usual to dirty them up a little (as mentioned, Portishead in particular are renowned for recording all their instruments to old analogue tapes before submitting them to digital media for editing and mixing). This helps to retain the 'old' feel that is often crucial to the production of the genre.

When it comes to the equipment used by the artists, many of them are particularly nonchalant about what's used, but Portishead are notoriously cagey about both their production techniques and equipment. Nevertheless, producing this style of music does require you to use the 'right' type of instrumentation and effects to produce the atypical feel of the genre. The first of these is access to a large collection of old vinyl records, particularly jazz, a decent record player and a sampler. Due to legal reasons it's immoral for me to suggest that you should rely on sampling from records to produce music but it is important to note that much of this genre relies so much on sampling that many record companies now ask which records have been sampled when the music is submitted.

Of course, these samples are very rarely used 'as is' and it's common to manipulate them in wave editors, or more specifically, sample slicing programmes such as Wavesurgeon. This is predominantly the case with the drum rhythms as these, more than any other aspect, are commonly sampled to help attain the feel of the genre. Generally, it's the break beat of the record that is sampled (i.e. the middle eight of the original record where the drummer goes off on one for a couple of bars) which is then cut up, rearranged. If required, the tempo is reduced or increased to between 100 and 120 BPM to form the basis of the requisite relaxed feel. That said, it should be noted that while sampling and rearranging breaks is the most common form of creating a loop, it is not the only method employed. The rhythms can also be programmed in drum machines or sequencers. It all depends on the artists working on the project and their methodology, so we'll examine the principles behind generating new grooves with both applications.

The most popular process, as touched upon, is to sample breakbeats from old jazz records which can then be sliced, diced and rearranged sample slicing programmes, such as Wavesurgeon or they can be sliced and have each individual sample mapped to a key in the sampler to be played live from a keyboard. Although sampling from vinyl will often produce quite dirty timbres, it's often the case that they're not filthy enough, so when first sampling the breakbeats it's advisable to sample at a much lower rate than CD quality (most trip-hop musicians use the E-Mu SP1200 or the Casio FZ due to the poor sampling quality). The typical resolution is 12-bit 22 kHz, but reducing the bit rate further to 8 bit may provide even better results, depending on the source. Alternatively, when using sample CDs in Wav or AIFF format you can often lower the rate in most wave editors.

While sampling at a lower quality will grit up the beats to a good degree, it's worth manipulating the sounds further using EQ, filters, transient designers and distortion to acquire the sounds you need. The use of these effects is purely down to experimentation, but it's often worth inserting a compressor directly after them so you can go as mad as you like without the fear of overloading the mixers inputs/outputs.

Of course, sampling beats from old records brings up numerous copyright issues, so it's sometimes worthwhile programming your own drum timbres and contorting them with effects into sounds that are more suited to the genre. These timbres can be sourced from almost anywhere, including the jazz kit from GM modules or more commonly a mix of the Roland TR808, TR909 and the E-Mu Drumulator. As always, however, you can program all these timbres in any capable synthesizer and then experiment with the various parameters on offer before applying effects to construct the atypical 'trip-hop' kit.

The trip-hop kick drum can range from being quite low and boomy to quite bright and snappy: it depends on how you want the loop to appear. As a good starting point, the kick can be produced with a 100 Hz sine wave that's modulated with a positive pitch envelope set to an immediate attack and a medium decay. If you need it to be a little tighter, then reduce the decay and place a small transient click at the beginning of the waveform. A good way to produce this click is to sample any of the GM drum kicks and, using a wave editor, remove the tail end of the sample before dropping it over the top of the programmed kick. Alternatively, it can be produced using a square wave with an immediate attack and very short release for the amp envelope. Once programmed, this can be laid over the top of the sine wave and can be tuned up or down to produce the basic timbre. It's important to note here that you're only after a fairly basic kick drum timbre at this point, since after the rhythm has been programmed it will be affected further to produce the finished article.

Moving onto the snares, these tend to be quite bright and snappy, which is often accomplished by filtering and augmenting the timbre with small amounts of controlled distortion, but the basic sound can be produced with nothing more than a triangle wave mixed with a white noise waveform. An amp envelope set to an immediate attack with no sustain or release and a very short decay is used to shape the triangle, while a second amp envelope is used to shape the noise waveform. This uses the same attack, sustain and release parameters but, by setting the decay a little longer than the triangle, the noise will ring out beyond the triangle and gives you independent control over the triangle and noise to create the timbre you're after. As the final icing on the cake, a band-pass filter can be used to remove the low-end frequencies along with some of the high-end noise to produce a typical trip-hop snare. On the other hand, on occasion once the snares have been created, the tail is removed so that only the initial transient is left, producing a click rather than a thwack. Massive Attack and Portishead have both used this technique.

This same sample cutting technique is also commonly employed on the high hats to produce a timbre that ticks rather than hisses. In fact, some trip-hop tracks have actually used samples of the ticking of a clock to produce the timbre. These can of course be synthesized quite easily using any synth. Although ring modulation does produce better hi-hat timbres, for trip-hop white noise is perfectly sufficient. To create this simply select white noise as an oscillator and set the filter envelope to a fast attack, sustain and release with a medium to short decay. Once this is done, set the filter to a high pass and use it to roll-off any frequencies that are too low to create a high hat. Once this basic timbre is down, lengthening the decay parameter can create open hi-hats, and shortening it will produce closed hats. By setting this very short you can also produce the typical ticking timbre. Alternatively, if samples are used it's prudent to either use the amplifier's decay envelope in the sampler or use a wave editor to remove the subsequent tail of the hats.

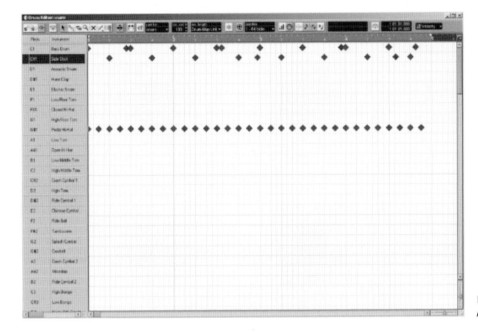

FIGURE 16.1
A trip-hop rhythm

Whether the grooves are sampled or programmed, the most fundamental aspect of creating the typical groove lies with the pattern. All trip-hop derives its feel from hip-hop and jazz and so relies on a relatively sparse drum arrangement with the laid-back feel produced by the interplay between the kick and the snare. As a consequence, the often-used kick/snare/kick/snare pattern is often (but not always) avoided. Generally speaking they commonly feature more rhythmic kick patterns, which are augmented with the occasional snare. Of course, closed hi-hats, ride cymbals and pedal hi-hats are also employed, along with the occasional open hat or crash cymbal, all of which add some syncopation and steadiness to the rhythm (Figure 16.1).

As mentioned, snares are not always used in the production of trip-hop and any 'clicky' percussive timbre can be used in its place. What is important, however, is that the entire loop is quite reserved in terms of instruments and patterns. In many cases these are kept incredibly simple to allow room for the delicate vocals, chords and lead instruments to play over. Also, it's important to bear in mind that many of these elements will sit just off the beat rather than strictly quantized on it, so that the rhythm appears to be played live. In fact, the real key to producing these rhythms is to keep the drums as simple as possible and play them live from a workstation or keyboard, and then use functions such as iterative quantize to prevent the pattern from becoming too wayward.

Velocity also plays a large role in the creation of the rhythm to keep the realism. The most common method of producing a live beat is to implement the *strong–weak–medium–weak* syncopation on the kick drums, although the *actual* velocities may stray wildly from this guideline depending on the sound *you*

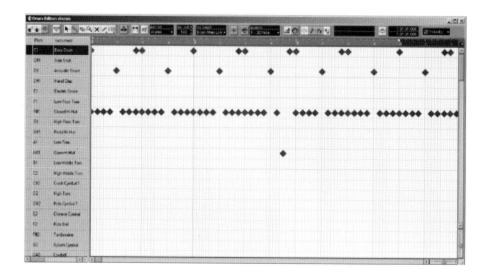

FIGURE 16.2
A trip-hop rhythm
using snares

wish to achieve. The snares too are often subjected to different velocities, with the first and final snares using the highest values and those in between using a mixture of different values (Figure 16.2).

On top of this, many trip-hop tracks also often employ small snare rolls in the rhythms to add to the flow of the groove. These are often played monophonically, however, and very close together so that the consecutive snares remove the tails of those that precede them. This can be accomplished by setting the synth or sampler to monophonic operation. Alternatively, they can be programmed in audio by physically cutting off the tails of the snares and positioning them one after the other. For added interest these can also be subjected to pitch bend so that the snares roll downwards in pitch. This downwards roll is often preferred to pitching upwards since a movement down the scale tends to sound less uplifting (Figure 16.3).

As always, these are only generalizations and are fully open to further experimentation. Indeed, after programming a typical loop it's quite usual to use a series of effects and processors to seriously grunge up the audio. The first of these techniques involves speeding up the tempo of the rhythm from the usual 100–120 BPM to 150 or 160 BPM. This loop is then sampled at a low-bit rate, typically 12- or 8-bit, set across the keyrange of the sampler and then played in the lower registers so that not only does the tempo slow down but the pitch also becomes significantly lower. Alternatively, the loop could be sampled at the original tempo and then time-stretched numerous times by extreme amounts each time. The more this is applied, the more the loop will begin to degrade as each subsequent time-stretching algorithm will impart some degradation into the audio.

Another technique involves playing the loop through a speaker system and miking up the speakers using a poor-quality microphone and pre-amp. This

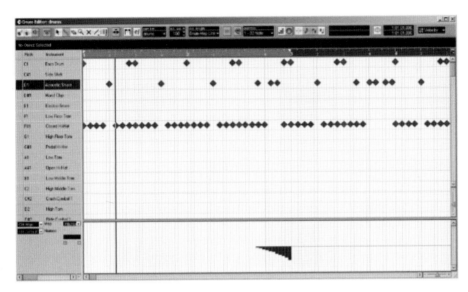

FIGURE 16.3
A typical trip-hop snare roll augmented with pitch bend

can often add sufficient dirt to anything that is recorded, and you can experiment by moving the placement of the microphone. You may also capture small amounts of incidental sounds such as traffic outside using this method which may actually enhance the overall lo-fi effect. Similarly, recording the loop down to an old (and hopefully knackered) analogue tape machine will introduce adequate hiss and sound degradation that can then be re-recorded into the audio editor. So long as any of these techniques are followed by a compressor, you should feel free to experiment wildly.

On the subject of compression, trip-hop drum loops benefit hugely from incredibly heavy compression. In many instances this contributes a great deal to the overall sound. Although there are no definitive settings – depends on the loop and the timbres being used – a good starting point is to set a very low threshold with a ratio of approximately 10:1. Once these are set, experiment with the attack and release settings. While generally the idea is not to create a loop that pumps, you need to compress the loop as hard as possible by trying out various settings so that the loop becomes as 'heavy' as possible.

With the basis for the drums laid down, it's quite usual to follow these with the chords or leads rather than with the bass. Since trip-hop draws its inspiration for bass lines from dub (and in some cases hip-hop), they remain relatively simple and only act as a basic underpinning for the chords and lead to interweave with. As a consequence, it's usually prudent to lay down the chord structure first, followed by the leads and vocals before finally developing the bass to sit around them.

Generally speaking the chords and most of the overlaying instruments are usually written in a minor rather than major key since, as we've already touched upon in music theory, these provide a more serious and sometimes dark feel to the music. Naturally, the actual chord structure is up to your own artistic license, but generally chord structure does not involve large amounts of movement through the pitch range and tends to stay rather static, using simple inversions rather than jumping manically from one key to another. A good starting point for these is to write a chord in A minor and then create some inversions based around this. That said, it is often worth moving between consonant and dissonant chords to produce a cacophonous feel to the music. Portamento is also useful here, so experiment by overlapping notes slightly and using it so that the pitch sweeps up and down between the notes of the consecutive chords. This will invariably produce more laid-back results than a chord sequence that triggers suddenly on each new note.

The timbres used for the chords can range from samples of a real string section (ensure that you write these in the right key – violins, for example, cannot play any lower than the G below middle C!) to synthesized pads that are swept with a low- or high-pass filter. If you decide to take the latter approach, the timbre to use is down to your discretion, but we can nonetheless look at the construction of a basic pad which can then be contorted further to suit your music.

Typically, the pad in trip-hop is fairly rich in harmonics, allowing it to be swept with a low-pass filter if that is required to add some weight, or alternatively swept with a high-pass or band-pass filter if the idea is to produce a ghostly backing. A good starting point for this is to use either two sawtooth waveforms or a sawtooth and a pulse wave. Detune one oscillator from the other by 5 or 7 cents and then set the amp envelope to use a fast attack with a short decay, medium sustain and a fast release. If you want to sweep the filter (keep in mind that a swept low-pass filter will use a proportionate amount of frequencies in the mix) then set the filter's envelope to a slow attack and decay with no sustain or release and set this to positively modulate the filter's cut-off. Otherwise, use a high-pass or band-pass filter or, to keep the timbre static (at this point), use the same settings as the amp envelope with a slightly longer decay on the filter and set the filter key follow to positively track the pitch. If the pad is not being swept and consists of long sustained notes in the chord structure then it's prudent to add some movement to the timbre by using an LFO to augment the pulse width or the resonance of the filter. The waveform, rate and depth to use on the LFO will obviously depend on what you wish to achieve, but generally a square, triangle or sine wave will produce the most 'natural' results.

Although any effects should generally be avoided this early since they tend to occupy a significant amount of space within a mix, if the pad is to play a large role in the music then it may be worth applying effects now to produce the finished timbre. Typical effects for this include heavy chorus or light reverb, but it's worth experimenting with other effects to produce the results you want. One of the classic effects to use on a pad in trip-hop is a vocoder. If your employ the tracks drum loop as a modulator and use the pad as a carrier wave, the drums will impose a new character onto the pad, forcing it to flow in and out of the mix with the drums. Alternatively, you can make a copy of the drum track, move it forward in time and use this to modulate the pad through the vocoder.

This results in the pad 'pulsing' between the kicks of the original loop creating a forced syncopation. And of course, as a final stage, the pad will need to be sampled at 8- or 12-bit to produce the archetypal sound of the genre.

With the chord structures down, the vocals and lead melodies can be laid over the top. Roughly speaking, the vocals are usually recorded before any lead instrumentation since these instruments are often kept very simple to prevent detracting the attention from the vocals. In fact, the vocals play such a fundamental role in trip-hop that it's rare to hear of a track that doesn't feature any. Usually, mellow, laid-back female vocals are used, but occasionally they are performed by male singers, although if this is the case they tend to be more rap (i.e. poetic) based than actually sung.

The typical microphone for recording trip-hop is a large diaphragm model such as the AKG C414, which is amplified with a valve pre-amp to add the requisite warmth. It should be noted that this is not necessarily a requirement. Indeed, since this genre relies on a gritty, dirty feel the vocals can be recorded with pretty much anything so long as they're decipherable. Even a cheap dynamic microphone from the local electronics store recorded direct into a portable cassette recorder such as a Walkman can produce great results.

If this latter approach is used, then generally you will not need a compressor while recording as the cassette tapes will reach magnetic saturation before clipping is introduced. But if you're recording into a digital system, compression is vital to prevent any clipping. At this stage it should be applied lightly since it cannot be undone later. A good starting point is to set a threshold of −8 dB with a 3:1 ratio and a fast attack and moderately fast release. Then, once the vocalist has started to practice, reduce the threshold so that the reduction meters are only lit on the strongest part of the performance.

Once the vocals have been recorded (whether to analogue tape or hard disk) they will invariably require more compression to even out the levels so that they don't disappear behind other instruments in the mix. As always, the settings to use will depend on the initial performance, but to begin with use a threshold setting so that you have −7 dB on the reduction meter, along with a ratio of 3:1 and a fast attack and moderately fast release (approximately 200 ms). If you're after a more commercial sound, then try increasing the ratio to 5:1 but exercise caution and use your ears to decide whether it's right or not. Keep in mind that trip-hop is all about generating atmosphere, and much of this is derived from the vocals. If these are too heavily compressed, the ratio between the peak and average signal level will be seriously compromised, which can remove all emotion from a recording. If they sound a little thin in the context of the mix, then small amounts of reverb may help, but similar to compression this should be kept to a minimum to retain the feel of the music. In fact, reverb should be applied so lightly that its application is only noticeable when it's removed. Alternatively, making a second copy of the vocal track and pitching this up or down by 5 semitones can produce harmonies which can help fill out the vocals.

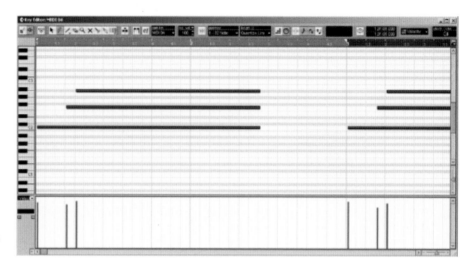

FIGURE 16.4
A typical Rhodes piano
lick

With the vocals in the track, it's much easier to hear gaps in both frequency spectrum and arrangement, and these can be filled out with the lead instrumentation. As already mentioned, lead instrumentation very rarely includes complicated riffs so as not to draw attention, but simply adds to the atmosphere of the music. As a result, in the majority of tracks leads consist of single hits at the beginning of the bar or extremely simple melodies which appear as though they've been played ad-lib over the chords and vocals. The old adage of 'less is more' is certainly the case with most trip-hop. Fundamentally, these tend to be played with real rather than synthetic instruments and a proportionate amount of tracks will use clavinets, organs, horns, theremins or flutes but more commonly Rhodes pianos and electric guitars that are treated heavily to tremolo effects (i.e. pitch modulation), phasing, flanging, distortion and a whole host of effects. In actual fact, as these leads play such simple riffs it's the effects applied that add to the atmosphere of the music and so experimentation is the key (Figure 16.4).

Most artists rely on guitar pedals to produce most of the effects and ideally, it's wise to follow suit. Since these are essentially produced to be used live on stage, they don't have a particularly low noise floor, so not only are they powerful effects, they're also pretty noisy. You can pick up guitar pedals new for around £50 and they can change hands for as little as £20 on the used market. Generally, you'll need tremolo, flanger, reverb and delay pedals to begin with, as these are all suited and used in the production of the genre. You can, of course, use digital or plug-in effects units in their place, but these tend to be too clean so you'll need to grunge them up a little.

Due to the sparseness of trip-hop, delay is the most commonly used of these effects. This can be subsequently soiled by sending the audio out to a delay unit set to a high feedback and sampling the subsequent delays. Once you have these, import them into a separate track in an audio sequencer and, using a mix of EQ and distortion, automate the EQ to gradually remove both the high and low frequencies on each delay while increasing the distortion applied to

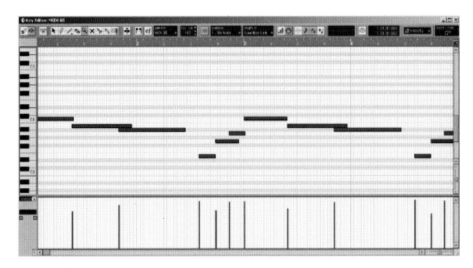

FIGURE 16.5
A typical trip-hop bass
rhythm

each. Export this as another audio file and then use the scissors tool (or similar) to cut up the delays and place them where required in the arrangement. If the sequencer does not allow automation (or you're using hardware), you can simply record the EQ and distortion tweaks to audio cassette (or DAT) and then edit the results in the sequencer.

The final element of trip-hop is the bass, but before proceeding any further it should be noted that not all tracks will feature a bass line, and many that do will use an upright bass that can also act as a lead instrument. Trip-hop is a descendant of hip-hop mixed with elements of jazz and ambient, and as such it is totally unnecessary to fill up the entire frequency spectrum and arrangement with sounds. In fact, the relatively large spaces that are left between instruments help in creating the all-important atmosphere of the music. This is important to keep in mind when writing the bass since it can be quite easy to get carried away and produce a bass with far too much movement.

Fundamentally, trip-hop draws the basses from dub, so on the whole basses are kept relatively simple as they merely act as the underpinning of the track and should not form a major melodic element in themselves. This is mainly to prevent the bass from drawing attention away from the vocals. Consequently, a bass line will often consist of nothing more than a few 1/8th or perhaps 1/4th notes per bar that will tend to stay at one pitch or move by no more than 3, 5 or 7 semitones. Simplicity is the solution here. More importantly, since the bass is often played live, when programmed in MIDI, the notes should very rarely occur at the beginning of each bar. It's much more natural for the bass to begin just a few ticks *before* or *after* to offset it from the rest of the record. This not only emulates the nuances of a real bassist but it also makes the drums appear to pull or push the record forward, helping to create a more flowing groove. Creating this relationship can form an essential part of the music but whether to sit it before or after the beat is entirely open to artistic license. After programming it's worth experimenting by moving it before or after the beat to see which produces the best results (Figure 16.5).

For the timbre, real bass guitars, particularly upright and acoustic, are usually chosen due to the resonant sounding strings which help it cut through the mix. These sounds are best sourced from a sample CD or bass-specific module such as *Spectrasonics Trilogy* since recording live guitars is difficult without good equipment. This subject is beyond the scope of this book as it requires a whole new set of production ethics. Nevertheless, provided that you have a good bass module it is perfectly possible to emulate a real guitar with MIDI. While it does involve a lot of work, it is only as much as is involved with recording a real bass guitar – it just doesn't take as much equipment. The details of emulating a bass with MIDI have already been discussed in Chapter 15, but for the benefit of those who've jumped straight to this chapter you can play the book- style version of Dungeons and Dragons by jumping back and forth, depending on what information you want, or just read on.

The key to programming any real instrument is to take note of how it is played and then emulate this action with MIDI and a series of CC commands. In this case, most bass guitars use the first four strings of a normal guitar E–A–D–G, but these are tuned an octave lower, resulting in the E being close to three octaves below middle C. Also they are monophonic, not polyphonic, so the only time notes will actually overlap is when the resonance of the previous string is still dying away as the next note is plucked. This effect can be emulated by leaving the preceding note playing for a few ticks while the next note in the sequence has started. The strings can either be plucked or struck and the two techniques produce different results. If the string is plucked, the sound is much brighter and has a longer resonance than if it were simply struck. To copy this, the velocity will need to be mapped to the filter cut-off of the bass module so that higher values open the filter more. Not all notes will be struck at the same velocity, though, and if the bassist is playing a fast rhythm the consecutive notes will commonly have less velocity since he has to move his hand and pluck the next string quickly. Naturally, this is only a guideline and you should edit each velocity value until it produces a realistic feel.

'Bassist's may also use a technique known as 'hammer on' whereby they play a string and then hit a different pitch on the fret. This results in the pitch changing without actually being accompanied with another pluck of the string. To emulate this, you'll need to make use of pitch bend. First set it to a maximum bend limit of two semitones, since guitars don't easily 'bend' any further than this. Begin by programming two notes – for instance an E0 followed by an A0 – and leave the E0 playing underneath the successive A0 for around a hundred ticks. At the very beginning of the bass track, drop in a pitch bend message to ensure that it's set midway (i.e. no pitch bend) and just before the second note drop in another pitch-bend message to bend the tone up to A0. If this is programmed correctly, on playback you'll notice that as the E0 ends the pitch will bend upwards to A0, simulating the effect. Although this could be left as is, it's sensible to drop in a CC11 message (expression) directly after the pitch bend, as this will reduce the overall volume of the second note so that it doesn't sound like it has been plucked.

In addition to this, it's also worthwhile employing some fret noise and finger slides. Most good tone modules will include fret noise that can be dropped in between the notes to emulate the bassist's fingers sliding along the fret board. The pitch bending is best emulated by first programming the notes to overlap slightly and then recording movements of the pitch-bend wheel live and editing them in the sequencer. Again, since the pitch-bend range is set to 2 semitones only the MSB value will need editing if the live recording didn't go according to plan.

Of course, you don't particularly have to use a real bass guitar and, as ever, it's open to artistic license. Pretty much any low frequency sound can be used provided that it sounds dark, moody and atmospheric. As we all have different ideas of what constitutes a moody, atmospheric bass, there are no real generic timbres and practically any synthesizer can be used to programme the basic sound. As a recommendation, the *Access Virus*, *Novation Supernova* or the *Novation Bass Station* (VSTi) are particularly suited towards creating deep, atmospheric bass sounds.

Fundamentally, a synthesized bass for trip-hop should exhibit plenty of low-end presence (dub influence) yet at the same time not be so powerful that it takes up too much of the mix (hip-hop influence). To accomplish this, a good starting point is to use three oscillators: one set to a sawtooth waveform, the second set to a sine wave and the third using a triangle wave. Transpose the triangle wave up an octave and then adjust the amp's EG to a fast attack, decay and release but a high sustain. This will allow the bass note to sustain while the key is depressed (many trip-hop rhythms depend on long notes rather than short, stabby ones). If a resonant pluck is needed for the beginning of the note, using a low-pass filter, set the cut-off quite low with a high resonance and use a filter EG with a zero attack, sustain and release with a medium decay. Finally, set the filter envelope to full positive modulation and then experiment by detuning the sine wave downwards from the saw. The further away this is detuned the deeper and more 'full' the bass will become. If the notes are sustained for a lengthy period of time then it may also be worth applying very small amounts of LFO modulation to the filter or pitch of one or all of the oscillators, depending on the type of sound you want to achieve.

As with most other instruments in this genre, the bass can also benefit from effects, but these must be chosen and applied carefully. Any stereo effects such as flangers and phasers will often spread the sound across the image (which will in turn move the bass from the centre of the mix and consequently destroy the stereo perspective) so generally only mono effects should be applied. This can include distortion, EQ and filters, but if you are using any of these it's useful to place a compressor directly after so that you can experiment by overdriving the signal without fear of actually overdriving the desk (although this can be put to creative uses if it's an analogue desk).

As touched upon throughout, the most fundamental aspect of creating trip-hop is the dirty/gritty character that the whole mix exhibits. While this is certainly not an excuse for poor mixing, it is a good reason to be experimental and push things further to attain the character of the sounds. For instance, record everything to analogue tape rather than direct to hard disk, or create wild feedbacks by sending an audio signal to one channel of a stereo effect then return the signal back into two inputs of the desk and feed one of these back out to the other channel of the effect. Alternatively, feed an effect as usual but return the outputs into a normal mixing channel and then feed these down another aux send back into the effect. Also feel free to experiment with very heavy EQ to thin sounds right down, or use filters such as the Sherman Filter Bank to warp sounds beyond comprehension. So long as each of these processes is followed by a compressor to keep the levels under some control there should be no restrictions. What's more, even if the elements of the record are *not* sampled, they should nevertheless sound as though they are. This means gratuitous use of bit reduction, sample reduction along with sampling the vinyl crackle from a record and applying this over the top of the mix. For added hiss, a popular method is to record a source at a relatively low level and then increase the gain so it comes up to nominal level, thus also increasing the noise floor. Furthermore, you may wish to play your final track through a pair of poor-quality speakers and record the sound output with a poor-quality microphone (as suggested for individual sounds). By importing this sound back into your sequencer you will have further material to edit and add back to your original track, creating areas of lo-fi double-tracked sound.

Once the basic elements are programmed, you can begin to lay the arrangement down. Since trip-hop relies heavily on vocals, the music tends to be structured in a similar manner to most popular music songs consisting of a verse/chorus structure. As such, it can roughly be broken down into four distinct parts consisting of the verse, chorus, bridge and middle eight. If we break this down in a number of bars we can derive that the order of a song and subsequent bars are as follows:

- Verse 1 – *Commonly* 16 bars
- Chorus – *Commonly* 8 bars
- Verse 2 – *Commonly* 16 bars
- Chorus – *Commonly* 8 bars
- Verse 3 – *Commonly* 16 bars
- Chorus – *Commonly* 8 bars
- Bridge – *Commonly* 1 bar
- Middle eight – *Commonly* 8 bars
- Double Chorus – *Commonly* 16 bars

This is, of course, open to artistic license and some artists will play the first two verses before hitting the chorus. This approach can often help to add excitement to a track as the listener is expecting to hear the chorus after the first verse. If this route is taken, even though there may be different vocals employed in the verses, if they are played one after the other it can become tiresome, so a popular trick is to add a motif into the second verse to differentiate it from the first. Also, keep in mind that this is only a general guideline and as always the best way to gain a good representation of how to arrange a trip-hop track is to actively listen to what are considered the classics and then copy them. This isn't stealing, it's research and nearly every musician on the planet follows this same route.

> The data CD contains a full mix of a typical Trip Hop track with narration on how it was constructed.

RECOMMENDED LISTENING

Ultimately, the purpose of this chapter has been to give some insight into the production of trip-hop. There is, of course, no one definitive way to produce the genre. Indeed, the best way to learn new techniques and production ethics is to actively listen to the current market leaders and be creative with processing, effects and synthesis. With this in mind, what follows is a short list of some of the artists who, at the time of writing, are considered the most influential in this area:

- Tricky
- Massive Attack
- DJ Shadow
- Portishead
- Red Snapper
- DJ Food
- DJ Cam
- DJ Kicks
- Thievery Corporation
- DJ Krush
- Coldcut
- Moloko
- DJ Vadim
- Herbalizer
- 9 Lazy 9

CHAPTER 17
Ambient/Chill Out

'Ambient is music that envelops the listener without drawing any attention to itself...'

Brian Eno

Ambient music has enjoyed a long, if somewhat diverse, history and its subsequent offshoots have formed an important role in dance music since 1989. However, it only recently re-established itself to many as part of the dance music scene when beats were again dropped over the atmospheric content and it was relabelled by record companies and the media as 'chill out' music.

The roots of ambient music are nothing if not unusual. It's believed that it first came about in the mid-1970s when Brian Eno was run over by a taxi. While he was recovering in hospital, a friend gave him a tape machine with a few dodgy tapes of harp music. The result was that music didn't remain at a constant volume and on occasion dropped considerably in gain whereby it mixed with the rain hitting the hospital windows. This second accident formed the beginning of ambient as Eno began to experiment by mixing in real-world sounds such as whale song and wind chimes with constantly changing synthesized textures. Described by Eno himself as music that didn't draw attention to itself (go figure), it enjoyed some success but was soon relabelled as 'Muzak' and subsequently poor imitations began to appear as background music for shopping centres and elevators, and to many the genre was soon written off as 'music suitable only for hippies'.

When the rave generation emerged in the late 1980s and early 1990s, a DJ named Alex Patterson began to experiment with Eno's previous works, playing it back to clubbers in small side rooms who needed a rest from the fast, hard-hitting beats of the main room. These side rooms were soon labelled by clubbers as 'chill out' rooms, a place where you could go and take a break from the hectic four to the floor beats. As these 'chill out' rooms began to gain more popularity with clubbers, Patterson teamed up with fellow musician Jimmy Cauty to form The Orb and they jointly released what many believe to be the first ever ambient house music for clubbers, somewhat strangely named *A Huge Ever Growing Pulsating Brain That Rules from the Centre of the Ultraworld*. Soon after the release of the album, Patterson and Cauty went their separate ways. While Cauty teamed up with Bill Drummond to from KLF, Patterson continued to write under the moniker of The Orb and to DJ in the chill out rooms.

This form – ambient house – began to grow into its own genre and chill out rooms became a fundamental part of the rave scene. Some DJs became VJs (video jockeys), mixing not only real-world sounds with slow, drawn-out drum loops but also projecting and mixing images onto large screens to accompany the music. This was soon followed by a series of ambient compilations hitting the public market, and artists such as William Orbit and Aphex Twin soon released their own ambient house albums.

In 1992, the genre was in full flow. As different artists adopted the scene, each putting their own twist on the music, it began to diversify into subgenres such as ambient dub (ambient music with a bass), conventional (ambient with a 4/4 backbeat), beatless (no backbeat but following the same repetition as dance music) and soundscape (essentially pop music with a slow, laid-back beat). By 1995, ambient music was everywhere, as the larger record companies took the sound on board and saturated the market with countless ambient compilations (although thankfully there was never any *Now That's What I Call Ambient Music* Volume 390). Even artists who had ignored the genre before began to hop on board in the hopes of making a quick buck out of the new fad.

As with most music that is encapsulated and bastardized in this way, eventually ambient house became a victim of its own success. The general public became tired of the sound. To the joy of many a clubber, it was no longer the new fashion and it returned to only being played where it had originated from – the chill out rooms.

Fast forward to the year 2000. A small Balearic island in the middle of the Mediterranean began to revive the public's and record companies' interest in ambient house. DJs in Ibiza's Café Del Mar began to tailor music to suit the beautiful sunsets by mixing Jazz, Classical, Hispanic and New Age together to produce laid-back beats for clubbers to once again chill out to. Now repackaged and relabelled 'chill out music' it's enjoying renewed interest and has become a genre in its own right. However, while chill out certainly has it roots deeply embedded in ambient music, they have over time become two very different genres. Thus, as the purpose of this book is to cover dance music, for this chapter we'll concentrate on chill out rather than ambient.

MUSICAL ANALYSIS

As always, possibly the best way to begin writing any dance music is to find the most popular tracks around at the moment and break them down to their basic elements. Once this is accomplished, we can begin to examine the similarities between tracks and determine exactly what it is that differentiates them from other musical styles. In fact, all music that can be placed into a genre-specific category will share similarities in terms of the arrangement and/or sonic elements.

Generally, chill out is music that incorporates elements from a number of different styles such as Electronica, New Age, Classical, Hispanic and Jazz. However, it's this very mix of different styles that makes an *exact* definition impossible. Indeed, it could be said that as long as the tempo remains below 120 BPM and it employs a laid-back groove, it could be classed as chill out. In fact, the only way to analyse the music is to take Brian Eno's advice in that it shouldn't draw too much attention and ideally be the type of inoffensive music that most people could easily sit back and relax to. (Sometimes the best chill out can be derived from just rolling up a fat err… cigarette and playing about.)

Defining exactly what this means is difficult, but we can settle for saying that chill out commonly has a slow rhythmic or trance-like nature, combining both electronic and real instruments which are often backed up with drop-out beats and occasionally smooth, haunting vocals. Generally, many of these real instruments will be recorded live or sampled from sample CDs, but in some instances they can also be 'borrowed' from other records. Consequently, chill out can utilize almost any time signature, from the four to the floor, to a more swing orientated 3/4, and be produced adagio, andante or moderato. In terms of physical tempo, this can range from 80 BPM and move towards the upper limits of 120 BPM, but most chillout tends to use a happy medium, staying around the 110–120 BPM mark. This, of course, isn't a demanding rule but it is important not to go too fast as you need to relax to it!

To better explain how this theory is applied in practice, we'll look at the creation behind a typical chill out track. As always, any music is an entirely individual, artistic endeavour, and the purpose of this analysis is not to lay down a series of 'ground rules' on how it should be created. Rather, my intention is to describe some of the general principles behind the characteristic arrangements, sounds and processing. Indeed, in the end it's up to the individual (i.e. you) to experiment and put a twist on the music that suits your own particular style.

THE RHYTHMS

There are no particular distinctions as to what makes a good chill out drum loop. The only guide that can be offered is that the loop should remain relatively simple and exhibit a laid-back feel. The kick drum can lie on the beat, every beat, or it can be less common, appearing on the second and fourth, or first and third beat, or any division thereof. This is simply because the laid-back feel is often derived from the pattern and position of the snares in relation to the kick. Indeed, it's the rhythmic interplay between these two elements that creates the 'feel' of the drum rhythm. Additionally, closed and open hi-hat

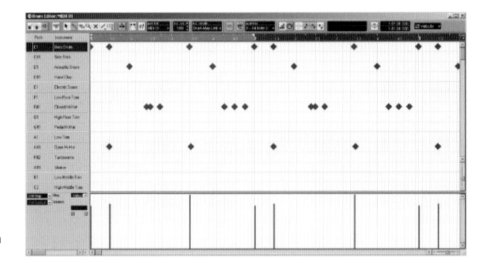

FIGURE 17.1
Typical chill out drum loop

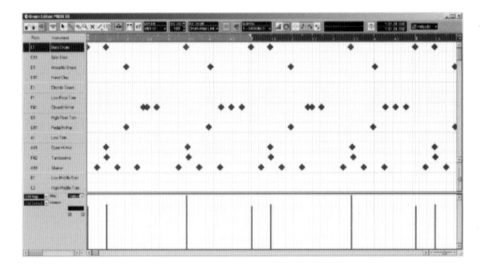

FIGURE 17.2
A more complex chill out drum loop

patterns are invariably employed in these rhythms to act as a form of rhythmic syncopation (i.e. not on the beat) to help the groove flow along (Figure 17.1).

As the diagram in Figure 17.1 shows, the closed hi-hats, unlike most other dance genres, do not necessarily have to sit on every 1/16th division, and along with the open hats they can play a pattern of their own. It is wise, however, to remove any closed hi-hats that occur on the same position as the open hi-hat, as this results in a frequency clash and can often loose a rhythm to sound too 'programmed'. Auxiliary instruments such as congas, bongos, toms, cabassas, triangles, shakers and tambourines also often make an appearance in chill out, and similar to the hi-hat patterns these act as a form of syncopation to help the rhythm flow (Figure 17.2).

This, of course, is only a general guideline to the drum patterns used and it's fully open to artistic license. The key is to produce a loop that does not sound rigid or programmmed through experimenting by moving the kicks in relation to the snares to adjust the interplay between the two. If the tempo appears too fast, reducing the amount of snares or auxiliary instruments employed in the rhythm will often help slow it down and is preferable to physically slowing down the tempo, since this will affect the rest of the instrumentation laid on top. Additionally, as this genre occasionally follows a verse chorus structure similar to most 'pop' music to denote the change between the two, it's common to employ a couple of extra snare hits positioned together at the end of the bar to create a skip in the rhythm.

Of particular note, the velocities of each percussive instrument should be manipulated to help create a more flowing groove. Although this is entirely open to interpretation, the kick commonly follows the typical *strong–weak–medium–weak* syncopation. This is, of course, provided that there are four kicks per bar. If there are three, then it could follow a *strong–weak–medium* or perhaps a *strong–medium–strong* syncopation. What is common, however, is that the first kick of the bar remains the strongest so as to denote the beginning of a new bar. Even then, this is simply convention and convention shouldn't always play a part in your music, so you should feel free to experiment by placing the accents at different divisions on the beat. When you do this, the rhythm of the piece can change quite severely, so it is worth experimenting.

With chill out, the sounds used for the rhythm often play a vital role in obtaining the laid-back feel. Once the basic sounds have been laid down it's quite usual to apply effects to the individual elements. The hits can be sourced from anywhere, including other records, most drum machines (including the TR909 and TR808) or even the standard GM drum kits (particularly Jazz kits). It is, of course, possible to synthesize a kit. This can often prove the best method since the available parameters will allow you to modify each sound to suit any particular rhythm. As we've already discussed the basics behind creating and synthesizing a drum kit, what follows is a short overview of the three most important elements (kicks, snares, hi-hats) with some tips on how the parameters can be used to create typical 'laid-back' timbres. Keep in mind that these are only tips and above all, however you do it, if it sounds right then it most probably is.

RHYTHM TIMBRES

Generally, the kick is quite deep and 'boomy' as tight kicks tend to add some small urgency to the music. Thus a good starting point is a 40 Hz sine wave with positive pitch modulation using a fast attack and a medium-to-longish decay period. There is usually little need to employ a 'click' at the beginning of the waveform since this makes it appear sharper, which isn't *generally* required but, as always, experimentation is essential. If possible, it's also worthwhile experimenting by setting the decay's transition to exponential so that the pitch curves outwards as it falls, creating a 'boomier' sound. Additionally, it's often worth sending the kick to a reverb unit set to a medium pre-delay and short tail to help lengthen the timbre and prevent it from dropping off too sharply.

If you decide to use a kick sample or a GM module and there is too much of an attack on the sound, a compressor set so that it clamps down on the transient may produce the sound you require. That said, in the long run a preferred option would be to use a transient designer such as featured in the Waves *Transform Bundle* plug-in or SPL's hardware *transient designer*. These can be used to remove the front of the kick to produce a more relaxed-sounding timbre.

In direct contrast to the kicks, the snares often use a sharp transient stage, as these dictate the rhythmic flow of the loop and need to be apparent in the loop – more so if it's a particularly busy loop packed with auxiliary instruments. The typical snare can be created by using a triangle oscillator mixed with pink noise. A high-pass filter across these generally produces the most common chill out snares, but it's also worth trying a notch or band-pass filter to see which you prefer. The amps attack should be set fast while the decay can be set anywhere from medium to long depending on the sounds that are already in the mix and how slurry you want the snare to appear.

Alternatively, if it's possible with the chosen synthesizer, it's prudent to employ a different amp EG for both the noise and the triangle wave. The triangle wave can be kept quite short and swift by using a fast decay, while the noise can be made to ring a little further by increasing its decay parameter. The further this rings out, the more slurred and 'ambient' it will become. If the snares are too bright, brightness can be removed with some EQ or, preferably, try replacing the triangle wave with a pulse and experiment with the pulse width.

The snares also often benefit from small amounts of reverb by sending, not inserting, them to a reverb unit. A typical snare room preset that's available on nearly every reverb unit around will usually suffice, but depending on the loop it may be worth reducing the decay slightly. Again, if samples are used in place of synthesis, the Jazz snares on most GM modules produce a fairly good rendition which may be more suited to your track. Using a transient designer along with sending the snare to a reverb unit will often help you achieve the typical 'chilled' feel.

Some chill out will benefit from a snare with a less prominent attack phase which can be accomplished by using a transient designer to remove the attack phase. On top of this, it may also be worthwhile introducing a small amount of pink noise over the top of the snare to give the impression of a brush stick used to play. Alternatively, you could just use a brush kit from the GM standard or sample one from a Jazz record. For obvious legal reasons I can't condone this latter approach but that's not to say that a proportionate amount of dance artists don't do it.

The hi-hats can be created in two ways: with noise or with ring modulation. Typically for chill out, ring modulation produces better results. This is accomplished by modulating a high-pitched triangle wave with a lower pitched triangle to produce a high-frequency noise. As a starting point, set the amp envelope to a zero attack, sustain and release with a short-to-medium decay. If there isn't enough of a noise present, it can be augmented with a white noise waveform using the same envelope. Once this basic timbre is constructed, shortening the decay creates a closed hi-hat, while lengthening it will produce an open hat.

As touched upon, not all chill out will use MIDI programmed loops and some artists will borrow loops or at least elements of loops from previous records. For obvious legal reasons I can't condone this behaviour but nevertheless it is a popular technique. These sampled loops are often imported into a sample slic-ing programme and messed around until it sounds nothing like the original. Of course, you shouldn't just settle with rearranging parts. After they're remod-elled, it's worthwhile importing them into a wave editor and using a transient designer to model the transients and releases of the drum hits. Alternatively, another approach is to use two sampled loops and mix them together in a sequencer. Time-stretch the loops so they share the same tempo, and play the loops simultaneously, and any conflicting elements can be cut-out from one of the loops (Figure 17.3).

This technique is useful if you wish to create particular poly-rhythms that are difficult to accomplish any other way. On top of this, it's worth experimenting by moving one of the loops back in time to make some of the elements of the loop occur in different places.

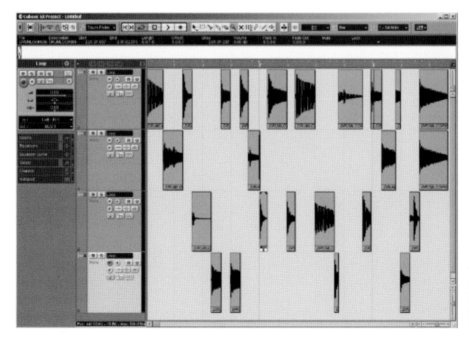

FIGURE 17.3
Lining up a series of
loops and removing
element to create a
basic rhythm

With the loop laid down, you can then look towards effecting and processing the rhythm, although roughly speaking, chill-out loops are not compressed for numerous reasons, including:

- They've already been sampled from a record and are already compressed.
- They are from a synth and are already compressed.
- They don't need to 'pump' musically.
- Second-order harmonic distortion from valve compression isn't usually required.

However, if the loop has been composed of numerous elements from different sources, compression may help to bring some of the levels under control. Typically, as a starting point, set a relatively slow attack with a medium release and a ratio of about 4:1. Then adjust the threshold so that the loop registers approximately 4 or 5 dB on the gain reduction meter. It's usually wise to ensure that the loop doesn't pump, so always experiment with the compressor's release to make its action as transparent as possible.

Although in many genres it's usually worth avoiding applying any effects to drum timbres, chill out is often an exception to the rule. While it's unwise to wash the entire loop in cavernous reverb, if it's lightly applied to the snares and kicks, it can often present a more laid-back feel to the timbres. That said, any reverb or any effect for that matter must be applied cautiously. The drums form the backbone of the record, and if an effect is applied too heavily it will often destroy the transients, reducing the bottom-end groove to mush.

MELODIC LEADS

Although it isn't unusual for a chill out track to be based entirely around synthetic instruments, most do tend to contain real instruments. These can range from acoustic guitars playing Spanish rhythms, Classical strings, pianos, wind or a combination thereof. Roughly speaking, these are best laid down before the bass, as much of the attention of the track is commonly directed towards the lead and the bass simply acts as the underpinning. Since most real instrument leads are commonly sampled from sample CDs or other records, it's much easier to form the bass around their progression, rather than to try to re-programme a lead to sit around a bass. Although it is possible to cut up and rearrange a lead in a sample splicing programme to suit a pre-programmed bass, this can often result in the lead behaving unnaturally, so it's prudent to rearrange the bass instead.

Samples of real instruments do not necessarily have to be used (although in the majority they do sound better when used for leads), and provided that you have a good tone module it's possible to programme realistic instruments through MIDI. If you take this approach, however, it's important to take note of the key and the range you write the song in because every acoustic instrument has a limit to how high and low it can play. For instance, with an acoustic guitar the (standard) tuning is E, A, D, G, B, E with this latter E a major third above middle C. This is incredibly important to keep in mind while programming since if you exceed the limitations of an acoustic instrument the human ear will instinctively know that it's been programmed rather than played.

ACOUSTIC GUITARS

The key to programming acoustic guitars is to take note of how they're physically played and then (provided the tone module produces an authentic sound) emulate this with MIDI. Firstly, you need to take into account the way that a guitarist's hand moves while playing. If the style is Hispanic, then the strings are commonly plucked rather than strummed; so the velocity of each note will be relatively high throughout since this will be mapped to control the cut-off (i.e. the harder a string is plucked, the brighter the timbre becomes).

Not all of these velocity plucks will be the same, however, since the performer will want to play in time with the track, and if a note occurs directly after the first, it will take a finite amount of time to move the fingers and pluck the next string. This often results in the string not being accentuated as much as the preceding one due to 'time restraints'. Conversely, if there is a larger distance between the notes, then there is higher likelihood that the string will be plucked at a higher velocity.

In many instances, particularly if the string has been plucked hard, the resonance may still be dying away as the next note starts, so this needs to be taken into account when programming the MIDI. Similarly, it's also worth bearing in mind that a typical acoustic guitar will not bend more than an octave, so it's prudent to set the receiving MIDI device to an octave so you can use the pitch bend wheel to create slides. In between notes that are very close together, it may also be worth adding the occasional fret squeal for added realism.

If the guitar is strummed, then you need to take the action of the hand into account. Commonly, a guitarist will begin by strumming downwards rather than upwards, and if the rhythm is quite fast on an upwards stroke it's rare that all the strings will be hit. Indeed, it's quite unusual for the bottom string to be struck, as this tends to make the guitar sound too 'thick', so this will need to be emulated while programming.

Additionally, all the strings will not be struck at exactly the same time due to the strumming action. Obviously, this means that each note on message will occur a few ticks later than the previous, which will depend on whether it's strummed upwards or downwards and the speed of the rhythm. To keep time, guitarists also tend to continue moving their hands upwards and downwards when not playing; so this 'rhythm' will need to be employed if there is a break in the strumming to allow it to return at the 'right' position in the arrangement. Finally, and somewhat strangely, if guitarists start on a downward stroke they tend to come in a little earlier than the beat, while if they start on an upward stroke they tend to start *on* the beat. The reason behind this is too complex to explain plus it would also involve me knowing why.

If you want to go even further and replicate trills or tremolo, it's worthwhile recording pitch bend movements onto a separate MIDI track and imposing it onto the longer notes of the original guitar track by setting it to the same MIDI channel. These pitch bend movements are recorded onto a separate channel. Since once it's imposed onto the notes it's likely that the pitch bend information will need further editing.

WIND INSTRUMENTS

Although fundamentally there are two types of wind instruments – brass and reed—in the concept of MIDI programming they're both programmed following similar principles since they both rely on variations in air pressure to produce the timbre. As a result, if you're planning on making music that contains wind instruments, it's prudent to invest in a breath controller. These are small devices that connect directly to the MIDI IN port and act similar to any wind instrument, albeit they do not produce any sound of their own. Rather they measure changes in air pressure as you blow into them. They convert this into CC2 (breath controller) messages that can then be used to control the connected MIDI synthesizer or sampler. These can be expensive, though, so if the use of wind instruments is only occasional it's worth programming the typical nuances by hand into the sequencer.

Firstly, the volume and brightness of the notes are proportional to the amount of air pressure in the instrument. All good MIDI instruments will set the velocity to control the filter cut-off of the instrument, so the brightness can be controlled with judicious use of MIDI velocity. The volume of the note, however, is a little more complicated to emulate since many reed instrumentalists will deliberately begin most notes by blowing softly before increasing the pressure to force the instrument to become louder. Essentially, this means that the notes begin softly before quickly rising in volume, and occasionally pitch. The best way to emulate this feel is to carefully programme a series of breath controller (CC2) messages while using a mix of expression controllers (CC11) to control the volume adjustments. Alternatively, brass instruments will often start below the required pitch and slide up to it; this is best emulated by recording live pitch bend movements into a sequencer and editing them to suit later.

On top of this, many wind instruments also introduce vibrato if the note is held for a prolonged period of time due to the variations in air pressure through the instrument. While it is possible to emulate this response with expression (CC11) controllers, generally speaking, introducing small pitch spikes at the later stages of the sustain can produce better results. It should be noted that these pitch 'spikes' only appear in the later stages of the notes sustain, though, and should not be introduced at the beginning of a note. Possibly the best way to determine where these spikes should appear is to physically emulate playing a wind instrument by beginning to blow at the start of the MIDI note on and when you begin to draw short of breath, and insert some pitch bend messages. Alternatively, if the synth allows you to fade in the LFO, you can use this to modulate the volume by setting it to a sine wave on a slow rate, modulating the volume lightly. As long as the LFO fade-in time is set quite long, it will only begin to appear towards the end of a note.

On the subject of breathing, bear in mind that all musicians are human and as such need to breathe occasionally. In the context of MIDI this means that you should avoid playing a single note over a large number of bars. Similarly, if a series of notes are played consecutively, remember that the musician needs enough time to take a deep breath for the next note (wind instruments are not polyphonic!). If there isn't enough time, the next note will generally be played softer due to less air velocity from a short breath, but if there is too short a space the instrument will sound emulated rather than real.

Finally, you need to consider how the notes will end. Neither reed nor brass instruments will simply stop at the end of the note; instead they will fade down in volume while also lowering in pitch as the air velocity reduces (an effect known musically as diminuendo). This can be emulated with a series of expression (CC11) messages and some pitch bend.

As touched upon, some tracks do not employ real instruments and rely solely on synthetic. If this is the case it's often worth avoiding any sharp aggressive synthesizer patches such as distorted saw-based leads since these give the impression of a 'cutting' track, whereas softer sounds will tend to sound more laid back. This is an important aspect to bear in mind, especially when mastering the track after it's complete. Slow, relaxed songs will invariably have the mid-range cut to emphasize the low and high end, while more aggressive songs with have the bass and high hats cut to produce more mid-range and make it appear insistent.

In MIDI programming, if synthetic timbres are used, then many tracks tend to refrain from using notes less than 1/4th in length since shorter notes will make the track appear faster. Additionally, with longer notes many of the tones can utilize a long attack, which will invariably create a perception of more relaxed music. In fact, long attack times can form an important aspect of the music. It's good to experiment with all the timbres used in the creation of this genre by lengthening the amps/filters attack and release times and taking note of the effect it has on the track.

Vocal chants are sometimes used in the place of leads as these can be used as instruments in themselves. Typically, in many popular tracks these are sampled from other records or sample CDs, but it's sometimes worthwhile recording your own. Characteristically, condenser microphones are the preferred choice over dynamic as these produce more accurate results, but the diaphragm size will depend on the effect you wish to achieve. As touched upon in the chapter on recording real instruments, many producers will use a large diaphragm for vocals, but if you're after 'ghostly' chants, a small diaphragm mic such as the Rode NT1 will often produce better results due to the more precise frequency response. Of course, this is simply the conventional approach and, if at all possible, it's worthwhile experimenting with different microphones to see which produces the best results.

Vocals will also benefit from compression during the recording stage to prevent any clipping in the recording, but this must be applied lightly. Unlike most other genres where the vocals are often squashed to death to suit the compressed nature of the music, chill out relies on a live feel with the high frequencies intact. A good starting point is to set a threshold of -9 dB with a 2:1 ratio and a fast attack and moderately fast release. Then, once the vocalist has started to practice, reduce the threshold so that the reduction meters are only lit on the strongest part of the performance.

Similar to the drum rhythms, effects can also play a large part in attaining the right lead sound and you should feel free to experiment. Reverb is the main suspect here, but phasers and flangers can also work especially well if they are lightly applied. A good (if often overused) technique to produce ghostly vocal chants is to reverse the vocal sample, apply a large cavernous reverb and then reverse them back the right way around again. This way, the reverb tail-off will act as a build-up to the vocal chant. The key is, as always, to experiment.

BASS

With the lead laid down, the bass is often programmed next to form the basic groove of the song. Generally speaking, this tends to be quite minimal, consisting mostly of notes over 1/8th in length to prevent the groove from appearing too fast. These are often derived from the principles encapsulated in the dub scene, staying quite minimal to allow space for the lead to breathe without having to compete with the bass for prominence. Indeed, it's incredibly important that the bass is not too busy either rhythmically or melodically and many chill out tracks borrow heavily from the pentatonic scale used in dub and R 'n' B. That is, they use no more than a five-note scale playing a single harmony restricting the movements to a maximum five-semitone shift.

Naturally, the bass should interact with the lead and, as touched upon in the chapter on music theory, this can be accomplished through using parallel, oblique or contrary motions. In many cases, an oblique motion is the preferred option since a majority of this music is driven by the melody and you don't want the bass to detract from it. Having said that, you should feel free to try out both contrary and parallel motions to see which produces the best results. What is important, however, is that the bass coincides with the drum rhythm. Bear in mind that if the drums have been affected with reverb, the subsequent tail-off between each hit will leave less room for the bass, so it will need to be less melodic or rhythmic. Conversely, if the drums are left dry, it's possible to employ a more rhythmic or melodic bass without creating any conflicts (Figure 17.4).

Typically, most chill out basses are synthesized rather than real since the timbre is often quite deep and doesn't want to attract as much attention as the lead. Analogue synthesizers usually provide the best results due to the uncontrollable

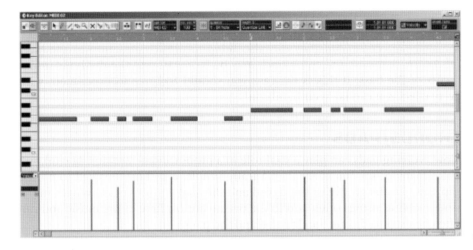

FIGURE 17.4
A typical chill out bass

yet pleasing phasing of the oscillators. Nonetheless, the beginnings of a typical timbre can be programmed on most synthesizers. A good starting point is to use a sine wave with a sawtooth transposed up from it by 5 or 7 cents. Unlike most of the timbres used throughout the production of chill out, the attack should be well defined to prevent the bottom-end groove turning to mush, but there is usually very little decay, if any, to prevent it from becoming too plucky. Indeed, in much of the music of this genre, the bass tends to hum rather than pluck to prevent it from drawing attention away from the chilled feeling. The best way to accomplish this type of sound is to use a simple on/off envelope for the amp envelope. This is basically an envelope with no attack, decay or release but a quite high sustain. This forces the sound to jump directly into the sustain stage which produces a constant bass tone for as long as the key is depressed. If this results in a bass with little or no sonic definition, a small pluck can be added by using a low-pass filter with a low cut-off and high resonance that's controlled with an envelope using a zero attack, sustain and release but a medium decay. By adjusting the depth of the envelope modulating the filters or increasing/decreasing the decay stage, more or less of a pluck can be applied. If you use this process, however, it's prudent to employ filter key tracking so that the filter action follows the pitch.

At this stage, it's also prudent to experiment with the release parameters on the filter and amp envelope to help the bass sit comfortably in the drum loop. In fact, it's essential to accomplish this rhythmic and tonal interaction between the drum loop and the bass by playing around with both the amp and filters A/R EG before moving on. This is the underpinning of the entire track and it needs to be right for the style of music.

Since these waveforms are produced from synthesizers, there is little need to compress the results – they're already compressed at the source. Also, it's unwise to compress the drums against the bass to produce the archetypal 'dance music' pumping effect, as this will only accentuate the drum rhythm and can make the music appear too 'tight' rather than relaxed. Similarly, effects are usually avoided on the bass timbre because they can not only spread the timbre across the stereo spectrum but they can also bring too much attention to the bass. Of course, that said, the entire track may be based around the rhythmic movements of the bass. If this is the case you should feel free to experiment with every effect you can lay your hands on!

CHORDS/PADS

Slow-evolving pads can play a vital role in a proportionate amount of chill out. Since many of the tracks are thematically simple, a pad can be used to fill the gap between the groove of the record and the lead and/or vocals. What's more, slow-evolving strings often add to the overall atmosphere of the music and can often be used to dictate the drive behind the track.

We've already covered the principles behind creating chord structures in the chapter on compression, processing and effects, but as a quick refresher, the chords here will act as a harmony to the bass and lead. This means that they should fit in between the rhythmic interplay of the instruments so far without actually drawing too much attention (rather like the genre as a whole). For this, they need to be very closely related to the key of the track and not use a progression that is particularly dissonant. Generally, this can be accomplished by forming a chord progression from any notes used in the bass and experimenting with a progression that works. For instance, if the bass is in E, then C major would produce a good harmony because this contains an E (C–E–G). The real solution is to experiment with different chords and progressions until you come up with something that works.

Once the progression is down, it can then be used to add some drive to the track. Although many chill out tracks will meander along like a Sunday stroll, if you place a chord that plays consistently over a large number of bars, all of the feel can quickly vanish. This can sometimes be avoided by moving the chords back in time by a small amount so that they occur a little later than the bar. Alternatively, if they follow a faster progression, they can be moved forward in time to add some push to the music. It should be noted here, however, that if the pads employ a long attack stage, they may not actually become evident until much later in the bar, which can destroy the feel of the music. In this instance, you will need to counter this effect by moving the chords so that they occur much earlier (Figure 17.5).

The instruments used to create these are more often than not analogue in nature due to the constant phase discrepancies of the oscillators, which creates additional movement. Thus, analogue or analogue emulations will invariably produce much better results than an FM synth or one based around S&S. What's more, it's of vital importance that you take into account the current frequencies used in the mix so far.

With the drums, bass, lead and possibly guide vocals playing along, there will be a limited frequency range where you can fit the chords in. If the timbre used

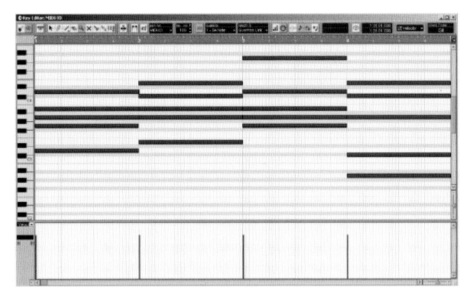

FIGURE 17.5
Chill out chord
structure

is too harmonically rich it will be incredibly difficult to fit them into the mix without having to resort to aggressive EQ cuts. This will change the entire character of the chords and in some instances may make it inappropriate for the music; so it's much better to programme the chords timbre to fit in now, rather than when approaching the mix. Ideally, you should look towards producing a 'mix' that sounds right during this construction stage and not rely on the mixing desks EQ later to cure a host of problems. Keep in mind that the EQ on a mixing desk is very subtle and usually kept aside for subtle tonal changes. This obviously means that there are two programming possibilities for the chords. They either need to be quite lush to fill out any noticeable gaps in the mix or they need to be quite thin so they do not collide with the rest of the instrumentation.

If the mix is particularly busy in terms of frequencies, then it's prudent to build a relatively thin pad that can then be 'thickened' out later, if required, with phasers, flangers, and reverb or chorus effects. This is accomplished by using pulse waves, as they have less harmonic content than saws and triangles and do not have the 'weight' of a sine wave (plus, of course, they sound more interesting!). Generally only one pulse wave is required with the amp envelope set to a medium attack, sustain and release but a fast decay. If this timbre is to sit in the upper midrange of the music, then it's best to use a 12 dB high-pass filter to remove the bottom end of the pulse; otherwise use a low-pass filter to remove the top end and then experiment with the resonance until it produces a general static tone that suits the track. Following this, set an LFO to positive modulation with a slow rate on the pulse width of the oscillator and the filter to add some movement. Which waveform you use for the LFO will depend on the track itself, but by first sculpting the static tone to fit into the track you'll have a much better idea of which wave to use and how much it should modulate the parameters. If the timbre appears too 'statically modulated' in that it still seems uninteresting, use a different rate and waveform for the oscillator and filter so that the two beat against each other. Alternatively, if the timbre still appears too thin even after applying effects, add a second pulse detuned from the first by 3 cents with the same amp envelope, but use a different LFO waveform to modulate the pulse width.

If the track has a 'hole' in the mix, then you'll need to construct a wider, thicker pad to fill this out. As a starting point for these types of pads, square waves mixed with triangles or saws often produce the best results. Detune the saw (or triangle) from the square wave by 5 or 7 cents depending on how thick you need to the pad to be and set the amp attack to a medium sustain and release, but no attack, and a short decay. Using a low-pass filter, set the cut-off to medium with a high resonance and set the filters EG to a short decay and a medium attack, sustain and release. This should modulate the filters positively so that the filters sweep through the attack and decay of the amp but meet at the sustain portion, although it is worth experimenting with negative modulation to see if this produces better results for the music. This will produce a basic pad timbre, but if it seems a little too static for the music, use an LFO set to a sine or triangle wave using a slow rate and medium depth to positively modulate the pitch of the second oscillator. For even more interest, you could also use an LFO to modulate the filter's cut-off.

Effects can also play an important role in creating interesting and evolving pads. Wide chorus effects, rotary speaker simulations, flangers, phasers and reverb can all add a sense of movement to help fill out any holes in the mix. Again, as with most instruments in this genre, compression is generally not required on pads since they'll already be compressed at the source, but a noise gate can be used creatively to model the sound. For instance, set a low threshold and use an immediate attack and release; the pad will start and stop abruptly, producing a 'sampled' effect that cannot be replicated through synthesis parameters. What's more, if you set the hold time to zero and adjust the threshold so that it lies just above the volume of the timbre, the gate will 'chatter', which can be used to interesting effect.

ARRANGEMENT

As touched upon in the musical analysis, chill out is incredibly diverse, and pretty much anything with a slow tempo and a laid-back feel could be classified as part of the genre. Consequently, the arrangement of the music can follow any structure you feel suits. In fact, as this genre is so varied, it's difficult to point out globally accepted arrangements, and the best way to obtain a good representation of how to arrange the track is to listen to what are considered the classics and then copy them. This isn't stealing, it's research and nearly every musician on the planet follows this same route. As a helping hand, however, we'll look at one of the popular structures used by this genre, based on the same principle as most popular music.

Most popular music is constructed of around five parts, consisting of the verse, chorus, bridge, middle eight and occasionally a key change near the end. The verse is the part of the song where the story is told and a good song will feature around three or four verses. The chorus is the more exciting part of the music which follows the verse and is where all the instruments move up a key and give the listener something to sing along with. The bridge is the break between the verse and chorus and usually consists of a drum fill which leads onto the middle eight of the track. This, as its name would suggest, is usually eight bars long and is the break of the track that is often the most exposed segment to the dance musician's sampler. Finally, there's the key change and although not all records will use this, it consists of shifting the entire song up a key to give the impression that it has reached its apex. If we break this down in a number of bars, we're left with:

- Verse 1 – *Commonly* 16 bars
- Chorus – *Commonly* 8 bars
- Verse 2 – *Commonly* 16 bars
- Chorus – *Commonly* 8 bars
- Verse 3 – *Commonly* 16 bars
- Chorus – *Commonly* 8 bars
- Bridge – *Commonly* 1 bar
- Middle eight – *Commonly* 8 bars
- Double chorus – *Commonly* 16 bars

This is, of course, open to artistic license and some artists will play the first two verses before hitting the chorus. This approach can often help to add excitement to a track as the listener is expecting to hear the chorus after the first verse. If this route is taken, even though there may be different vocals employed in the verses, if they are played one after the other it can become tiresome; so a popular trick is to add a motif into the second verse to differentiate it from the first.

> The data CD contains a full mix of a typical Chill Out track with narration on how it was constructed. Vocals by Tahlia Lewington.

RECOMMENDED LISTENING

Ultimately, the purpose of this chapter is to give some insight into the production of chill out; there is no one definitive way to produce the genre. Indeed, the best way to learn new techniques and production ethics is to actively listen to the current market leaders and be creative with processing, effects and synthesis. Chorus, delay, reverb, distortion, compression and noise gates are the most common processors and effects used within chill out, so experiment by placing these in different orders to create the 'sound' you want.

While the arrangement and general premise of each track is generally similar, it's this production that differentiates one from the other. With this in mind, what follows is a short list of some of the artists that, at the time of writing, are considered the most influential in this area:

- Aphex Twin (*Selected Ambient Works*)
- William Orbit (*Excursions in Ambience*)
- Future Sound of London
- Ministry of Sound (*Chillout Sessions*)
- Groove Armada

CHAPTER 18
Drum 'n' Bass

'It became the basis for drum-and-bass and jungle
music – a six-second clip that spawned several entire
subcultures...'

Nate Harrison

Pinpointing the foundations of where drum 'n' bass originated is difficult. In a simple overview, it could be traced back to as simply being a natural development of jungle; however, jungle was a complex infusion of breakbeat, reggae, dub, hardcore and artcore. Alternatively, we could look further back to 1969 and suggest that the very beginnings lay with a little known record by the Winston's. The B side of a record entitled *Colour Him Father* which featured a 6s drum break – the Amen Break, taken from the title of the record, *Amen Brother* – that became the staple basis for jungle and drum 'n' bass for many years. Or it could be traced back to the evolution of the sampler, offering the capability to cut and chop rhythmic material to create the resulting jungle and more refined drum 'n' bass.

Note: For those who are reading this book cover to cover, we have already discussed the history of jungle in Chapter 13 (UK Garage), so feel free to skip the history and move onto the analysis.

We can roughly trace the complex rhythms developed in jungle back to breakbeat, whose foundations can be traced back to the 1970s. Kool Herc, a hip-hop DJ began experimenting on turntables by playing only the exposed drum loops (breaks) and continually alternating between two records, spinning back one while the other played and vice versa. This created a continual loop of purely drum rhythms allowing the breakdancers to show off their skills. DJs such as Grand Wizard Theodore and Afrika Bambaataa began to copy this style, adding their own twists by playing two copies of the same record but delaying one against the other, resulting in more complex asynchronous rhythms.

It was early 1988 and the combined evolution of the sampler and the rave scene that really sparked the breakbeat revolution. Acid house artists began to sample the breaks in records, cutting and chopping the beats together to produce more complex breaks that were impossible for any real drummer to play naturally. As these breaks became more and more complex, a new genre evolved known as hardcore. Shifting away from the standard 4/4 loops of typical acid house, it featured lengthy complex breaks and harsh energetic sounds that were just too 'hardcore' for the other ravers.

Although initially scorned by the media and record companies as drug-induced rubbish that wouldn't last more than a few months, by 1992 the entire rave scene was being absorbed in the commercial media machines. Riding on this 'new wave for the kids', record companies no longer viewed it as rubbish but a cash cow and proceeded to dilute the market with a continuous flow of watered down rave music. Rave became commercialized and this was taking its toll on the nightclubs.

In response to the commercialization, in 1992 two resident DJ's – Fabio and Grooverider – pushed the hardcore sound to a new level by increasing the speed of the records from the usual 120–145 BPM. The influences of house and techno were dropped and quickly replaced with ragga and dancehall, resulting in mixes with fast complex beats and a deep bass. Although jungle didn't exist as a genre just yet, these faster rhythms mixed with deep basses inspired artists to push the boundaries further and up the tempo to a more staggering 160–180 BPM.

To some, the term jungle was attributed to racism but the name was derived from the 1920s. It was used on flyers to describe music produced by Duke Ellington. This featured exotic fast drum rhythms and when Rebel MC sampled an old dancehall track with the lyrics 'Alla the Junglists', jungle became synonymous with music that had a fast beat and a deep throbbing bass. This was further augmented with pioneers of the genre such as Moose and Danny Jungle.

Jungle enjoyed a good 3–4 years of popularity before it began to show a decline. In 1996 the genre diversified into drum 'n' bass as artists such as Goldie, Reprazent, Ed Rush and LTJ Bukem began to incorporate new sounds and cleaner production ethics into the music. Goldie and Rob Playford are often credited with initiating the move from the jungle sound to drum 'n' bass with the release of the album *Timeless*.

Of course, jungle still exists today and by some jungle and drum 'n' bass are viewed as one and the same but to most, jungle has been vastly overshadowed by the more controlled production and deep basses of 'pure' drum 'n' bass.

MUSICAL ANALYSIS

Examining drum 'n' bass in a musical style is more difficult than examining its roots since it has become a hugely diversified genre. On one end of the scale it can be heavily influenced with acoustic instruments while on the other it will only feature industrial or synthetic sounds. Nevertheless, there are some generalizations that all drum 'n' bass records share that can be examined in more detail.

Typically a drum 'n' bass tempo will remain between 165 and 185 BPM, although in more recent instances the tempo has sat around 175–180 BPM. Generally speaking, a snare will remain on beats 2 and 4 of the bar to help keep the listener or dancer in time, while the kick drum will dance around this snare. The bass often plays at one-quarter or half time of the drum tempo, again to keep the dancer 'in time'. It should be noted, however, that the more complex the drum rhythm is, the faster the music will appear to be. Therefore, it's prudent to keep the loop relatively simple if the tempo is higher or relatively complex if the tempo is lower.

Much of the progression of this genre comes from careful development of the drum rhythms, not necessarily a change in the rhythm itself but from applying pitch effects, flangers, phasers and filters as the mix progresses. Much in the way that techno uses the mixing desk as a creative tool so does drum 'n' bass with the EQ employed to enhance the rhythmic variations and harmonic content between them.

To the uninitiated, drum 'n' bass may be seen to only include drum rhythms and a bass, but very few tracks are constructed from just these elements. Indeed, although these play a significant role in the genre, it also features other instrumentation such as guitars or synthesis chords, sound effects and vocals.

PROGRAMMING

The obvious place to start is to begin by shaping the drum rhythm, and the timbres are more often than not sourced from sample CDs or other records. Single hits from sample CDs can be imported directly into the sequencer and arranged, while drum loops are often imported and cut into individual hits. It is important to note that not all the individual hits from a drum loop must be used, and even if it only results in a single snare hit, it has served its purpose.

For this genre it is preferable to make your own kick and snare, since these can be pitched up and down to help achieve the sound typical of the genre. Being able to pitch these instruments up and down when the rest of the instrumentation is in place will also help to clear the mix. Nevertheless, you will need to sample from a record (legally) or from a sample CD to later complete the loop and give it the requisite feel.

Snares are possibly best laid down first, and these can be created in most synthesizers provided they feature white noise. Using a triangle wave in the first oscillator, select white noise for the second and set the amp EG to a zero attack, sustain and release and then use the decay parameter to set the length of the snare. Generally, in drum 'n' bass, the snare is particularly 'snappy' and often pitched up the keyboard to give the impression of a sampled sound having its frequency increased within the sampler.

If possible, employ a different amp EG for both the noise and the triangle wave. By doing so, the triangle wave can be kept quite short and swift with a fast decay, while the noise can be made to ring a little further by increasing its decay parameter. This, however, should never be made too long since the snares should remain snappy. If the snare has too much bottom end, employ a high- pass, band-pass or notch filter depending on the type of sound you require. Notching out the middle frequencies will create a clean snare sound that's commonly used in this genre. Further modification is possible using a pitch envelope to positively modulate both oscillators. This will result in the sound pitching upwards towards the end, giving a brighter snappier feel to the timbre.

The kick can again be sampled or created in any competent synthesizer. Kicks are initially made from a sine wave, but the frequency will determine the kick's body. Since the bass timbre is generally quite heavy in drum 'n' bass, the kick is best programmed around 100–150 Hz. Using an attack/decay EG, modulate the pitch of sine wave with a maximum positive depth setting and an immediate attack and short release. The timbre can be further modified by adding a square wave pitched down with a very fast amplifier attack and decay setting to produce a short sharp click. The amount that this wave is pitched down will depend entirely on the sound you want to produce, so it's sensible to layer it over the top of the sine wave and then pitch it up or down until the transient of the kick sounds right for the genre. The timbre will also probably benefit from some compression; experiment by setting the compressor so that the attack misses the transient but grips the decay stage. Increasing the gain of the compressor can make the timbre more powerful.

Whether sampled or synthesized, the snare is generally placed on bars 2 and 4 of the loop, and the kick can then be programmed to play around these snares. A good starting point is to place a kick on the first beat of the bar, followed by a beat just before the snare and two beats directly after. Adding two kicks on the next bar, followed by another kick before the snare and one after can produce a good place to start. It is a case of experimenting with placing kicks around the snares. But bear in mind that the more kicks that play around the snare positions the faster the rhythm will appear, so it is a case of experimentation until you produce an initial loop that capture the essence you're looking for (Figure 18.1).

Although this generally provides a good starting pattern, any further loops should generally be sampled for either a sample CD or another record (legally, of course). It is unusual to sample a loop from another drum 'n' bass record but rather from a record of a different genre such as rock music. The amen loop sampled by so many artists in this genre was never a drum and bass rhythm; it was simply employed and mixed with a pre-programmed loop to produce the feel of the music. If you do use a sample, you will not require the kick in from the sample. Contrary to popular belief, drum 'n' bass does not use multiple layered kicks but instead relies on just one kick and derives much of its complexity from differing hi-hat/percussion rhythms. Generally speaking, it is not unusual to layer 5–6 differing percussive elements under the kick and snare to produce the feel.

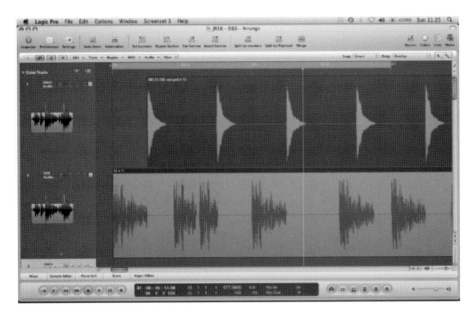

FIGURE 18.1
The beginnings of a loop

FIGURE 18.2
A drum 'n' bass loop

The kicks in any sampled loops can be removed via wave editing/sequencing software; however, a better approach is to employ an EQ unit to roll off the bottom end of the loop. This removes the body of kick and much of the body of the snare just leaving the higher percussive elements to the mix and allowing the originally programmed kick and snare to take prominence in the loop. A typical example of a drum loop is shown in Figure 18.2.

Note that in the above example, many of the sampled loops actually land on the beat, rather they are all offset slightly. This helps them to combine more freely with the initially created loop, creating thicker textures as the sounds play slightly earlier or later than the first loop, thus preventing the loops from becoming too metronomic and adding a more breakbeat feel to the rhythm. The principle is to create a series of loops, each of which focuses on a different timbre to carry the loop forwards. It is also worthwhile experimenting by mixing loops with different time signatures together. For instance, two 4/4 loops mixed with two 3/4 loops can produce results that cannot be easily acquired any other way.

Although for this example we've only discussed a limited number of loops, you should feel free to programme and layer as many loops together as you feel are necessary to accomplish the overall sound. This can range from using just three to layering over eight together and then progressively reducing the content of each until you are left with harmonically rich and interesting rhythm. The layering of these should also not just be restricted to dropping them dead atop one another, and in many cases moving one of two of the loops forward or back in time from the rest can be used to good effect.

While some producers frown on using compression on a drum 'n' bass rhythm, there is certainly no harm in applying it, either creatively or naturally. Applying compression how it is 'supposed' to be used can often aid in controlling the rhythms and result in a more well-produced loop, aiding the various elements to sit together as a whole. Generally compression of this nature should be applied lightly, with a threshold that just captures the transients, a low ratio and a quick attack and release. If applied in a more creative way, it can often breathe new life into the rhythm. A trick sometimes employed is to access the compressor as a send effect.

With a medium threshold setting and a high ratio, the returned signal can be added to the uncompressed signal. You can then experiment with the attack and release parameters to produce rhythm that gels with the rest of the instruments. If you do not have access to a valve compressor, then sending the signal to the PSP Vintage Warmer or Sonalksis TBK3 Uber Compressor can often add more harmonics to the signal.

THE BASS

The second, vital element of drum 'n' bass is the bass. Generally speaking, the bass consists of notes that play at either one quarter or half the tempo of the drum rhythm. This is accomplished through using length bass notes, set at one quarter, half or full notes and sometimes straddling over the bars of the drum loop. The notes of the bass usually remain with an octave and rarely move further than this to prevent the music from becoming too active and deterring the listener from the rhythmic interaction of the drums (Figure 18.3).

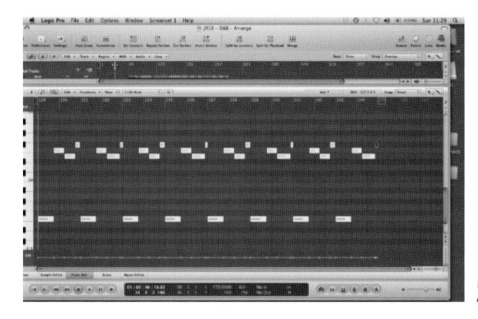

FIGURE 18.3
A typical bass line

In the above example, the bass remains within an octave, exhibiting a slow movement to counteract the speed set by the drum rhythm. Note also in the above example a secondary note overlaps the first. Using pitch bend, this creates a timbre that rises into the next note, and it's this type of movement that creates some interest in the bass.

Indeed, since the bass notes are kept quite lengthy, it is not uncommon to employ some movement in the bass to maintain interest. This is often accomplished through filters or pitch modulation. Since the bass in this genre is supposed to remain deep and earth shaking, a sine wave makes the perfect starting point for this timbre.

If you programmed your own kick drum, as described earlier, try copying the preset over to another bank and lengthen the decay and release parameter of the amp envelope. If you sampled the drum kick instead, use a single oscillator set to a sine wave and positively modulate its pitch with an attack/decay envelope. Then experiment with the synthesizer's amplitude envelopes. This should only be viewed as a starting point and, as always, you should experiment with the various modulation options on the synthesizer to create some movement in the sound.

As an example of a typical drum 'n' bass timbre, using a sine wave, set the amplifiers attack to zero and increase the decay setting while listening back to the programmed motif until it produces an interesting rhythm. Next, modulate the pitch by a couple of cents using an envelope set to a slow attack and medium decay. This will create a bass timbre where the note bends slightly as it's played. Alternatively, you can use an LFO set to a sine wave with a slow rate and set it to start at the beginning of every note.

Experimentation is the key, changing the attack, decay and release of the amp or/and filter EG from linear to convex or concave will also create new variations. For example, the decay to a convex slope setting will produce a more rounded bass timbre. Similarly, small amounts of controlled distortion or very light flanging can also add movement.

Alongside the synthetic basses, many drum 'n' bass tracks will employ a real bass. More often than not these are sampled from other records – commonly dancehall or raga, but they are also occasionally taken from sample CDs rather than programmed in MIDI. Provided that you have (a) plenty of patience, (b) a willingness to learn MIDI programming and (c) a good tone or bass module (such as Spectrasonics Trilogy, EW QL Collosus or any professional MIDI tone module), it is possible to programme a realistic bass that could fool most listeners. This not only prevents any problems with clearing copyright but it also allows you to model the bass to your needs.

Programming a realistic bass has already been covered in detail within the chapter on house, but if you've jumped directly to this chapter what follows is a rundown on how to programme a realistic sounding bass instrument.

The key to programming any real instrument is to take note of how they're played and then emulate this action with MIDI and a series of CC commands. In this case, most bass guitars use the first four strings of a normal guitar E–A–D–G, which are tuned an octave lower, resulting in the E being close to three octaves below middle C. Also they are monophonic, not polyphonic, so the only time notes will actually overlap is when the resonance of the previous string is still dying away as the next note is plucked. This effect can be emulated by leaving the preceding note playing for a few ticks while the next note in the sequence has started. The strings can either be plucked or struck and the two techniques produce different results. If the string is plucked, the sound is much brighter and has a longer resonance than if it were simply struck. To copy this, the velocity will need to be mapped to the filter cut-off of the bass module so that higher values open the filter more. Not all notes will be struck at the same velocity, though. If the bassist is playing a fast rhythm the consecutive notes will commonly have less velocity since he has to move his hand and pluck the next string quickly. Naturally, this is only a guideline and you should edit each velocity value until it produces a realistic feel.

Depending on the 'bassist', they may also use a technique known as 'hammer on' whereby they play a string and then hit a different pitch on the fret. This results in the pitch changing without actually being accompanied with another pluck of the string. To emulate this, you'll need to make use of pitch bend. First set the pitch bend to a maximum limit of 2 semitones, since guitars don't 'bend' any further than this. Begin by programming 2 notes, for instance an E0 followed by an A0 and leave the E0 playing underneath the successive A0 for around a hundred ticks. At the very beginning of the bass track, drop in a pitch bend message to ensure that it's set midway (i.e. no pitch bend) and just before where the second note occurs, drop in another pitch bend message to bend the tone up to A0. If this is programmed correctly, on play back you'll notice that as the E0 ends the pitch will bend upwards to A0 simulating the effect. Although this could be left as is, it's sensible to drop in a CC11 message (expression) directly after the pitch bend as this will reduce the overall volume of the second note so that it doesn't sound like it has been plucked. In addition to this, it's also worthwhile employing some fret noise and finger slides. Most good tone modules will include fret noise that can be dropped in between the notes to emulate the bassist's fingers sliding along the fret board.

As the rhythmic movement and interaction with the bass and rhythms provide the basis for this genre, it's recommended that you also experiment by applying effects to the bass timbre. While most effects should be avoided since they tend to spread the sound across the image, in this genre the bass is one of the most important parts of the music. Small amounts of delay can create interesting fluctuations, as can flangers, phasers and distortion.

As with the drum rhythms, creative compression can also help in attaining an interesting bass timbre. As before, try accessing the compressor as a send effect with a medium threshold setting and a high ratio. The returned signal can be added to the uncompressed signal. You can then experiment with the attack and release parameters to produce an interesting bass tone. Alternatively, try pumping the bass with one of the rhythms. Set the compressor to capture the transients of a percussive loop and use it pump the bass by experimenting with the attack, ratio and release parameters.

CHORDS

Many aficionados of the genre recommend only using minor chords to act as a harmony, although more recently major chords have been making an appearance. This, as always, depends upon your artistic interpretation, but it is generally accepted that minor chords produce better results than major which can tend to lift the music too much.

Generally speaking, chords in A minor can produce good results. The movement of the chords can often work well when contrasted with the bass line. This can be accomplished by copying the bass line down to another sequencer track and converting this new track into a chord. Once created, when the bass moves up in pitch, move the chords down in pitch and vice versa (Figure 18.4).

If you're struggling understanding how to build a minor chord, what follows are some typical minor chords to get you started:

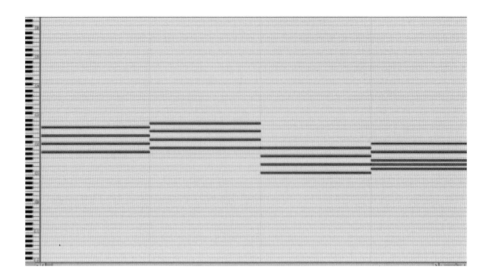

FIGURE 18.4
A minor chord progression

G–Bb–D
A–C–E
B–D–F#
C#–E–G#
Eb–Gb–Bb
Gb–A–Db
Ab–B–Eb
Bb–Db–F

With a general idea of the chord structure down, you can programme (or sample) a string to fit. A good starting point for programming a drum 'n' bass string can be created by mixing a triangle and square wave together and detuning one of the oscillators from the other by 3–5 cents. Set the amps attack to zero with a medium sustain and release. Set the filter envelope to a long attack, sustain with a medium release and short decay. Set the filter cut-off quite low and the resonance about midway, then modulate the pitch of either the triangle or, square wave with a sine wave set to a slow rate with a medium depth. If the string is going to continue for a length of time it's worthwhile employing a sine, pulse or triangle wave LFO to modulate the filters cut-off to help maintain interest. As always, this should only be considered as a starting point, and experimentation is the key to gaining good results.

Effects can also play an important role in creating interesting strings for the genre, although these should be used conservatively so as not to detract from the bass rhythm. Often, wide chorus effects, rotary speaker simulations, flangers, phasers and reverb can all help to add a sense of interest. A compressor can also be used creatively to model the sound. For instance, if you put a threshold to capture the kick drums transient and use an immediate attack and release, the pad will pump in volume on every kick, producing an interesting 'sampled' effect that cannot be replicated through synthesis parameters.

VOCALS AND SOUND FX

One final aspect yet to cover is the addition of sound effects and vocals. The vocals within drum 'n' bass more often than not consist of little more than a short vocal snippets; however, there have been some more commercial drum 'n' bass mixes that have featured a verse/chorus progression. It seems to be a point of contention among many drum 'n' bass producers as to whether a verse/chorus is actually part of the genre or is in fact diversifying again to produce a new genre of music. Others, however, believe that it's simply a watered down, commercialized version of the music made solely for the purpose of profit margins. Nevertheless, whether you choose to use a few snippets of vocals, some ragga or MC'ing, or a more commercialized vocal performance is entirely up to you. It's musicians pushing boundaries that reaps the greatest rewards.

The sound effects can obviously be generated by whatever means necessary, from sampling and contorting sounds or samples with effects and EQ. For contorting audio, the *Mutronics Mutator*, *Sherman Filterbank 2*, the *Camelspace* range of plug-ins and Steinbergs *GRM Tools* are almost a requisite for creating strange evolving timbres. The effects and processing applied are, of course, entirely open to artistic license as the end result is to create anything that sounds good and fits within the mix. Transient designers can be especially useful in this genre as they permit you to remove the transients of the percussive rhythms which can evolve throughout the track with some thoughtful automation. Similarly, heavy compression can be used to squash the transient of the sounds, and with the aid of a spectral analyser you can identify the frequencies that contribute to the sound while removing those surrounding it. Alternatively, pitch-shift individual notes up and by extreme amounts, or apply heavy chorus or flangers/phasers to singular hi-hats or snares, or try time stretching followed by time compression to add some digital clutter and then mix this with the other loops.

ARRANGEMENT

Drum 'n' bass arrangement can follow a number of diverse paths, from the typical verse/chorus structure of the more commercialized music to the adaptation and interrelationship between all of the elements together to create the arrangement. The latter approach consists of dropping different rhythms in and out of the mix, along with the bass, vocals and sound effects. The ideals are not so much to create a track that builds to a crescendo or climax but rather stays on one constant rhythmical level that warps from one rhythmically interesting collective to another. This is accomplished through filters and effects and not by introducing new melodic elements.

The more commercialized approach is to use five-part arrangement typical of most popular music tracks: verse, chorus, bridge, middle eight and occasionally a key change near the end. The verse is the part of the song where the story is told, and a good song will feature around three or four verses. The chorus is the more exciting part of the music which follows the verse and is where all the instruments move up a key and give the listener something to sing along with. The bridge is the break between the verse and chorus and usually consists of a drum fill, which leads onto the middle eight of the track. This, as its name would suggest, is usually 8 bars long and is the break of the track that is often the most exposed segment to the dance musician's sampler. Finally, there's the key change and although not all records will use this, it consists of shifting the entire song up a key to give the impression that it has reached its apex.

This is, of course, open to artistic license, and some artists will play the first two verses before hitting the chorus. However, with so much diversity in the genre and arrangements, the best solution is to listen to what are considered the classics and then copy them. This isn't stealing, its research and nearly every musician on the planet follows this same route.

> The data CD contains a full mix of a typical Drum 'n' Bass track with narration on how it was constructed.

RECOMMENDED LISTENING

Ultimately, as with all the chapters in this section, its purpose is not to tell you how to write or produce a specific genre. Rather it should be seen as offering a few basic starting ideas that you can evolve from. There is no one definitive way to produce any genre, and the best way to learn new techniques and production ethics is to actively listen to the current market leaders and experiment. With this in mind, what follows is a short list (and by no means exhaustive) of artists that are considered influential in this area:

- Golide
- Grooverider
- 4 Hero
- Aphex Twin
- Roni Size
- LTJ Bukem
- Photek
- Lamb
- Fabio

PART 3
Mixing & Promotion

CHAPTER 19

Mixing

'The ideas for "A Huge Ever Growing Pulsating Brain That Rules From the Centre of the Ultra World" were there weeks before, but the mix itself took twenty minutes...'

Alex Patterson

Mixing is viewed by many as the most complex procedure involved in producing a dance record, but the truth of the matter is 99.9% of the time a poor mix isn't a result of the mixing process but a culmination of factors before anything even reaches the mixing desk.

If you've taken care in producing the track throughout, then when it eventually comes to mixing you'll find that the music has more or less mixed itself and all that is required is gentle use of EQ, volume and effects to polish the results.

It is incredibly important to bear in mind that the mixing desk is simply a tool used to position instruments in the mix and allow you to add a *very* small final sparkle. So, before you even touch the desk the only problem should be that you can't hear some of the instruments too clearly. Apart from that, the mix should otherwise sound just as you imagined it would. Thus, before we even touch upon the use of faders, EQ and effects in mixing, you need to go back and examining what you have so far.

Generally, a poor 'pre-mix' is created by one or all of the following:

- Poor recording, programming or choice of timbre/sample.
- Poor-quality effects, or use of effects when programming.
- Poor arrangement or MIDI programming.

We've touched upon the importance of all of these in earlier chapters; at this stage you should go back and check everything one final time to ensure that everything sounds the best it possibly can. If you're unhappy with any of the timbres, re-program or replace them. It is vital at this stage that you not settle for anything less than the very best – it may be an overused analogy but you wouldn't expect to make a great tasting cake if some of the ingredients were out of date!

Perhaps, most importantly, before you approach the mix you need to ask yourself one final yet vitally significant question about your track. Close your eyes, listen back to the music and ask yourself 'can you feel it?'

Above the arrangements, sounds, processing and effects this is the ultimate question and the answer should be a resounding (and honest!) yes. Dance music is ultimately about 'groove', 'vibe' and 'feel' and without these fundamental elements it isn't dance music. A track may be beautifully programmed and arranged but if it has no 'feel' then the whole endeavour is pointless.

MIXING THEORY

At the outset, it's important to understand that mixing is a creative art: there are no right or wrong ways to go about it. All engineers will approach a mix in their own distinctive way – it's their music and they know how they want it to sound. A unique style of mixing, provided that it sounds transparent, will define your creative style as much as the sounds and arrangement do. Despite this, there are some practices that you can employ that will put you on the right track to producing transparent and detailed mixes.

The first step to creating any great dance mix is to understand the theory behind mixing and how it can be adapted in a more practical sense to suit your own particular style. This means that you need to comprehend how the soundstage of a mix can be created and adapted, how frequency ranges are grouped together and how to come to terms with our natural hearing limitations. We'll look at each of these in detail.

HEARING LIMITATIONS

Our hearing is far from perfect because not only do we perceive different frequencies to be at different volumes but the overall volume at which we listen to a mix also determines the dominant frequencies. This inaccurate response can be traced to a time when we lived in caves. Since talking (or more likely grunting) to each other in a cave resulted in vast amounts of reverberation, our hearing adapted itself to concentrate on the frequencies where speech is most evident and decipherable in this situation – approximately 3–4 kHz. This means that at conversation level, our ears are most sensitive to sounds occupying the mid-range and any frequencies higher or lower than this must be physically louder in order for us to perceive them to be at the same volume.

At normal conversation levels it's sixty-four times more difficult to hear bass frequencies than it is to hear the mid-range and eighteen times more difficult to perceive the high range.

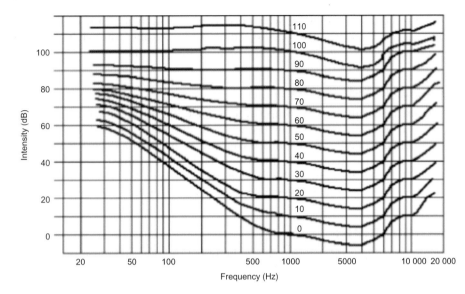

FIGURE 19.1
The Fletcher Munson
contour control curve

If, however, the volume is increased beyond normal conversation level, the lower and higher frequencies gradually become (albeit perceivably) louder than the mid-range. This again is related to when we lived in caves and it played a part in our survival. It allowed us to understand when something was shouted rather than spoken – not many people are in the habit of casually telling you there's a hungry-looking bear behind you.

In the 1930s, two researchers, Fletcher and Munson from Bell Laboratories, were the first to accurately measure the uneven response of the ear and so we refer to the *Fletcher Munson contour control curve*.

It's absolutely vital that you keep this contour control in mind while mixing since it means that the bass energy produced will depend entirely on the mix volume. For example, if you balance the bass elements at a low monitoring level, there will be a huge bass increase at higher volumes. Conversely, if you mix at high levels, there will be too little bass at lower volumes.

Mixing appropriately for all three volumes is something of a trade-off as you'll never find the perfect balance for all listening levels, but whenever mixing dance music you should check that it sounds okay at low and medium volumes but great at high volume. That is, of course, assuming that clubs are your market – they don't play music at low or medium levels (Figure 19.1).

FREQUENCY BANDS

The second step is to become familiar with how frequencies are generally banded together as these bands play a large role whenever we speak about sub-bass, bass, mid-range, etc. This isn't as difficult as it may initially sound: we've been subjected to these frequencies since we were born; you just need to understand how they are grouped together. To help you decipher this, what follows is a list of how frequencies are generally banded and described.

Sub-Bass: Under 50 Hz

At frequencies this low, it's impossible to determine pitch; nonetheless this range is commonly occupied by the very lowest area of a kick drum and bass instruments. Notably, most loudspeaker and near-field loudspeaker monitors cannot reproduce frequencies this low reliably, and in most genres of dance music all signals this low will be rolled off to prevent loss of volume in hi-fi systems or to prevent damaging the bass response in club PA systems.

Bass: 50–250 Hz

This range is typically adjusted when the bass boost is applied on most home stereos, and where most of the bass is contained in all dance music mixes. EQ cuts (or boosts) around this area can add definition and presence to the bass and kick drum.

Mid-Range Muddiness Area: 200–800 Hz

This frequency range is the main culprit for mixes that are described as sounding muddy or ill-defined. It is often the reason why a mix is fatiguing to the ear. If too many sounds are dominating in this area, a track can quickly become tiring and irritating.

True Mid-Range: 800–5000 Hz

As previously touched upon, the true mid-range is where loudspeakers produce most of their energy; human hearing is particularly sensitive to these frequencies. This is because the human voice is centred here, as are TVs and radios; thus, even small boosts of a decibel will be perceived to be the same as boosting 10 dB at any other frequency. Subsequently, you should exercise extreme caution when adjusting frequencies in this area.

High Range: 5000–8000 Hz

This range is typically adjusted when the treble boost is applied on most home stereos and is where the 'body' of hi-hats and cymbals often reside. This area is sometimes boosted by a couple of decibels to make sounds artificially brighter.

Hi-High Range: 8000–20 000 Hz

This final frequency area often contains the higher frequency elements of cymbals and hi-hats. Some engineers also apply a small shelving boost at 12 000 Hz to make the music appear more hi-fidelity. This often adds extra detail and sheen without introducing any aural fatigue. It does, however, require plenty of care as boosting anywhere in this region can intensify any background hiss or high-frequency noise.

There's much more to applying EQ to a mix than understanding the separate frequency bands and to become truly competent at engineering. It's also vital to train your hearing so that you can identify timbres and the corresponding frequency in hertz. Unfortunately, this isn't something that you can pick up just from reading about it but will only come with practical experience and careful listening (Figure 19.2).

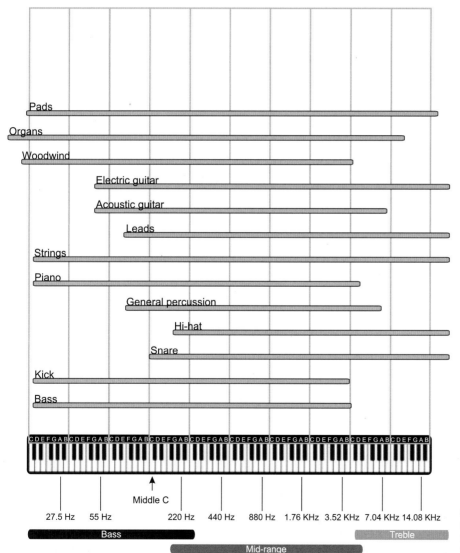

FIGURE 19.2
A chart defining the general ranges taken up by instruments in dance music

CREATING A SOUNDSTAGE

Finally, you need to be able to envisage the 'stage' on which the instruments will be placed. When approaching any mix it's usually best to imagine a three-dimensional room – the soundstage – on which you can place the various instruments.

Sounds placed on this stage can be positioned anywhere between the left or right 'walls' using a pan pot; they can be positioned at the front or back, or anywhere in between, using volume; and the frequency content of the sound will determine whether it sits at the top of the stage (high frequencies), the middle (mid-range frequencies) or the bottom (low frequencies).

The concept behind mixing is to ensure that each sound occupies its own unique space within this room so that it can not only be heard but also fits in well with everything else. To do this we can break the soundstage into three distinct areas: front to back, horizontal and vertical. We'll begin by examining the front to back area.

FRONT–TO–BACK PERSPECTIVE

One of the primary auditory clues we receive about how far away we are from a sound source is through the intensity of air pressure that reaches our ear drums. As we touched upon in Chapter 1, sound waves spread spherically outwards in all directions from any sound source. The further these have to travel the less intense the sound becomes. In other words, the further we are from a source of sound, the more the sound waves will have dissipated, resulting in a drop of volume.

The intensity of sound is governed by the inverse law which states that:

> 'Sound pressure decreases proportionally to the square of the distance from the source'

Roughly translated, this means that each time the distance from the sound doubles it'll become roughly 6 dB quieter.

If we interpret this in the context of a mix, we can say that the louder an instrument is the more 'in your face' it will appear to be. However, although many dance mixes appear to have everything right in your face and at the front of the soundstage, this is far from the case. Indeed, depth perception is the first aspect to take into consideration when producing a good mix.

If every instrument were placed at the forefront, all the volumes would be at equal gain; this would produce a cluttered image as every instrument fights to be at the front. What's more, the mix would appear just two-dimensional because when listening to music our minds always work on the basis of comparison. That is, for us to gain some perception of depth there must be some sounds in the background so we can determine that some sounds are at the front of the mix and vice versa.

This means that you need to determine which sounds should be up front and which should be progressively further back in the mix but this isn't as difficult as it may sound. Dance music, by its very nature, is based on rhythm and groove because that's what we dance to. Thus, in all genres of dance it makes sense that both the drums and bass should take soundstage priority over every other instrument.

The most obvious way to provide this depth perspective is with volume adjustment, as it's natural for us to believe that the louder a signal is the closer it must be. This only applies to sounds that we are familiar with, though, and for dance music it's quite usual to use unnatural or unrealistic timbres. We can overcome this problem by taking advantage of the fact that how our ears and mind perceive a sound depends on the acoustic properties it projects.

As touched upon in the very first chapter of this book, the higher the frequency of a sound the shorter the wavelength becomes; therefore, we can assume that if a high-frequency sound has to travel over a long distance, many of these frequencies will dissipate along the way. This means that if a high-frequency sound is playing a good distance away from a listener, some of the high-frequency content will be reduced.

This effect can be particularly evident with passing cars whose stereo systems take up more space than the engine and shift enough air to blow out a candle at 50 paces. You hear the low frequencies while the car is at a distance. As it approaches, the higher frequencies become more pronounced until it passes by, whereby the higher frequencies begin to decay as it moves off into the distance. We can therefore emulate this effect with some creative EQ. Cutting just a few decibels at the higher frequency ranges of a sound makes the sound appear more distant when it is heard in association with other sounds with low-frequency content. Alternatively, increasing the higher frequencies using enhancers or exciters can make a timbre appear much more up front.

Notably, this frequency-dependent effect is an important aspect to consider when working with a compressor since if a fast attack is used the transient will be clamped down, thus reducing some of the high-frequency content and pushing it to the rear of the mix. This is often the cause of numerous problems, as most usually you would like to compress to prevent clipping or to introduce 'punch' into a full mix. While you could always increase the gain to bring the timbre to the front of the mix, it would not work as well as if they all had the high-frequency content intact. The only suitable way to avoid this is to use a long attack on the compressor or to employ a multi-band compressor and set it to compress only the lower frequency content, leaving the higher frequencies unaffected.

Another characteristic of the front-to-back perspective is derived from the amount of reverberation the signal has associated with it. All sound has some natural reverberation as the reflections emanate from surrounding surfaces, but the amount and stereo width of these reflections depends on how far away the source of the sound is. If a sound is far away, then the stereo width of the reverberations will dissipate as they travel through the air, and they will be subjected to more reverberation. This is important to bear in mind since many artists wash a sound in stereo reverb to push it into the background and then wonder why it doesn't sound quite 'right' in context with the rest of the mix.

From this we can also determine that if you need to place an instrument at the rear of a mix, it's wise to use a mono reverb signal with a long tail and to remove some of the higher frequencies. This will emulate the natural response we've come to expect from the real world, even if that particular timbre doesn't occur in the real world. Most good reverb units will allow you to remove higher frequencies with an on-board filter. If not, by returning the reverb effect into a normal mixing channel you can use the channels EQ to remove the high frequencies this way.

Of course, sounds that are up close or at the front of a mix will have little or no reverb associated with them but in some instances you may wish to apply reverb to thicken the timbre. On these occasions you should apply it in stereo; you should also aim to keep the stereo width of the reverb contained to prevent it from occupying too much of the left and right perspective. Also, use a short tail and a pre-delay of approximately 50–90 ms to separate the timbre from the effect to prevent it from washing over the attack stage; otherwise it may force the instrument to the back of the mix.

Always keep in mind that applying effects too heavily can make sounds very difficult to localize and it's absolutely vital that each instrument be pinpointed to a specific area within a mix. Otherwise it will appear indistinct and muddy.

THE HORIZONTAL PERSPECTIVE

The next consideration in a mix is the horizontal plane – the distance and positioning of sounds between the left and right walls of the virtual room. The major audio clue that helps us derive the impression of panning and stereo can be determined by (a) the volume intensity between sounds or (b) the timing between sounds.

The principle behind altering the volume of sounds to produce a stereo image was first realized by Alan Blumlein in the early 1930s. An inventor at EMI's Central Research Laboratories, he researched the various ways in which the ears detect the direction from which a sound comes. Along with deriving the technique for creating a stereo image in gramophone records, he also figured that to maintain realism in a film, the sound should follow the moving image.

This technique was first employed in Walt Disney's *Fantasia*, when sound engineers asked Harvey Fletcher (of the same Fletcher Munson curve) if he could create the impression of sound moving from left to right for the movie. Drawing on Alan Blumlein's previous work, Fletcher came to the conclusion that if a sound source is gradually faded in volume in one speaker and increased in the other it will gradually move from one to the other. The engineers at Disney put this idea into practice by using a potentiometer to vary the volume between two speakers and labelled the process 'panoramic potentiometer', hence the term pan pot.

Although this volume-intensity difference between two speakers is still the most commonly used method for panning a sound around the image, we can also receive directional clues from the timing between sounds. Otherwise known as the directional cues, Precedence or Haas effect, this process takes advantage of the Law of the First Wavefront, which states that:

> 'If two coherent sound waves are separated in time by intervals of less than 30 ms, the first signal to reach our ears will provide the directional information'.

In layman's terms, if a direct sound reaches our ears anywhere up to 30 ms before the subsequent reflections, we can determine the position of a sound. For example, if you were facing the central position of the mix, any sound leaving the left speaker would be delayed in reaching the right ear and vice versa for the left ear, an effect known as interaural time delay (ITD). Considering that sound travels at approximately 340 m/s, this effect can be emulated by setting up a delay unit on a mono signal and delaying it by a couple of milliseconds, producing the impression that the sound has been panned.

For this effect to work accurately, we also need to consider that our ears are on the side of our head and therefore our head gets in the way of the frequencies from opposite speakers (an effect known as head-related transfer function). This means that, provided we are facing the centre of the stereo image, some of the higher frequencies emanating from the left speaker will be reduced before they reach the right ear, simply because sound doesn't travel through our heads, it has to travel around it. This effect can be simulated by cutting a decibel at around 8 kHz from the delayed signal.

Naturally, to accomplish these types of effects you need to use mono sounds within the mix. Here is where many practicing musicians make one of their biggest mistakes. With all of today's tone modules, keyboards and samplers featuring stereo outputs, it's quite easy to fall into the trap of using stereo sounds throughout an arrangement, but this only leads to a mix that lacks any real definition.

A stereo image is formed by spreading any sound to both the left and right speaker using one of the two methods: effects or layering. The results are then panned left and right so that they play across both speakers. While this always makes them sound much more interesting in isolation, it's only to persuade you to part with your money for the synthesizer.

If you made a mix completely from stereo files, they would collate in the mix and all occupy the same area of the soundstage. Of course, you could narrow the stereo width of each file with the pan pots so that they don't all occupy the same position, but this approach is not particularly suitable for making a transparent mix.

The soundstage for any mix should be transparent enough that you can picture the mix in three dimensions with your mind's eye and pinpoint the exact position of each instrument. Many dance mixes are quite busy, encompassing anything from 8 to 12 different elements playing at once. If these were all stereo it would be difficult, if not impossible, to find a pan placement for each. This often leads to an inexperienced engineer resorting to unnecessary EQ in a futile attempt to carve out some space for the instruments or start to 'creep the faders'.

Creeping faders is typical of most inexperienced engineers and is the result of gradually increasing the volume of each track so that it can be heard above other instruments. For instance, they may increase the gain of vocals so that they can be heard above the bass, and then increase the drums so they can be heard above the vocals, and then increase the bass so it can be heard with the drums and then move back to the vocals and, well, you get the picture. Eventually, the mixer is pushed to its maximum headroom and the mix turns into loud, incomprehensible rubbish.

Unnecessary EQ and volume creep can be avoided by utilizing mono sound sources, since if two instruments share the same frequency range they can be separated by panning one sound to the left and the other to the right. There's little need to spatially separate them by panning them as far as possible in each direction and, in many cases, a panning space of approximately 'two hours' will allow both instruments to be heard clearly. By 'hours' we mean the physical position of a pan pot on a mixing (i.e. 1 o'clock, 2 o'clock, etc.) Thus, for a 2-hour panning the right pan would read 1 o'clock and the left would read 11 o'clock.

This physical pan pot measurement is only mentioned in a theoretical sense. You should always pan with your ears and not your eyes. All mixing desks are not equal and where one may place the sound at a certain position at a certain setting, the same setting on a different mixer may place the sound in a different position.

There is yet another reason why you should avoid using stereo files throughout and this is based upon our perception of stereo. Just as different volumes aid us in gaining some perception of depth within a mix, the same is true of stereo.

It's in our nature to turn and face the loudest, most energetic part of any mix, which in many mixes is the kick drum. If this happened to be a stereo file, it would be spread between the left and right speakers and the main energy would be spread across the soundstage. If, however, mono samples were used and the kick drums were placed centrally, we would perceive the position of other sounds much more clearly, which would increase our overall sense of space within the mix.

On top of this, when working with mono files on a stereo system we also perceive the volume of a sound by its position in the soundstage. This means that if a signal is placed dead centre, the energy is shared by both speakers and it will appear louder than if the same signal, at the same gain, were placed in the left or right speaker alone.

Typically, this perceived volume difference can be as much as 3–6 dB, so many professional and semi-professional mixing desks (and some audio sequencers) will implement the panning law. According to the panning law, any sounds that are placed centrally are subjected to either 3 or 6 dB of attenuation, which can be set by the user. It should be noted that not all mixing desks will implement the panning law, so after panning the instruments to their respective positions you may need to readjust the respective volumes again.

Perhaps, most important of all, if you wish to pan instruments faithfully, your speaker system must be configured correctly in relation to your monitoring position. Ideally, you should be positioned at one point of an equilateral triangle with the loudspeakers positioned at the other two points. This means that speakers should be positioned an equal distance apart from one another and your listening position to ensure that the signal from each reaches your ears at the same time.

If there is just a few inches difference between these three points, the sound from one speaker could be delayed in reaching your ears by a couple of milliseconds which results in the stereo image moving to the left or right. What's more, you also need to ensure that the volume from each speaker is the same when it reaches your current listening position – even differences as small as 1 or 2 dB can shift the image considerably to one side.

A possible solution to producing an accurate soundstage would be to mix through headphones, but this should be avoided at all costs. While they are suitable for listening to a mix without disturbing the neighbours, they overstate the left and right perspective because they're positioned on either side of your head. Keep in mind that when listening to a mix from loudspeakers, sound from the left speaker reaches your right ear and vice versa for the left speaker. If you mix using headphones alone, you can easily over- or understate the stereo field.

Finally, any stereo mix must also work well in mono. Most mixing desks will offer a mono button that, when used, sums the channels together into a mono signal. This prevents any phasing that, while not immediately evident in stereo playback, can result in a comb filtering effect. While this may seem worthless when all of today's hi-fi systems are stereo, most TV and radio stations still broadcast in mono, most clock/transistor radios are also mono, and so are many club PA systems.

This means that if you've relied entirely on stereo while mixing, then parts of the mix can disappear altogether. This doesn't mean you have to destroy the stereo mix. When you switch the mix to mono, the image will simply be collated into the centre. When in mono, the general tone of the instrument and their volumes should remain constant; if not, you will need to re-examine the mix and check for what is causing mono compatibility problems. In the majority of cases this is caused by using stereo files or the application of too many stereo effects.

THE VERTICAL PERSPECTIVE

The final perspective of a mix is the top to bottom of the soundstage, with higher frequency elements sitting towards the top of the stage and lower frequencies towards the bottom. Much of this positioning will already be dictated by the timbres used in the mix. For example, basses will tend to sit towards the bottom of the mix and hi-hats will naturally sit towards the top. However, it should be noted that these, and any other timbres, will contain frequencies that do not necessarily contribute to the sound when placed into a mix.

A typical case in point is that the basses used in dance music do not consist entirely of low-frequency elements but also some mid-range elements and possibly some high frequencies too. If you were to tonally adjust this type of sound by removing all of the mid-range frequencies, the bass would move more towards the bottom of the soundstage, making some space for instruments to sit in the mid-range. This 'corrective' adjustment is one of the most fundamental aspects of mixing and is accomplished with the EQ on a mixing desks channel.

EQ is a frequency volume control that allows you to reduce or increase the gain of a band of frequencies contained within a sound, but it can be used for much more than simply making space in a mix. Applied cautiously it can be used to make insipid timbres much more interesting, add more definition to the attack portion of a sound to pull it to the front of a mix or prevent the bass from overpowering and 'muddying' the entire lower frequency range of the mix. However, if you apply it without knowing exactly what affect it will have on a sound and those surrounding it, a mix can fall apart within minutes and you end up with nothing more than a muddy, ill-defined mess that doesn't translate at all well on a typical club PA system.

To understand the concept behind using EQ, we need to revisit the first chapter which explained that all sounds are made up of a number of frequencies that are a direct result of the fundamental and its associated harmonics at differing amplitudes. Also, that it's this predetermined mix of harmonics that helps us establish the type of sound we perceive, whether it's an earth-shaking bass, a screaming lead or a drum kick.

We only require all these frequencies present if the sound is played in isolation. If it's mixed in with a number of other instruments, we only need to hear a few basic frequencies because our hearing/perception will persuade itself that the others are present – they're just masked behind the other instruments.

This natural occurrence is one of the main keys to mixing. If frequencies that we don't necessarily need are removed, well have more room for the important frequencies of other instruments that we do need to hear. This means that we need a way of pinpointing a specific group of frequencies within a sound that we can remove or enhance. In 'professional' desks this is accomplished with a parametric EQ.

Fundamentally, a parametric EQ consists of three controls: a frequency pot, a bandwidth pot and a gain pot. The frequency pot is used to select the centre frequency that you want to enhance or remove, while the bandwidth pot determines how much of the frequencies either side of the centre frequency should be affected. Finally, the gain pot allows you to increase or reduce the volume of the chosen frequencies.

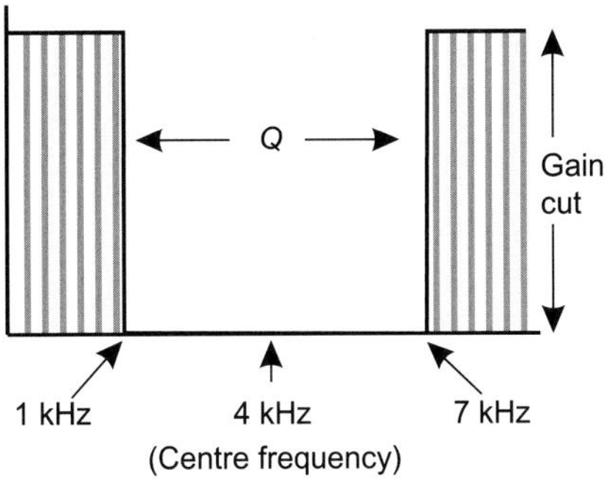

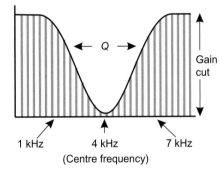

FIGURE 19.3
Theoretical EQ

Suppose you had a lead sound that contained frequencies ranging from 600 Hz to 9 kHz but the frequencies between 1 and 7 kHz were not required. Using the frequency pot, you could home in on the centre of these (4 kHz) and then set the bandwidth wide enough to affect 3 kHz either side of the centre frequency. If you then reduce the gain of this centre frequency, the frequencies 3 kHz either side would be attenuated, too.

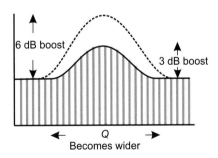

FIGURE 19.4
Practical EQ

The size of this bandwidth is often referred to as the Q (quality). The smaller the Q number, the larger the 'width' of the bandwidth (Figure 19.3).

Such a precise bandwidth as shown in the figure is entirely theoretical. In the real world the two extremes of the bandwidth are not attenuated at right angles. This is because, similar to synthesizer's

FIGURE 19.5
Non-constant Q

filters, EQ has a transition period or 'response curve' that allows it to appear more natural. Typically, this is around 3 dB per octave but can sometimes be 6 dB per octave (Figure 19.4).

The problem with this approach is that if you apply a particularly heavy cut or boost and the Q does not remain constant, the bandwidth will increase in size as you boost or cut further. Typically, the width of the Q is measured at 3 dB from the floor, so if the bandwidth were set to remove 2 kHz either side of the centre frequency at an EQ boost of 3 dB and you were to then boost by another 3 dB, the bandwidth would affect more of the frequencies either side of the centre frequency (Figure 19.5).

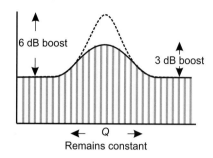

FIGURE 19.6
Constant EQ

This action is known as a 'non-constant Q' because the Q changes as you increase or decrease the gain at the centre frequency. This creates an unnatural sound, so many professional desks and external EQ units will vary the response curve as the gain is increased or decreased to maintain a constant bandwidth. This produces a sound that appears much more natural to our ears and is the sign of a well-designed EQ unit (Figure 19.6).

Q should only be used when discussing the theory behind EQ, and I wouldn't advise taking it any further than this book. In a studio situation you're communicating with musicians, not techno freaks, so it's usual to talk about the Q in octaves. Otherwise you could find yourself laughed out of the studios. For those with an unhealthy academic interest into how the Q value can be transformed into octaves, you can use the following math theorem:

$$(4 \times Q^{2+1})^5 = B$$

$$\log_2\left(\frac{B+1}{B-1}\right) = \text{Octave}$$

But for those who simply can't be bothered, it's worth memorizing this table as it covers the most commonly used Q settings while mixing:

Q Setting	Octave Range
0.7	2 Octaves
1.0	$1\frac{1}{3}$ Octaves
1.4	1 Octave
2.9	½ Octave
5.6	¼ Octave

Generally speaking, a bandwidth of 1/4th of an octave or less is used for removing problematic frequencies, while a width of a couple of octaves is suitable for shaping a large area of frequencies. For most work, though, and especially if you're new to mixing, you should look at always maintaining a non-constant Q of 1.0. This is sometimes referred to by many engineers as the 'magic Q' since it commonly avoids any masking problems (we'll look at this in a moment) and sounds much more natural to the ears.

As of yet, we've only concerned ourselves with one form of EQ – parametric – but there are shelving EQs too. These will, or at least should, be available on most semi-professional and all professional desks and consist of low- and high-pass filters along with low- and high-shelf filters. These two different forms of filters are often confused with one another due to their similar nature, but it is important to differentiate between them.

Both low-pass and high-pass EQs are based on exactly the same principle as the low- and high-pass filters used in synthesizers. The low-pass (sometimes referred to as high-cut) will attenuate the higher frequencies that are above the cut-off point. A high-pass (sometimes referred to as low-cut) will attenuate all the lower frequencies that are below the cut-off point. Conversely, a low-shelf filter will boost (or cut) any frequencies that are below the cut-off point, while a high-shelf will boost (or cut) any frequencies above the cut-off point.

Of course, this does mean that if only low- and high-pass filters are available on the desk, it is possible to emulate the action of a low- and high-shelf and this is the reason why the two types are often confused. For instance, if you need to emulate the action of a high-shelf, you could use the high-pass filter and increase the EQ overall output level.

The specifications of the mixer will depend on how much control you have over the filters and shelves but, generally, a shelving filter will have two controls: one to select the frequency (often called the knee frequency) and one to adjust the gain of the frequencies above (or below) the shelf. Some of the more expensive desks may also offer the opportunity to adjust the transition slope of all these filters, and this can sometimes be important when we need to 'roll off' frequencies.

Since many speaker systems are incapable of producing frequencies as low as 40 Hz, they should be removed altogether and the attenuation of frequencies above or below the cut-off point is often termed as 'rolling off'. Thus, if you were asked to roll off all frequencies below 40 Hz, you would employ a high-cut filter and position the cut-off point at 40 Hz. Or at least you would if the transition slope of the EQ filter were more or less vertical. As we've seen, EQ units will use a transition slope to make the effect appear more natural to our ears, but we don't necessarily need this when rolling off frequencies that are not required; so it is useful to reduce this if it happens to be particularly long (6 dB per octave and over) (Figure 19.7).

Naturally, all this control requires dedicated pots on a mixer and additional circuitry to implement it; so many of the cheaper mixing desks will not employ shelving or low/high-pass filters and in some instances may not even offer any parametric EQ. Indeed, it's fair to say that the cheaper the desk, the less EQ control you'll be offered. EQ can broadly be categorized into three areas: fixed frequency, semi-parametric and graphic.

Fixed Frequency

Fixed-frequency EQ is possibly the worst type to have on any mixing desk as it's the least flexible, but as it is the cheapest to implement, many cheap consumer desks will employ it. Fixed-frequency EQ consists of two, sometimes

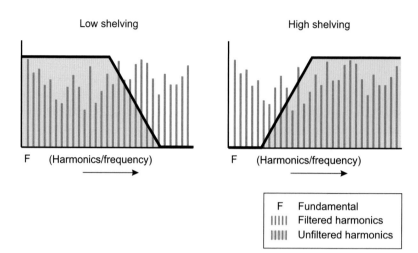

FIGURE 19.7
The action of shelving
filters

three, rotary pots that control the low, mid and high frequencies. To accomplish this, they commonly use low/high-pass filters with a fixed cut-off frequency at 12 kHz and 75 Hz for the high and low band, respectively. If there is a third pot, then this controls the mid-band and is generally a notch filter with a centre frequency that's fixed at around 2 kHz and has a fixed bandwidth of approximately two octaves (see the chart for the respective Q!). This type of EQ makes precise, frequency-specific shaping impossible, and unless you have an external parametric EQ unit it is far from suitable for creating a good mix.

Semi-Parametric

Semi-parametric EQs (sometimes known as sweep EQs) utilize a similar design as fixed-frequency EQs with a fixed low/high-pass filter at 75 Hz and 12 kHz, respectively. Rather than having a fixed mid-band, only the bandwidth is fixed and you are free to sweep the centre frequency. This type of control is typical of mid-priced consumer mixing desks, but while offering more freedom to select frequencies, the fixed bandwidth means that it isn't versatile enough for creative EQ. It can be particularly difficult to produce a good mix with this limitation.

Graphic

Graphic EQs do not appear on mixing desks but are sometimes used by mastering engineers to set the overall tonal colour of a full mix. They're also possibly the most instantly recognizable of all EQ units. Most home stereos now have them fitted as standard, appearing as a number of faders, each of which offers control over a specific frequency band.

The size of the bandwidth covered by each fader obviously depends on how many faders it has. So the more faders available, the better the EQ, since each bandwidth will be smaller. The reason these are referred to as graphic EQs is because you can often tell what type of music people listen to by the way the equalizer is configured. Hence it's rather graphic in nature.

With the intervention of computers and audio sequencers, this graphic nature is not just limited to graphic EQs anymore, as many software mixing desks and some of the latest digital desks will mix a parametric EQ with the visuals of a graphic EQ to produce a paragraphic EQ. This gives all the freedom of a parametric EQ with the visual ease of using a graphic EQ. The immediate benefit of this is obvious. Rather than having to determine the current EQ settings through a series of often cryptic pot positions, you can actually see the frequency range being affected.

That said, keep in mind that your potential audience does not see EQ settings, they only hear them and so you should do the same. One of the biggest problems with mixing on any 'visually enhanced' software desk is that you have the tendency to use your eyes more than your ears and this can be disastrous. If it looks right, it may not sound right; if it sounds right, it most probably is.

PRACTICAL MIXING

Armed with the theory of mixing, we can approach a mix on a more practical sense, and it should go without saying that the first step to creating a good mix is to ensure that you can actually hear what you're doing reliably. If you don't hear the frequencies within a mix accurately, then you certainly can't EQ or affect them accurately. This means that you *must* use loudspeaker studio monitors to scrutinize a mix rather than placing your trust in everyday hi-fi speakers.

No matter how 'hyped' these are by manufacturers for having an excellent frequency response, all hi-fi speakers deliberately employ frequency-specific boosts and cuts throughout the sonic range to make the sound produced by them appear full and rounded. As a result, if you rely entirely on these while mixing, you'll produce mixes that will not translate well on any other hi-fi system. For instance, if your particular choice of hi-fi speakers features a bass boost at 70 Hz and a cut in the mid-range, you're obviously going to produce a mix based on this. If your mix is then played on a system that has a flatter response there will be less bass and an increased mid-range.

It's impossible to recommend any particular make or model of studio monitor because we all interpret sound differently, but generally you should look for a model that makes any professionally produced music you like sound moderately lifeless. If you can make your mixes sound good on these, then it's likely to sound great on any other system. Then again, you shouldn't aim for a totally flat response, and it's usual to look for monitors that are a happy medium between coloured and flat. Indeed, most dance musicians will deliberately use monitors with a vaguely coloured response, since 99.9% of people who are going to listen to it will do so in a club, in a car or on a typical home stereo system.

Even with a good monitoring system, it's important to also note that the actual positioning of them will have an effect on the frequencies they produce.

Placing any monitor close to a corner or wall will increase the perceived bass by 3, 6 or 9 dB due to the boundary effect. This is similar in some respects to the proximity effect that's exhibited when singing too close to a microphone – there's an artificial rise in the bass frequencies.

With monitors this can be attributed to the low-frequency waveforms emanating from the rear of the monitor onto the wall or corner directly behind it. This low-frequency energy combines with the energy produced by the monitor itself and doubles the perceived bass. As a result, the general rule of thumb is that unless they're specifically manufactured to be used close to a wall (i.e. they have a limited bass response and use the wall to produce the low-frequency energy), they should be placed at least 3 feet away from any wall. Of course, this may not always be possible, and if it isn't then you'll have to learn to live with the increased response by listening to as many commercial mixes you can and training your ears to accept the response they produce.

On that note, before you even approach any mix you should take time out to listen to some of your favourite commercial mixes. We hear music everyday but we very rarely actually *listen* to it, and if you want to become competent at mixing, you need to differentiate between hearing and listening. This means that you need to sit down and play back your favourite commercial mixes at a moderate volume (just above conversation level) and try to pick out the action of the bass, the drums, vocals, leads and any sound effects. Listen for how the melodies have been programmed and ask questions such as:

- Is there any pitch bend in the notes?
- Does the drum loop stay the same throughout the track or does it change in any way?
- What pattern are the hi-hats playing?
- Do they remain constant throughout?
- What effects have been used?
- Where is each instrument placed in the mix?

Although many of these may not seem related to mixing, listening this closely for small nuances in drum loops, hi-hat patterns and basses, etc., will train your ears into listening closely to any music and will help you to not only begin identifying frequencies but also increase your awareness of arrangement and programming techniques.

Most important of all, though, despite the huge array of audio tools available today, the most significant ones are those that are stuck to the side of your head. Continually monitoring your mixes at excessive volume or listening to any music loud will damage the ears' sensitivity to certain frequencies. A recent survey conducted showed that those in their late teens and early twenties who constantly listened to loud music had more hearing irregularities than people twice their age.

Also, keep in mind that dabbling in pharmaceuticals for recreational use leads to a heightened sensory perception. Although they may have played a large role in the development of the dance scene, they don't help in the studios. Apart from losing hours from fits of laughter about terms such as ring modulation, the heightened sensory perception makes (a) everything sound great or (b) everything sound wrong, so you spend the next 10 h twiddling EQ on something that doesn't need it.

VOLUME ADJUSTMENTS

The first step in creating a mix is to begin by setting the relative volume levels of each track, and this means taking the musical style into account. You need to identify the defining elements of the genre and make sure that these are placed at the very front of the mix. Typically for all dance music this means that the drums and bass should sit at the front of the mix with everything else that must be located centrally sitting behind them. This includes the vocals as more often than not they will not be the truly defining part of the music.

It's generally best to begin mixing by starting with these defining points, so it's prudent to mute every channel bar the drum tracks and commence mixing these. As the kick drum is the most prominent part of a drum loop, set this so it's at unity gain and then introduce the snare, claps, hi-hats, cymbals and other percussion instruments at their relative levels to the kick.

At this point don't concern yourself with panning, EQ or effects but try to find a happy medium so that the drum levels are close to how you imagine the finished product to sound. Follow this by adding the bass, then the vocals (or the next important part of the mix) and gradually add each audio track in order of priority, adjusting the volume levels until all the instruments are playing together in the mix. It's vital that you're not concerned with masking, EQ or effects at this stage. You need to make the mix work quickly before you start jumping in with compressors, EQ or effects as these can quickly send you off on a tangent and you can lose the direction of the mix. If you're experiencing troubles setting the volume levels, then a quick trick is to switch the mix to mono as this allows you to hear the relative volume levels more accurately.

This form of 'additive' mixing is only one way to approach a mix, and some users feel more at ease using the 'subtractive' method. This means that all the faders are set at unity gain, and each is reduced to obtain the appropriate levels required. Which to use is entirely up to your own digression and depends on which you feel most comfortable with.

Whichever method you feel at ease with, an approach I often recommend is that when adjusting any volume fader, do not listen to the instrument you are adjusting the volume of, rather listen to the whole mix and how adjusting the volume is affecting the mix. For instance, as you increase the volume of the strings (if they are present), do not listen to the strings, rather listen to how the other instruments are being affected by the strings' increase in gain.

PANNING

With all the relative volume levels set, you can then begin to pan the instruments to their appropriate positions. This is where your own creativity and scrutinizing of previous mixes of the same genre will come into play, as there are no 'strict' rules for creating a good soundstage. Of course, this isn't going to be much help to those new to mixing, so what follows is a *very* generalized guide to where instruments are usually positioned in most dance genres.

Instrument	Pan Position/Description
Kick	Positioned central so that both speakers share the energy
Snare	Positioned from 1 o'clock to 2 o'clock or central
Hi-hats	Positioned at the far left of the soundstage with a delayed version in the right
Cymbals	Positioned central or from 1 to 2 o'clock
Percussion	Positioned so that the different timbres are in the left and/or right of the soundstage
Bass	Positioned central so that both speakers share the energy
Vocals	Positioned central since you always expect the vocalist to be centre stage
Backing vocals	Occasionally stereo so they're spread from 2 to 4 o'clock
Synth leads (trance)	A stereo file positioned fully left and right with perhaps a mono version sat central
Synthesized strings	Positioned at 4 o'clock or stereo spread at 9 and 3 o'clock
Guitars	Positioned at 3 o'clock
Pianos	Commonly stereo with the high notes in the left speaker and low notes in the right
Wind instruments	Positioned in the left or right speaker (or both) at 3 and 9 o'clock
Sound effects	Wherever there is space left in the mix!

Apart from placing the kick, bass and vocals in the centre of the soundstage, note that this is a *very* general guide and the panning of instruments will define your own particular style of mixing. Indeed, the best solution to panning instruments is to have a damn good reason as to why you're placing a timbre there in the first place. Always have a plan and don't just throw sounds around the image regardless.

In direct contradiction, in many dance mixes the positioning of instruments is rarely natural, so you should feel free to experiment. In fact, in many cases it may be important to exaggerate the positions of each instrument to make a mix appear clearer and more defined. Nevertheless, you shouldn't position a sound in a different area of the soundstage just to avoid any small frequency clashes with other instruments. In these circumstances you should try to EQ the sound beforehand; if it still doesn't fit, consider panning it out of the way.

After the instruments have been panned it's highly likely that the relative volumes will have adjusted. You'll need to go back and readjust these so that the instruments are all at their appropriate volumes. Again, while panning the instruments, listen to how the rest of the instruments are being affected, rather than listening to where the instrument being panned is moving.

EQ

Although there are no rules on how to use EQ when it comes to mixing, some generalizations can be made. Firstly, any EQ is essentially acting like a filter and this means that like any other type of filter EQ introduces resonance at the cut-off point and phase shifting. In other words, it applies small amounts of comb filtering into the audio, which sounds like small amounts of distortion. This is where the quality of the EQ unit being used plays a major role: less competent/cheaper EQ units will introduce more comb filtering than the more expensive units.

As a result, you need to exercise care while EQ'ing sounds so that it doesn't become too apparent; this is why many engineers will advise that you should look to cutting frequencies rather than boosting them. This is because boosting any frequency will make the phase distortion much more noticeable. To support this theory further, in the 'real' world our ears are more used to hearing reductions in frequencies than boosts since frequencies are generally reduced by walls, objects and materials.

Yet another reason to cut rather than boost lies with the recording of the audio itself. To increase the signal-to-noise ratio, any audio should be captured as 'hot' as possible. If an audio file is already close to clipping, EQ boosts may push the audio to distortion.

As mentioned, these are only generalizations and so in direct contradiction, on occasion, boosting the frequencies that are lower in volume than the fundamental can be used as a creative sound design tool. Indeed, one of the most common misconceptions about EQ is that you should never really have to use it if the sound is recorded correctly in the first place.

The viewpoint stems from when EQ was first used as a means of correcting the frequency response of a recording to prevent it from compromising the recorded sound. This was because recording in the early days was carried out with poor-quality gear when compared with today's standards and, therefore, frequencies had to be adjusted to reproduce the original recording.

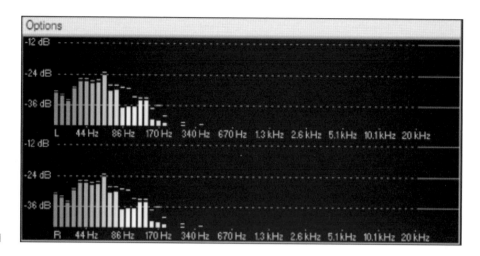

FIGURE 19.8

A spectral analysis of the house bass sound

While many older engineers still stand by this view, it means absolutely nothing to the dance musician since a proportionate amount of the timbres used will be synthetic anyway. Thus, apart from using EQ to help instruments sit together and create space within a mix, EQ should also be seen as a way of creatively distorting timbres.

Since EQ is essentially a filter, it can be used to reduce certain frequencies of a sound, which will have the immediate effect of increasing the frequencies that have not been touched. As an example of this, we'll use a spectral analyzer to examine the frequency content of the bass sound that was created for the house track (Figure 19.8).

Notice how there are a number of peaks in the timbre. If you were to listen to this sound, you would primarily hear these four loudest frequencies. However, using EQ creatively you can turn down these peaks to hear more of the entire frequency spectrum of the bass. It then becomes much more harmonically complex and can sustain repeated listening because the closer you listen to it the more you can hear. Making a sound appear as complex as possible by reducing the peaks is a common practice for many big-name producers. This could be further accentuated by applying small EQ boosting to the lower peaks of a sound.

On the other hand, simple ones can work as well as complex ones. This can be accomplished by cutting all but these four peaks. This cutting can also offer another advantage: as it is unlikely that you would be able to perceive the lowest frequencies (harmonics) contained in the sound, there is little need to leave them in; by removing them you would make more room for instruments in the mix.

You shouldn't simply grab a pot and start creatively boosting aimlessly since you need to consider the loudspeaker system that the mix is to be played back on. All loudspeaker systems have a limited frequency response and produce more energy in the mid-range than anywhere else. This means that if you decide to become 'creative' on the bass or mid-range, you could end up increasing energy that the speaker cannot reproduce faithfully, resulting in an over-zealous or muddy mid-range, or worse still, a quieter mix overall!

The main use for EQ within mixing, however, is to prevent any frequency masking between instruments. As touched upon in Chapter 1, the fundamental and subsequent harmonics contribute to making any sound what it is, but if two timbres of similar frequencies are mixed together, some of the harmonics are 'masked', resulting in the instruments sounding different in the mix than in isolation.

While this can have its uses in sound design (hocketing, layering etc.), during mixing it can cause serious problems since not only is it impossible to mix sounds that you can't hear, the frequencies that mix together can make both sounds lose their overall structure, become indistinct or, worse still, the confliction will result in unwanted gain increases.

Naturally, knowing if you have a problem with frequency masking is the first step towards curing it; so if you're unsure whether some instruments are being masked, you can check by raising the volume of every track to unity gain and then panning each track in turn to the left of the stereo field and then to right. If you hear the sound moving from the left speaker through the centre of the mix and then off to the right, it's unlikely that there are any conflicts. This, however, is about the only generalization you can make for mixing. The rest is entirely up to your own artistic preferences. Because of this, what follows can only be a very rough guide to EQ and to the frequencies that you may need to adjust to avoid any masking problems. As such, they are open to your own interpretations because, after all, it's your mix and only you know how you want it to sound.

Drums

DRUM KICK

A typical dance drum kick consists of two major components: the attack and the low-frequency impact. The attack usually resides around 3–6 kHz and the low-end impact resides between 40 and 120 Hz. If the kick seems very low without a prominent attack stage, then it's worthwhile setting a very high Q and a large gain reduction to create a notch filter. Once created, use the frequency control to sweep around 3–6 kHz and place a cut just below the attack, as this has the effect of increasing the frequencies located directly above.

If this approach doesn't produce results, set a very thin Q as before, but this time apply 5 dB of gain and sweep the frequencies again to see if this helps it pull out of the mix. Taking this latter approach can push the track into distortion, so remember to reduce the gain of the channel if necessary.

If the kick's attack is prominent but it doesn't seem to have any 'punch', the low-frequency energy may be missing. You can try small gain boosts of around 40–120 Hz but this will rarely produce the timbre. The problem more likely resides with your choice of drum timbre. A fast attack on a compressor may help to introduce more punch as it'll clamp down on the transient, reducing the high-frequency content. But this could change the perception of the mix, so it may be more prudent to replace the kick with a more significant timbre.

SNARE DRUM

Generally, snare drums contain plenty of energy at low frequencies that can often cloud a mix and are not necessary; so the first step should be to employ a shelving (or high-pass) filter to remove all the frequency content below 150 Hz.

The 'snap' of most snares usually resides around 2–10 kHz, while the main body can reside anywhere between 400 Hz and 1 kHz. Applying cuts or boosts and sweeping between these ranges should help you find the elements that you need to bring out or remove but, roughly speaking, cuts at 400 and 800 Hz will help it sit better while a small decibel boost (or a notch cut before) at 8 or 10 kHz will help to brighten its 'snap'.

HI-HATS AND CYMBALS

Obviously, these instruments contain very little low-end information that's of any use and, if left in, can cloud some of the mid-range. Consequently, they benefit from a high-pass filter to remove all the frequencies below 300 Hz.

Typically, the presence of these instruments lies between 1 and 6 kHz while the brightness can reside as high as 8–12 kHz. A shelving filter set to boost all frequencies above 8 kHz can bring out the brightness but it's advisable to roll off all frequencies above 15 kHz at the same time to prevent any hiss from breaking through into the track. If there is a lack of presence, then small decibel boosts with a Q of about an octave at 600 Hz should add some presence.

TOMS AND CONGAS

Both these instruments have frequencies as low as 100 Hz but are not required for us to recognize the sound and, if left, can cloud the low and low mids; thus, it's advisable to shelve off all frequencies below 200 Hz. They should not require any boosts in the mix as they rarely play such a large part in a loop but a Q of approximately 1/2 of an octave applied between 300 and 800 Hz can often increase the higher end, making them appear more significant.

Bass

Bass is the most difficult instrument to fit into any dance mix since its interaction with the kick drum produces much of the essential groove – but it can be fraught with problems.

The main problems with mixing bass can derive from the choice of timbres and the arrangement of the mix. While dance music is, by its nature, loud and 'in your face', this is not attributed to using big, exuberant, harmonically rich sounds throughout. As we've touched upon, our minds can only work in contrast, so for one sound to appear big, the rest should be smaller. Of course, this presents a problem if you're working with a large kick and large bass, as the two occupy similar frequencies which can result in a muddied bottom end.

This can be particularly evident if the bass notes are quite long as there will be little or no low-frequency 'silence' between the bass and the kicks, making it difficult for the listener to perceive a difference between the two sounds. Consequently, if the genre requires a huge, deep bass timbre, the kick should be made tighter by rolling off some of the conflicting lower frequencies and the higher frequency elements should be boosted with EQ to make it appear more 'snappy'. Alternatively, if the kick should be felt in the chest, the bass can be made lighter by rolling off the conflicting lower frequencies and boosting the higher elements.

Naturally, there will be occasions whereby you need both heavy kick and bass elements in the mix, and in this instance, the arrangement should be configured so that the bass and the kick do not occur at the same point in time. In fact, most genres of dance will employ this technique by offsetting the bass so that it occurs on the offbeat. For instance, trance music almost always uses a 4/4 kick pattern with the bass sat in between each kick on the eighth of the bar.

If this isn't a feasible solution and both bass and kick must sit on the same beat, then you will have to resort to aggressive EQ adjustments on the bass. Similar to most instruments in dance music, we have no expectations of how a bass should actually sound, so if it's overlapping with the kick making for a muddy bottom end, you shouldn't be afraid to make some forceful tonal adjustments.

Typically for synthetic instruments, small decibel boosts with a thin Q at 60–80 Hz will often fatten up a wimpy bass that's hiding behind the kick. If the bass still appears weak after these boosts, you should look towards replacing the timbre; it's a dangerous practice to boost frequencies below these as it's impossible to accurately judge frequencies any lower on the near-fields. In fact, for accurate playback on most hi-fi systems it's prudent to use a shelving filter to roll off all frequencies below 60 Hz.

Of course, this isn't much help if you're planning on releasing a mix on vinyl for club play as many PA systems will produce energy as low as 30 Hz. If this is the case, you should continue to mix the bass but avoid boosting or cutting anything below 40 Hz. This should be left to the mastering engineer who will be able to accurately judge just how much low-end presence is required. As a very rough guide for *theoretical* purposes alone, a graphic equalizer set to −6 dB at around 20 Hz, gently sloping upwards to 0 dB at 90 Hz, can sometimes prove sufficient enough for club play.

If the problem is that the bass has no punch, then a Q of approximately half of an octave with a small cut or boost and sweeping the frequency range between 120 and 180 Hz may increase the punch to help it to pull through the kick. Alternatively, small boosts of half of an octave at 200–300 Hz may pronounce the rasp, helping it to become more definable in the mix. Notably, in some mixes the highest frequencies of the rasp may begin to conflict with the mid-range instruments, and if this is the case then it's prudent to employ a shelving filter to remove the conflicting higher frequencies.

Provided that the bass frequencies are not creating a conflict with the kick, another common problem is the volume in the mix. While the bass timbre may sound fine, there may not be enough volume to allow it to pull to the front of the mix. The best way to overcome this is to introduce small amounts of controlled distortion, but rather than reach for the atypical distortion unit, it's much better to use an amp or speaker simulator.

Amp simulators are designed to emulate the response of a typical cabinet, so they roll off the higher frequency elements that are otherwise introduced through distortion units. As a result, not only are more harmonics introduced into the bass timbre without it sounding particularly distorted but you can use small EQ cuts to mould the sound into the mix without having to worry about higher frequency elements, creating conflicts with instruments sitting in the mid-range.

If the bass is still being sequenced from a tone module or sampler, then before applying any effects you should attempt to correct the sound in the module itself. As covered in an earlier chapter, we can perceive the loudness of a sound from the shape of its amplitude envelope and harmonic content, so simple actions such as opening the filter cut-off or reducing the attack and release stage on both the amplitude and the filter envelopes can make it appear more prominent. If both these envelopes are already set at a fast attack and release, then layering the kick from a typical rock kit over the initial transient of the bass followed by some EQ sculpting can help to increase the attack stage, but at the same time, be cautious not to overpower the track's original kick.

Although most genres of dance will employ synthetic timbres, on occasion they do utilize a real bass guitar, and if so, a slightly different approach is required when mixing. As previously mentioned, bass cabs will roll off most high frequencies, reducing most high-range conflicts, but they can also lack any real bottom-end presence. As a result, for dance music it's quite usual to layer a synthesizer's sine wave underneath the guitar to add some more bottom-end weight.

This technique can be especially useful if there are severe finger or string noises evident, since the best way to remove these is to roll off everything above 300 Hz. More importantly, though, unlike tone modules where the output is compressed to even level, bass guitars will fluctuate widely in dynamics and these must be brought under control with compression. Keep in mind that no matter what the genre of dance, the bass should remain even throughout the mix and remain consistent. If it fluctuates in level, the whole groove of the record can be undermined.

If after trying all these techniques there is still a 'marriage' problem between kick and bass, then it's worth compressing the kick drum with a short attack stage so that its transient is captured by the compressor. This will not only make the kick appear more 'punchy' but also allow the initial pluck of the bass to pull through the mix. That said, you should avoid compressing the kick so heavily that it begins to ring, as this will only muddy up the bottom end of the mix.

As ever, the compression settings to be used are entirely dependent on the kick in question but generally a good starting point is a ratio of 8:1 with a fast attack and release and the threshold set such that every kick activates the compressor. Ultimately, as all bass sounds are different the only real solution is to experiment by EQ boosting or cutting around 120–350 Hz or by applying heavy compression to conflicting instruments.

Vocals

Although vocals take priority over every other instrument in a pop mix, in most genres of dance they will take a back seat to the rhythmic elements of a mix. Having said that, they must be mixed coherently since while they may sit behind the beat, the groove relationship and syncopation between the vocals and the rhythm is what makes us want to dance. Subsequently, you should exercise great care in getting the vocals to sit properly in the mix.

Firstly, it should go without saying that the vocals should be compressed so that they maintain a constant level throughout the mix without disappearing behind instruments. Generally speaking, a good starting point is to set the threshold so that most of the vocal range is compressed with a ratio of 9:1 and an attack to allow the initial transient to pull through unmolested.

The choice of compressor used for this is absolutely vital and you must choose one that adds some 'character' to the vocals. Most of the compressors that are preferred for vocal mix compression are hardware units such as the LA 2A and the UREI 1176, but the Waves RComp plug-in followed by the PSP Vintage Warmer can produce good results. This is all down to personal choice, though, and it's prudent to try a series of compressors to see which produces the best results for your vocals.

It's worth noting that even when compressed it isn't unusual to automate the volume faders to help the vocals stay prominent throughout. It's quite common to increase the volume on the tail end of words to prevent them from disappearing into the background mix. If you take this approach, you may need to duck any heavy breath noises but you should avoid removing them between phrases, otherwise they could sound false.

Once compressed, vocals will or should rarely require any EQ as they should have been captured correctly at the recording stage and any boosts now can make them appear unnatural. That said, small decibel boosts with a 1/2 octave Q at 10 kHz can help to make the vocals appear much more defined as the consonants become more comprehensible. If you take this 'clarity' approach, however, it's prudent to use it in conjunction with a good de-esser. This will remove the sibilance boosting at 10 kHz but will leave the body of the vocals untouched. This must be applied cautiously, otherwise it could introduce lisps.

Alternatively, if the vocals appear particularly muddy in the mix, an octave Q placing a 2 dB cut at a centre frequency of approximately 400 Hz should remove any problems. If, however, they seem to lack any real energy while sitting in the mix, a popular technique used by many dance artists is to speed (and pitch) the vocals up by a couple of cents. While the resulting sound may appear 'off the mark' to pitch-perfect musicians, it produces higher energy levels that are perfectly suited towards dance vocals; besides, not many clubbers are pitch-perfect.

It's also sensible to add any of the effects you have planned for the vocals. As these play an important part in the music, they must fit now rather than later as any instruments that are introduced after them should have their frequencies reduced if they conflict with the vocals or effects. For instance, it's quite typical to apply a light smear of reverb to the vocals (remember the golden rule, though!) to help them produce a more natural tone, but if this were applied later in the mix you may find yourself EQ'ing the effect to make room for other instruments, which can produce unnatural results.

Synthesizers/Pianos/Guitars

The rest of the instruments in a mix will (or should) all have fundamentals in the mid-range. Generally speaking, you should mix and EQ the next most important aspect of the mix and follow this with progressively less important sounds. If vocals have been employed, there will undoubtedly be some frequency masking where the vocals and mid-range instruments meet, so you should look towards leaving the vocals alone and applying EQ cuts to adjust the instrument. Alternatively, the mid-range can benefit from being inserted into a compressor or noise gate, with the vocals entering the side chain so that the mid-range dips whenever the vocals are present. This must be applied cautiously and a 'duck' of 1 dB is usually sufficient; any more and the vocals may become detached from the music.

Most mid-range instruments will contain frequencies lower than necessary, and while you may not actually be able to physically hear them in the mix, they will still have an effect of the lower mid-range and bass frequencies. Thus, it's prudent to employ a shelving filter to remove any frequencies that are not contributing to the sound within the mix. The best way to accomplish this is to set up a high-shelf filter with maximum cut and, starting from the lower frequencies, sweep up the range until the effect is noticeable on the instrument. From this point sweep back down the range until the 'missing' frequencies return and stop. This same process can also be applied to the higher frequencies if some are present and do not contribute to the sound when it's sitting in the mix.

Generally, keyboard leads and guitars will need to be towards the front of the mix but the exact frequencies to adjust will be entirely dependent on the instrument and mix in question. Nevertheless, for most mid-range instruments, it's worth setting the Q at an octave and applying a cut of 2–3 dB while sweeping across 400–800 Hz and 1–5 kHz. This often removes the muddy frequencies and can increase the presence of most mid-range instruments.

Above all, try to keep the instruments in perspective by asking yourself questions such as:

- Is the instrument brighter than the hi-hats?
- Is the instrument brighter than a vocal?
- Is the instrument brighter than a piano?
- Is the instrument brighter than a guitar?
- Is the instrument brighter than a bass?

What follows is a general guide to the frequencies of most sounds that sit in the mid-range along with the frequencies that contribute to the sound. Of course, whether to boost or cut will depend entirely on the effect you wish to achieve.

PIANOS

- 50–100 Hz: Add weight to the sound
- 100–250 Hz: Add roundness
- 250–1000 Hz: Muddy frequencies
- 1–6000 Hz: Add presence
- 6–8000 Hz: Add clarity
- 8–12 000 Hz: Reduce hiss

ELECTRIC GUITARS

- 100–250 Hz: Add body
- 250–800 Hz: Muddy frequencies but may add roundness
- 1–6000 Hz: Allow it to cut through the mix
- 6–8000 Hz: Add clarity
- 8–12 000 Hz: Reduce hiss

ACOUSTIC GUITARS

- 100–250 Hz: Add body
- 6–8000 Hz: Add clarity
- 8–12 000 Hz: Add brightness

SYNTH LEADS/STRINGS/PADS

- 50–100 Hz: Add bottom-end weight
- 100–250 Hz: Add body
- 250–800 Hz: Muddy frequencies
- 1–6000 Hz: Enhance digital crunch
- 6–8000 Hz: Add clarity
- 8–12 000 Hz: Add brightness

WIND INSTRUMENTS

- 100–250 Hz: Add body
- 250–800 Hz: Muddy frequencies
- 800–1000 Hz: Add roundness
- 6–8000 Hz: Add clarity
- 8–12 000 Hz: Add brightness

PROCESSING AND EFFECTS

We've looked at the various effects in previous chapters, so rather than reiterating it all we'll just say that you should refrain from using any effects during the mixing process (bar the vocals) and only once the mix is together should you consider adding any effects. Even then, you should only apply them if they are truly necessary.

Always keep in mind that empty spaces in a mix do not have to be filled with reverb or echo decays: a good mix works on contrast. When it comes to effects and mixing, less is invariably more. Any effects, but particularly reverb, can quickly clutter up a mix, resulting in a loss of clarity. Since there will probably be plenty going on already, adding effects will only make the mix busier than it already is and it's important to keep some space in between the individual instruments. Indeed, one of the biggest mistakes made is to employ effects on every instrument when, in reality, only one or two may be needed throughout. Therefore, before applying any effects it's prudent to ask yourself why you are applying them – to enhance the mix? or make a poorly programmed timbre sound better? If it's the latter, then you should look towards using an alternative timbre rather than trying to disguise it with effects.

Above all, remember the golden rules when using any effects:

1. Most effects should not be audible within a mix. Only when they're removed you should notice the difference.
2. If an effect is used delicately, it can be employed throughout the track, but if it's extreme, it will have a bigger impact if it's only used in short bursts. You can have too much of a good thing but it's better to leave the audience gasping for more than gasping for a break.

COMMON MIXING PROBLEMS

Frequency Masking

As touched upon earlier, frequency masking is one of the most common problems experienced when mixing, whereby the frequencies of two instruments are matched and compete for space in the soundstage. In this instance, you need to identify the most important instrument of the two and give this the priority while panning or aggressively EQ'ing the secondary sound to make it fit into the mix.

If this doesn't produce the results, then you should ask yourself if the conflicting instrument contributes enough to remain in the mix. Simply reducing the gain of the offending channel will not necessarily bring the problem under control as it will still contribute frequencies which can still muddy the mix; it's much more prudent to simply mute the channel altogether and listen to the difference that makes. Alternatively, if you wish to keep the two instruments in the mix, consider leaving one of them out and bringing it in later during another 'verse', or 'chorus' after the reprise.

Clarity and Energy

All good mixes work on the principle of contrast, that is, the ability to hear each instrument clearly. It's all too easy to get carried away by employing too many instruments at once in an effort to disguise weak timbres, but this will very rarely produce a great mix. A cluttered, dense mix lacks energy and cohesion, so you should aim to mix so that you can hear some silence behind the notes of each instrument; if you can't, start to remove the non-essential instruments until some of the energy returns.

If no instrument can be removed, then aim to remove the unneeded frequencies rather than the objectionable ones by notching out frequencies of the offending tracks either above or below where the instruments contribute most of their body. This may result in the instrument sounding 'odd' in solo, but if the instrument must play in an exposed part then the EQ can be automated or two different versions could be used.

Additionally, when working with the groove of the record, remember that the silence between the groove elements produces an effect that makes it appear not only louder (silence to full volume) but also more energetic, so in many instances it's worthwhile refraining from adding too many percussive elements. More importantly, though, good dance mixes do not bring attention to every part of the mix. Keep a good contrast by only making the important sounds big and upfront, and leave the rest in the background.

Prioritize the Mix

During the mixing stage, always prioritize the main elements of the mix and approach these first. It's all too easy to spend a full day 'twitching' the EQ on a hi-hat or cymbal without getting a good balance on the most important elements. Always mix the most important elements first and you'll find that the 'secondary' timbres tend to look after themselves.

Mixing for Vinyl and Clubs

When mixing a record down that will be pressed onto vinyl for club play, you'll need to exercise more care in the frequencies you adjust. To begin with, any frequencies above 6 kHz should not be boosted and all frequencies above 15 kHz should be shelved off. On top of this, it's also prudent to run a de-esser across each individual track.

These techniques will help to keep the high-frequency content under some control since if the turntable cartridges are old (as they usually are in many clubs) it will introduce sibilance into the higher frequencies. Additionally, the frequencies on the kick and bass should be rolled off at 40 Hz, while every other instrument in the mix should have any frequency below 150 Hz rolled off. While this may sound unnatural for CD, it makes for a much clearer mix when pressed to vinyl and played over a club PA system.

Relative Volume

Analytical listening can tire your ears quickly, so it's always advisable to avoid monitoring at loud volumes as this will only quicken the process. The ideal standard monitoring volume is around conversation level (85 dB), but you need to keep the Fletcher Munson contour control in mind during mixing. After every volume or EQ adjustment, reference the mix again at various gain levels. It can also be beneficial to monitor the mix in mono when setting and adjusting the volume levels, as this will reveal the overall balance of the instruments more clearly.

EQ

EQ can be used to shape all instruments but you can only apply so much before the instrument loses its characteristics, so be cautious with any EQ. Always bypass it every few minutes to make a note of the tonal adjustments you are making but remember that while an EQ'd instrument may not sound correct in isolation, what really matters is that it sounds right when run with the rest of the mix.

Cut EQ Rather Than Boost

Our ears are used to hearing a reduction in frequencies rather than boosts since frequencies are always reduced in the real world by walls, objects and materials. Consequently, while some boosts may be required for creative reasons you should look towards mostly cutting to prevent the mix from sounding too artificial. Keep in mind that you can effectively boost some frequencies of a sound by cutting others as the volume relationship between them will change. This will produce a mix that has clarity and detail.

Don't Use EQ as a Volume Control

If you find yourself having to boost frequencies for volume, you should not have to boost by more than 5 dB. If you have to go higher than this, then the chances are that the sound itself was poorly recorded or the wrong choice for the mix was made.

Remember the Magic *Q*

A Q setting of $1\frac{1}{3}$ octaves has a bandwidth that's generally suitable for EQ'ing most instruments and often produces the best results. That said, if the instrument is heavily melodic or you're working with vocals, wider Q settings are preferred and a typical starting point is about two octaves. Finally, drums and most percussion instruments will benefit from a Q of 1/2 an octave.

Shelf EQ

Shelf equalizers are generally used to cut rather than boost because they work at the extremes of the audio range. For instance, using a shelving filter to boost the low frequencies will only accentuate low-end rumble since there's very little sound this low. Similarly, using a shelf to boost the high range will increase all the frequencies above the cut-off point, and there's very little high-frequency energy above 16 kHz.

Fix it in the Mix

'Fix it in the mix' is the opinion of a poor engineer and is something that should never cross your mind. If a timbre is wrong, no matter how long it took to programme, admit that it's wrong and programme/sample one that is more suitable. The following chart indicates the frequencies for mixing:

Frequencies	Musical Effect	General Uses
30–60 Hz	These frequencies produce some of the bottom-end power but, if boosted too heavily, can cloud the harmonic content, introduce noise and make the mix appear muddy.	Boosts of a decibel or so may increase the weight of bass instruments for drum 'n' bass. Cuts of a few decibels may reduce any booming and will increase the perception of harmonic overtones, helping the bass become more defined.
60–125 Hz	These frequencies also contribute to the bottom end of the track but, if boosted too heavily, can result in the mix losing its bottom-end cohesion resulting in a mushy, 'boomy' sound.	Boosts of a decibel or so may increase the weight of kick drums and bass instruments and add weight to some snares, guitars, horns and pianos. Cuts of a few decibels may reduce the boom of bass instruments and guitars.
125–250 Hz	The fundamental of bass usually resides here. These frequencies contribute to the body of the mix but boosting too heavily will remove energy from the mix.	Small boosts here may produce tighter bass timbres and kicks, and add weight to snares and some vocals. Cuts of a few decibels can often tighten up the bottom-end weight and produce clarity.
250–450 Hz	The fundamentals of most string and percussion instruments reside here along with the lower end of some male vocalists.	Small boosts may add body to vocals, and kicks, and produce snappier snare timbres. It may also tighten up guitars and pianos. Cuts of a few decibels may decrease any muddiness from mid-range instruments and vocals.

(continued)

Frequencies	Musical Effect	General Uses
450–800 Hz	The fundamentals and harmonics of most string and keyboard instruments reside here, along with some frequencies of the human voice. Cuts are generally preferred here, as boosting can introduce fatigue.	Small boosts may add some weight to the bass elements of instruments at low volumes. Cuts of a few decibels will reduce a boxy sound, and may help to add clarity to the mix.
800–1.5 kHz	This area commonly consists of the harmonic content of most instruments, so small boosts can often add extra warmth. The 'pluck' of most bass instruments and click of the drum kicks attack also reside here.	Boosts of a decibel or so can add warmth to instruments, increase the clarity of bass, kick drums and some vocals, and help instruments pull out of the mix. Small decibel cuts can help electric and acoustic guitars sit better in a mix by reducing the dull tones.
1.5–4 kHz	This area also contains the harmonic structure of most instruments, so small boosts here may also add warmth. The body of most hi-hats and cymbals also reside here along with the vocals, BVs and pianos.	Boosts of a decibel or so can add warmth to instruments and increase the attack of pianos and electric/acoustic guitars. Small decibel cuts can hide any out of tune vocals (although they should be in tune!) and increase the breath aspects of most vocals.
4–10 kHz	Finger plucks/attacks from guitars, the attack of pianos and some kick drums, snares along with the harmonics and fundamentals of synthesizers and vocals reside here.	Boosts of a few decibels can increase the attack on kick drums, hi-hats, cymbals, finger plucks, synthesizer timbres and pianos. It can also make a snare appear more 'snappy' and increase vocal presence. Small decibel cuts may reduce sibilance on vocals, thin out guitars, synthesizers, cymbals and hi-hats, and make some sounds appear more transparent or distant.

Frequencies	Musical Effect	General Uses
10–15 kHz	This area consists of the higher range of vocals, acoustic guitars, hi-hats and cymbals, and can also contribute to the depth and air in a mix.	Boosts of a few decibels may increase the brightness of acoustic guitars, pianos, synthesizers, hi-hats, cymbals, string instruments and vocals.
15–20 kHz	These frequencies often define the overall 'air' of the mix but may also contain the highest elements of some synthesizers, hi-hats and cymbals.	Boosts here will generally only increase background noise such as hiss or make a mix appear harsh and penetrating. Nevertheless, some engineers may apply a shelving boost around this area (or at 10 kHz) to produce the bandaxall curve.

CHAPTER 20
Mastering

'A failure will not appear until it has passed final inspection...'

Albert Einstein

Mastering can be viewed as the final link in the music-making chain. After any track has been mixed and processed, it should be mastered. Mastering a record is important for any musical production since it allows your music to stand up to the competition in term of balance, EQ and – more importantly – loudness. If your record isn't as loud as other records it isn't going to sound as impressive.

However, it should be noted that mastering isn't something you should attempt yourself, especially if you plan on a commercial release. The importance of employing a professional mastering engineer cannot be stressed enough. Indeed, all record companies will have their music mastered professionally regardless of the proficiency of the original artist, and it is fully recommended that if you value your music, you should do the same. Simply because you produced and mixed the music does not mean you are the perfect choice to master it. It is essential to appreciate that there is a world of difference between mixing and mastering.

Whereas recording and mixing music is a definite skill that requires complete attention to every individual element of the track, mastering involves looking at the bigger picture subjectively. This subjective viewpoint is something that is practically impossible for the producer (i.e. you) to accomplish, due to the producer's involvement throughout the entire project.

The mastering process depends on the quality of the mix in question, but if it has been mixed competently then it will usually involve equalization to set the overall balance of the mix, compression to add more punch and presence, and loudness maximization to increase its overall volume. The result will be enhanced impact of the music and spectral balance that will compare well with every other record on the radio and CD.

This overall balance is vital if you plan to release music for commercial sale, as the public will expect the record to have a professional polish similar to other commercial releases. When the sound of a record is right, you'll sell more records. This is more complicated than it first appears, though, and accomplishing it proficiently requires experienced ears, technical accuracy, knowledgeable creativity and a very accurate monitoring environment.

Of course, some suggest that you could master it yourself and then test the music on a number of systems to check that it is correct, but this takes time and no matter how experienced you are, it is doubtful you would achieve the same results as a professional mastering engineer. That said, many artists will master their own music if it's only to be submitted to a website in MP3 format, to a magazine for a 'review' or to a professional internet record label. Consequently, some knowledge of what is involved in the mastering process and how it is employed is useful, so for this chapter, we'll look at the basic *theory* behind it all.

To master any record you'll need the appropriate tools for the job. These are available as all-in-one units in both hardware and software form. The most celebrated hardware mastering device is the TC Finalizer that has all the mastering tools you could require in one compact single rack unit. Alternatively, digital audio workstations can be used to master music, and this is usually accomplished with an established wave editor and a number of individual mastering plug-ins such as Steinberg's mastering tools, Waves tools or all-in-one units such as Izotope's Ozone.

More importantly, though, no matter how good the equipment it is extremely easy to misuse it. If the equipment is used incorrectly, the sound will deteriorate and the audio distort. This distortion will not be to the extent that you can hear it; rather it will make the mix sound more muddled or reduce the stereo perspective.

MASTERING THEORY

As we've previously discussed in earlier chapters, the order of processing can dramatically affect the overall sound. Thus in mastering, each processor should be connected in a sensible order. Although this can depend on the music being mastered and the mastering engineer, the most usual order consists of:

- Noise reduction
- Paragraphic equalization/harmonic balancing
- Mastering reverb (although this is uncommon)
- Dynamics
- Harmonic excitement
- Stereo imaging
- Loudness maximizer

We'll look at each of these processes in turn.

NOISE REDUCTION

Although any extraneous noise should have been removed during the recording stages, occasional clicks, pops, hiss or hum can slip through, so it is important to listen to the mix thoroughly and carefully on both monitors and headphones to see if there is any extraneous noise present. This is where mastering engineers will have an immediate advantage as they will use subwoofers and loudspeakers with an infrasonic response. Without these, microphone rumble, deep sub-basses or any jitter noise introduced with poorquality plug-ins can easily be missed.

Any audible noise present will need to be removed before the track is mastered. Stray transients such as clicks or pops will prevent you from increasing the overall volume, while any hum, hiss or noise present will also increase in the volume when you increase the gain.

Ideally, the best way to remove any noise problems in a recording is to re-record the offending part, as this will inevitably produce much better results than attempting to reduce them at the mastering stage. If this is not possible, then you will have no choice but to use noise reduction algorithms or close waveform editing. This latter method is especially suited for removing pops or clicks that are only a couple of samples long and involves using a wave editor to zoom right into the individual samples and then using a pencil tool, or similar, to reduce the amplitude of the offending click. This approach is only suitable for extraneous noises that occur over a very short period of a few samples, however. If the noise is longer than this it is advisable to use dedicated reduction algorithms. These have been designed to locate and reduce the amplitude of short instantaneous transients typical of clicks or pops and are available in most wave editors or as a third-party plug-in. Depending on the amount of pops and the mix in which they're contained, *Waves Click Remover* or *Waves Crackle Reduction* can be suitable for reducing or eliminating the problem.

Hum and/hiss are much more difficult to deal with and often require specialist plug-ins to remove. As the frequency of tape hiss (also the result of a poor soundcard A/D or low recording levels) consists of frequencies ranging from 8 to 14 KHz, removing these using a parametric EQ can result in the higher frequencies of the mix also being removed. Thus, it is important to use plug-ins specifically designed to remove it, such as *Steinberg's De-Noiser*. Again, the reliability of using these types of plug-ins will depend on the source, and if they are not used with some degree of caution they can flatten the overall sound of the mix.

A.C. hum is little easier to deal with than hiss, as much of the 'noise' is centred around its fundamental frequency that lies around 50 Hz (UK) to 60 Hz (USA). Using a parametric EQ set to a thin bandwidth with a maximum cut can reduce most hum, but there may also be additional harmonics at 120, 180, 200, 300, 350 and 420 Hz, and removing any of these can compromise the overall tonal balance of the music, muddying the lower- and mid-range frequencies. Thus, if a parametric EQ does not remove the problem, it's worth using a dedicated plug-in such as *Waves X-Hum*.

Clicks, pops, hiss and hum can also be removed, sometimes much more reliably, by using plug-ins or wave editing features that profile the noise. This is often referred to as fast Fourier transform (FFT) filtering. A section of the track that consists of just the noise it can be sampled by the FFT filter to create a noise profile that can then be applied over the entire mix to remove the extraneous noise.

Waves X-Noise and *Sonic Foundry's Noise Reduction* plug-ins are typical examples of this, but some wave editors may also offer this function. Notably, these do not always produce the desired results, and in some instances they can remove frequencies that are essential to the music, resulting in a mix that has no bottom-end weight or a flat top end.

Above all, any form of noise reduction is not something that should be approached lightly. Reducing any noise within a final stereo mix will often affect frequencies that you may wish to keep. This is especially the case with removing hiss or hum. In fact, if there is an obvious amount of noise that cannot be removed reliably using FFT noise profiling or plug-ins, then you should seriously consider re-recording the performance, seeing if a professional mastering engineer can resolve the problem or drop the project and start a new one. It is much better to write a new record than to attempt to release one that's below your best capabilities.

HARMONIC BALANCING AND EQ

Assuming that there is no noise present, or that it has been removed proficiently, the next step in the mastering process is harmonic balancing. Essentially, this consists of EQ'ing the record to achieve an overall pleasing tone and an overall tonal balance that is sympathetic to our ears.

The quality of the EQ being used obviously plays a large part in attaining this balance, and many mastering engineers will choose EQ units that have a pleasing resonant quality, helping to give an overall satisfying tone. These are not necessarily complex parametric or multi-band graphic EQ units, as any precise EQ adjustments should have been made during the mixing process. Rather, during the mastering process wide Q settings with gentle transition slopes (sometimes as low as 3 dB per octave) are used, as these are more pleasing to our ears.

When mastering, Q's of 0.4–0.9 are the most popular settings.

Generally speaking, mastering EQ can be broadly split into three different sections: the bass, the mid-range and the highs.

Bass

The only 'real' EQ that's required on the bass end of a mix is a low-shelf filter to remove any frequencies below 40 Hz. Frequencies lower than this cannot be reproduced by many loudspeakers. When a signal this low is sent to them, they will attempt to replicate the frequencies, resulting in the bandwidth of the speaker system being restricted and the overall volume of the mix dropping. When a high-pass shelving filter is set to remove everything below 40 Hz, the sub-harmonic content of the sound will be removed by opening up the bandwidth of the speakers for a higher volume and result in an improvement in the definition of the bass frequencies.

The bass frequencies may also appear quieter than on commercial records, but it is important not to use EQ boosts in an attempt to add anything that isn't already present. The same goes for any maximizers. Although bass-specific maximizers such as Waves MaxBass can be used to introduce additional harmonics based on those already present, these are best avoided at this stage as compression can be used to introduce additional low-end punch to the music. If, however, there is a distinct lack of definition in the low-end frequencies, they will need to be pulled out using EQ.

In most dance mixes the kicks main energy will lie around 100–200 Hz. Perceivably boosting this will only result in muddying the bottom end. Rather, it's the transient that needs to be more defined and this commonly resides around 1000–2500 Hz. Thus, you will need to notch out the frequency that lies just below the transient so that it brings it out of the mix slightly. Possibly the best way to find these transient frequencies are to set a very narrow Q with a gain boost of around 6 dB. By then sweeping through the frequency range you should be able to locate the required transient. Once you have located it, you can reduce the gain to around −4 dB and sweep to the frequencies just below this.

Mid-range

The mid-range is the most common source of problems during mastering, so you need to listen carefully to these frequencies and then identify any problems they may have. The best way to accomplish this is to compare your current mix with a professional CD of the same genre using your ears and a good spectral analyser. Take note of the frequency dips, if there are any, and then try to employ these to your mix to see if it makes a beneficial difference.

Similar to mixing, the mid-range problems can be divided into three categories: it's too muddy, too harsh or too nasal.

If the sound is too muddy, then, using a fairly wide Q, place a cut around 180–225 Hz of a few decibels so that the frequency range between 100 and 300 Hz is dipped with the largest cut centred around the current frequency setting. Then, listening carefully, reduce the width of the Q until the muddiness dissipates.

If the overall sound of the mix appears too harsh, using the previously described principle place a cut at around 1.8 KHz so that there is a transition slope down from 1 KHz to the centred frequency, and then a transition slope up to 3 KHz. Then, once again start reducing the width of the Q setting until the harshness is removed.

This same process applies when a mix that sounds too 'nasal'. In that case, place a cut at 500 Hz with a Q that will create a slope from 250 Hz back up to 1000 Hz and slowly reduce the width of the Q.

Above all, bear in mind that wide bands will sound more natural to our ears. Try to keep the Q quite wide unless you're making surgical corrections to a specific frequency. If you find that you have to use a very narrow Q or too much of a cut to remove the problem, then the trouble is a result of the mixing and you should return to the mixing stage to repair it.

High Range

Finally, carefully listen to the high-end frequencies in the mix. In the improbable event that these are too harsh or bright, wide cuts placed at a centre frequency of 14 KHz can reduce the problem, but it is more likely that they will not be bright enough when compared with commercial CDs.

Although brightness could be added to the signal by placing a wide Q with a boost of a few decibels at 12–16 KHz, as touched upon in the previous chapter, gain boosts are usually avoided, with many engineers preferring to use harmonic excitement for brightenness. Not all engineers will use harmonic excitement, preferring to leave the signal unaffected. In this instance they'll employ the Baxandall curve.

This is created by using a wide Q setting on a parametric EQ and boosting by 6 or 8 dB at 20 KHz. The width of the Q introduces a gradual transition curve from 10 KHz upwards which is not only similar to the ear's natural response but also introduces small amounts of harmonic distortion similar to valves. Both these factors contribute to a pleasing sound that is often preferrable to the sound acheived using harmonic enhancers.

While the majority of these adjustments should be made in stereo to maintain the stereo balance between the two channels, occasionally it may be more beneficial to make adjustments on only one channel. This is useful if the two channels have different frequency content from one another or there is a poorly recorded/mixed instrument in just one of the channels. In these instances, subtlety is the real key. The best approach is to start with extremely gentle settings, slowly progressing to more aggressive corrections until the problem is repaired.

Also, when it comes to this form of surgical EQ, the most important rule of thumb is to avoid 'itchy fingers' and to listen to the material carefully before making any adjustments. You need to identify exactly what the problem is before applying any adjustments and even then adjusting only the sections that require it. If the 'problem' frequency sounds fine while it plays along with the rest of the track but appears wrong in the breakdown, then it's more sensible to correct it during the breakdown than to attempt to change the frequencies throughout the mix. Ideally, there should be more things right with the mix than there are wrong, and if you concentrate too heavily on a fault you could end up messing up the rest of the frequencies while trying to repair just one small problem.

Like mixing, this type of EQ should also be applied decisively and rapidly. Although it is difficult to perceive on a home monitoring system, changes of less than 1/2 dB are audible and our ears can quickly become accustomed to the changes, so it's vital that you compare the recent EQ movements with the original mix.

As a general rule of thumb you should play the same piece of the track being mastered over and over for up to 10 min before switching to the un-EQ'd comparison and listening to that for another 10 min to hear the changes that are being made. On this same note, this should also be referenced against no-expense-spared professionally mastered music of the same genre, both aurally and visually, using a spectral analyser.

BOOSTS AND CUTS

Although frequency boosts are generally best avoided altogether to retain a natural sound, any EQ adjustments will affect the perception of the overall tonal balance of the whole mix and so must be applied carefully. For example, even applying a 1/2 dB boost at, say, 7 KHz could have the same result as dipping 300–400 Hz by 2 dB. Always bear in mind that all the frequency ranges on a full mix will interact with each other and even small changes can have an adverse affect on every other frequency, so whenever you make a change you will have to reconsider all the others too.

More importantly, the combination of all these EQ settings will determine the overall dynamics of the song. This is one of the most critical aspects of any mix – it must sound correct for the style of music. When anyone listens to a mix, the first thing they'll hear is the overall EQ; if this is wrong for the genre of music then the track will fall flat. This is where a mastering engineer's expertise and subjective viewpoint plays yet another role. Every genre of music is approached and balanced differently, and there is no definitive guide to how any one genre should be balanced.

As an example, consider two entirely different genres of music: techno and chill out. A techno record will generally need to sound sharp and aggressive to suit the style. This can be accomplished by cutting the lower- and higher-range frequencies, so the mid-range becomes more apparent. Conversely, chill out will be quite slow and nowhere near as aggressive, so a cut in the mid-range will accentuate the low and high frequencies, making it appear gentler to the ear. Similarly, euphoric trance is encapsulated by the filter swept lead motif, so this would need to be pulled to the forefront of the mix by dipping the low and high frequencies, whereas in hip-hop the bottom end is bigger and the top end is more aggressive, and thus, the mid-range is usually attenuated.

As with mixing, the key behind setting a good tonal balance with EQ is developing an ear for the different frequencies and the effects they have, but it's worth employing a spectral analyser so you can also see how the movements are affecting the overall balance of the music. Additionally, makes an analyser it possible to measure the frequencies of a professionally mastered track and then compare these with the results you have on your own mix.

A general guide to mastering EQ frequencies:

0–40 Hz

- All these frequencies should be removed from the mix using a high-shelving filter.

40–200 Hz

- These frequencies contain the bottom end of the mix and are not usually attenuated or boosted. If this range has no punch or depth, or appears 'boomy', then it is common practice to use compression to cure the problem.

100–400 Hz

- These frequencies are often responsible for a 'muddy' sounding mix. The effect can be reduced by cutting the gain.

400–1.5 KHz

- These frequencies can contribute to the 'body' of the sound, and both compression and EQ gain can be used to introduce punch and impact to the mix.

800–5 KHz

- These frequencies often define the clarity of the instruments. Perceived increases applied around here will occasionally pull out the fundamentals of many instruments, making them appear more defined.

5–7 KHz

- These frequencies are usually responsible for sibilance in a recording and can be removed with a parametric EQ set to a narrow bandwidth to cut the problem. Alternatively, compression can be used.

3–10 KHz

- These frequencies account for the presence and definition of instruments. Introducing perceived 1 or 2 dB increases here can augment the presence of some instruments, while cutting can reduce the effects of sibilance or high-frequency distortion.

6–15 KHz

- These can contribute to the overall presence of the instruments and the 'air' surrounding the track. Introducing small gain boosts here will increase the brightness but may also boost noise levels.

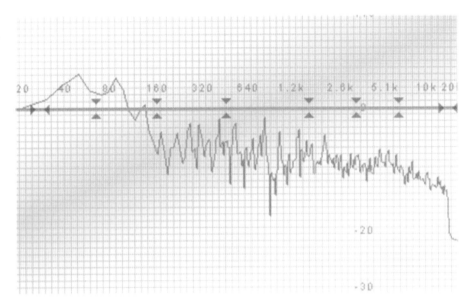

FIGURE 20.1
Spectral analysis of the results of mastering EQ on dance music

8–15 KHz

- Creating a boost at these frequencies will add sparkle and presence to a mix (although it will also boost hiss if there is any), while cutting them will help to smooth out any harsh frequencies.

20 KHz and above

- This is beyond the limit of human hearing, but setting a wide Q of around 1 and boosting (not perceived boosting!) by a few decibels at this frequency will create the Baxandall curve, creating a more ear-pleasing effect on the mix (Figure 20.1).

Naturally, these are just guidelines, and in the end you should rely on your ears and judgement as to what sounds right for your mix – provided of course, that you have a pair of monitors that give an accurate flat response, a room with a perfectly controlled response, an excellent EQ unit, an independent viewpoint on the mix and a pair of 'golden' ears.

Thankfully of developments in computers and associated processing power, more manufacturers are taking advantage and are releasing units that will aid with balancing a mix properly. One such product is HalBal, currently only available for the PC platform. This software analyses the music and shows the frequency response in a graphical form that you can then edit easily.

MASTERING REVERB

Mastering reverb (applying reverb across an entire mix) is rarely required in most mastering situations because it should have been applied properly during the mixing stages, but in certain cases it may still be required. Typically, this is when instruments don't seem to gel with one another, creating a series of holes in the mix, or when the mix seems to be lacking the fullness of other similar mixes.

Although from an ideal viewpoint you should return to the mixing stage and repair these problems, it may not always be possible, so lightly (and carefully!) applying reverb over the entire mix can help to blend the music together to produce a better result.

Of course, the amount of mastering reverb that should be applied is totally dependant on the mix, the spacing between the instruments, and the amount of reverb that has been applied to each during the mixing stage, but there are some general guidelines that can be followed.

First and foremost, the reverb must be of a very high quality as any tail noise created by poor algorithms will only serve to destroy the mix further. Indeed, if you don't have access to a good reverb unit or plug-in, it is not worth applying because it isn't worth turning an average mix into a poor one just for the sake introducing some cohesion between the instruments.

We've already looked at the uses and parameters of reverb in a previous chapter. Essentially the same methods and ideologies also apply during the mastering process, although a little more prudence should be exercised. Also, unlike all other effects in the mastering chain, reverb rather than is an insert as a send effect used because you don't want to wash over the mix completely, only add small amounts to the original signal. To start with a typical return is around 15% wet signal against 85% dry, but as always this depends entirely on the mix in question. Nevertheless, using this setting you can set the other reverb parameters and then back off the wet signal until it suits the mix.

Following the mix fader, it's worthwhile adjusting the decay time. Broadly speaking, you should aim to set this so that the tail is little longer than the reverb that's already been applied during the mixing stage so that the instruments mix together more coherently. Typically, for most styles of dance music this would equate to around 2–8 ms; any longer and the mix could appear too 'washy' and undefined. This, however, is related to the room size, and as a general starting point a square room (if the option is available) with a relatively small size to begin with will inevitably produce the best results. It's important to note that the size of the room will also determine the diffuse settings you should use; for a small room this should be set quite low so as not to create an inaccurate impression.

As you're working with a stereo reverb unit, you need to emulate the diffuse effect that results from the left and right reverberations reaching the ears at different times. If you set this high but use a small room, you would essentially be creating the effect of a small confined room with walls that are far apart. Although on some mixes this inaccurate impression has been known to work quite well, ideally it is best to keep small diffusion settings with small rooms and vice versa. More importantly, any adjustments with the diffusion can have a significant effect on the stereo aspect of the mix, so always check the mix in mono too.

Applying reverb to a signal may also introduce or enhance any sibilance or low-end artefacts. To remove these it's preferable to use the reverbs low- and high-pass filters. On average, for most dance mixes a high cut-off applied at 2–4 KHz should prevent the introduction of sibilance, while a low cut-off set at 80–120 Hz will prevent the low body of the mix from becoming too muddy.

Possibly the most important aspect of applying reverb, though, is to continually bypass the effect to check that you're not getting too carried away with its application. Ideally, as with mixing, reverb should be applied so that when it is present it isn't particularly noticeable but when removed the difference becomes immediately perceptible.

DYNAMICS

Following EQ (and perhaps mastering reverb), many engineers will compress the entire mix in an effort to add volume, punch, clarity and possibly emotion. This cannot be accomplished simply by strapping a standard compressor across the entire mix, however, since the loudest parts of the track will drive the compressor, forcing the entire mix to pump with each kick.

While all good dance mixes should pump with some energy, much better results can be gleaned from using multi-band compression. Using these, it's possible to alter the dynamics at different frequency bands, which can result in a clearer-sounding mix. For example, compressing the high end of a mix heavily can flatten the sound of a mix, so it's best applied lightly. On the other hand, the lower ranges of a mix are usually compressed more heavily, not only to even out the sound but also to help the mix pump along and add extra volume to some frequency bands of the mix.

This frequency-specific volume is an important aspect of most dance music, especially on the lower frequencies, and should not be underestimated. Almost every record company will want their music to measure up to the competition in terms of volume as it's a long-proven fact that if you play the same musical passage to an audience at different volume levels, most will prefer the louder playback. However, while it would seem sensible to reduce the dynamics of a mix as much as possible to increase the volume, there are limits to how far you should push it.

A typical home hi-fi system will reproduce about 80 or 90 dB of the dynamic range, but this is going to be pointless if the music is squeezed into just a few decibels. In fact we naturally expect to hear some dynamics when listening to music, and if they are eliminated it isn't going to sound real no matter how good the hi-fi system reproducing the music is.

Furthermore, while louder music will inevitably capture the attention of the listener, it will only be for the first minute or two. While the track may appear hard-hitting at first, if it's too 'hot' then it can quickly become fatiguing to listen to and most listeners will turn off even if they like the track. In fact, any ruthless dynamic restriction is only really considered useful for playback on radio stations.

All radio stations will limit the signal for two reasons:

(a) They need to keep the music's level at a specific maximum so as not to overmodulate the transmission.
(b) They want to be as loud as possible to hold an audience's attention.

This means that no matter what the input level is, the output level will always be constant. The inevitable result is that even if your mix happens to be a couple of decibels louder than every other mix, when it's played on the radio it will be at the same volume as the played before or after.

Incidentally, some artists will argue that by compressing a mix heavily at the mixdown/mastering stage the radio station cannot squash the signal any further so there is no possibility of any broadcasters destroying the signal. While there is some logic behind this, setting the average mix level too high through heavy compression can actually create more problems because the station's processors might confuse the energy of the music with the signal peaks and attempt to squash it even further, totally destroying what's left of the mix. This will seriously reduce the overall punch of the music, making it appear lacklustre when compared to other dance records.

Consequently, you will often have to settle for a compromise between the output volume and leaving some dynamic range in the mix. As a very general rule of thumb, the difference between the peak of the kick and the average mix level (known as the *peak-to-average ratio*) should never be less than 4 or 5 dB. This will enable you to increase the overall gain while also retaining enough dynamic variation to prevent the mix from sounding 'flat' on a home hi-fi system. Alternatively, if you really want a loud mix with less punch for radio airplay, then it may be worthwhile releasing two different versions of the record. The promotional records can be compressed heavily for radio airplay, while the record store version can be mastered with more punch for clubs or home equipment.

There's more to multi-band compression than simply using it to add some punch to a dance track, and a manipulation of dynamics around the vocals or lead melody can introduce additional impact. This is often what gives music that extra listenable edge, although acquiring this dynamic feel often requires a fresh pair of experienced ears, too, as it involves listening out for the mood, flow and emotion it conveys.

As an example consider this rhythmic passage:

'......I love to **party**......to dance the night **away**...............I just love **to party**'

Beats 1	2	3	4	1	2	3	4	1	2	3	4	1	2	3	4

Of course, this isn't a particularly complex vocal line, but notice that the accents occur on the downbeat of each bar. If you were to apply compression heavily throughout the mid-range (where the vocals reside) then this vocal line could easily become:

'……I love to party…to dance the night away…………I just love to party'

Beats 1	2	3	4	1	2	3	4	1	2	3	4	1	2	3	4

By compressing the vocal area hard throughout, you have lost the dynamics of the track and the resulting feel has been totally removed. However, by carefully compressing and only applying light amounts to specific parts of the vocals you could get this result:

'……I love to **party**……**to** dance the night **away**……………I just love **to party**'

Beats 1	2	3	4	1	2	3	4	1	2	3	4	1	2	3	4

As the accents show, you're not just working on the vocal but the mix that's also behind it, so you're also affecting the music. This can help to add a dynamic edge as the music will move dynamically with the vocals, enhancing the general feel and emotion of the song.

When you are working with such precise dynamic manipulation, you must set the compressors parameters accurately and carefully. If the release is set too long the compressor may not recover fast enough for the next accent or sub-accent, while if it's set too short, the sound could distort. Similarly, if the attack is too short, then the initial transients of the word and music will be softened, which defeats the whole point behind applying it in the first place.

More importantly, any compression will result in your having to increase the make-up gain on the passage being compressed, and as we've discussed, loudness can have an adverse affect on our judgement. Consequently, even applying accent compression in the wrong place will lead you to think it sounds better, so it's important to keep the gain the same as the uncompressed passages to ensure that it isn't just pure volume that's making the passage sound better.

When it comes to setting up a multi-band compressor for a mix, there is no set approach and there are certainly no generic settings. It depends entirely on the mix in question. All the same, there are some very general guidelines you can follow to initially set up a compressor across a mix.

- On a three-band compressor, start by setting each band of the compressor to 2:1. Then adjust the frequency of each band so that you're compressing the low range, mid-range and high range.
- On a four-band compressor, adjust the crossover frequencies so that one band is compressing the bass frequencies, two are compressing the mid-range and one is compressing the high range.
- On a five-band compressor, adjust the crossover so that two bands are concentrated on the bass, two are compressing the mid-range and the final is compressing the high range.

The crossover frequencies will depend on the mix being squeezed, but try to ensure that the vocals are not compressed with either the low- or high-frequency ranges of the compressor. By doing so, you will be able to adjust a single (or double on a five band) compressor concentrated on the vocals alone, to add the aforementioned dynamic movement to a mix.

Applying this multi-band compression to parts of the mix will upset the overall balance of mix. Rather than start EQ'ing again, adjust the output gain of each compressor to try to gain a more level mix. It's important at this stage to realize that overall track volume shouldn't be an immediate concern. Ideally, you should be aiming to produce a mix that has plenty of bottom-end energy with a well-balanced mid and high range.

If the bass end of the mix is short of any bottom-end weight, then it is worthwhile lowering the threshold further and raising the ratio until the bass end comes to the forefront. It should be noted, though, that the mid-range may simply need less compression to lift up the bass and it's a careful mix of all the bands that creates a fuller mix.

For instance, if the mid-range is compressed heavily it will be pushed to the front of the mix, overtaking both the high and bass ranges, so rather than compress the bass even more it may be more prudent to lower the compression on the mid-range. As a general guideline, for most dance mixes, it's usual to compress the bass heavily and use lighter settings on the mid-range because this can help the mix appear less muddied.

Both the attack and the release parameters can also affect the overall tonal balance of a mix, so these must be adjusted with caution. Usually, on the bass frequencies you can get away with using a fast attack time, as it consists purely of low-frequency waveforms with few or no transients. Also, the release setting should be set as fast as possible so the compressor can recover quickly, but you may find that it has to be set longer than both the mid and high range due to the waveform lengths of the bass frequencies.

Generally, the faster the release is set the more the bass end will pump, so it should be set so that it pumps slightly but not enough to destroy the feel of the mix. For the mid-range, it is sensible to use longer attack times than with the bass to help improve the transients that are contained here, while the release setting should be set as short as possible without introducing any gain pumping. Determining pumping at mid-range frequencies can be difficult, so it is preferable to use a compressor that allows you to solo the current band to hear the effect you're imparting onto the mix.

Finally, the high range should use a relatively fast attack to prevent the transients becoming too apparent, and as the waveform of these is often considerably shorter than lower frequencies you can also get away with using short release times. As with the mid-range, however, do not set this release too short, otherwise you may introduce pumping which will ruin the overall dynamics contained here.

As a very general starting point, what follows is a list of typical settings for three-, four- and five-band compressors.

Three-Band Compression

BAND 1: TO TIGHTEN UP THE BOTTOM END OF A MIX

- Frequency: 0–400 Hz
- Ratio: 4:1
- Threshold: 3 dB below the quietest notes so that the compressor is permanently triggered
- Attack: 10–20 ms
- Release: 140–180 ms
- Gain make-up: Increase the gain so that it is 3–5 dB above pre-compression level

BAND 2: TO TIGHTEN UP AND ADD SOME 'WEIGHT' TO THE MIX

- Frequency: 400 Hz–1.5 KHz
- Ratio: 2:1
- Threshold: Just above the quietest notes so that it is triggered often but not continually
- Attack: 10–20 ms
- Release: 100–160 ms
- Gain make-up: Increase the gain so that it is 1 dB above pre-compression level

BAND 3: TO INCREASE CLARITY OF INSTRUMENTS AND REDUCE 'HARSH' OR 'ROUGH' ARTEFACTS

- Frequency: 1.5–15 KHz
- Ratio: 2:1
- Threshold: Just above the quietest notes so that it is triggered often but not continually
- Attack: 5–20 ms
- Release: 100–130 ms
- Gain make-up: Increase the gain so that it is 1 dB above pre-compression level

Four-Band Compression

BAND 1: TO TIGHTEN UP THE BOTTOM END OF A MIX

- Frequency: 0–120 Hz
- Ratio: 4:1
- Threshold: 3 dB below the quietest notes so that the compressor is permanently triggered
- Attack: 10–20 ms
- Release: 150–180 ms

- Gain make-up: Increase the gain so that it is 3–5 dB above pre-compression level

BAND 2: TO TIGHTEN UP THE MIX IN GENERAL AND ADD WARMTH TO INSTRUMENTS AND VOCALS

- Frequency: 120 Hz–2 KHz
- Ratio: 2.5:1 or 3:1
- Threshold: Just above the quietest notes so that it is triggered often but not continually
- Attack: 20–30 ms
- Release: 100–160 ms
- Gain make-up: Increase the gain so that it is 1–3 dB above pre-compression level

BAND 3: TO INCREASE THE CLARITY OF INSTRUMENTS

- Frequency: 2–10 KHz
- Ratio: 3:1
- Threshold: 2 dB below the quietest notes so that the compressor is permanently triggered
- Attack: 10–20 ms
- Release: 100–180 ms
- Gain make-up: Increase the gain so that it is 2 dB above pre-compression level

BAND 4: TO INCREASE THE CLARITY OF MID- TO HIGH-FREQUENCY INSTRUMENTS

- Frequency: 10–16 KHz
- Ratio: 2:1 or 3:1
- Threshold: Just above the quietest notes so that it is triggered often but not continually
- Attack: 5–25 ms
- Release: 100–140 ms
- Gain make-up: Increase the gain so that it is 1 dB above pre-compression level

Five-Band Compression

BAND 1: TO TIGHTEN UP THE BOTTOM END OF A MIX

- Frequency: 0–180 Hz
- Ratio: 5:1
- Threshold: 3 dB below the quietest notes so that the compressor is permanently triggered
- Attack: 10–20 ms
- Release: 130–190 ms
- Gain make-up: Increase the gain so that it is 3–5 dB above pre-compression level

BAND 2: TO TIGHTEN UP THE RHYTHM SECTION AND THE MIX IN GENERAL

- Frequency: 180–650 Hz
- Ratio: 2.5:1 or 3:1
- Threshold: Just above the quietest notes so that it is triggered often but not continually
- Attack: 10–20 ms
- Release: 130–160 ms
- Gain make-up: Increase the gain so that it is 1–3 dB above pre-compression level

BAND 3: TO ADD SOME WEIGHT TO THE MIX AND INCREASE ITS ENERGY

- Frequency: 650 Hz–1.5 KHz
- Ratio: 3:1
- Threshold: 1 dB below the quietest notes so that the compressor is permanently triggered
- Attack: 15–25 ms
- Release: 100–130 ms
- Gain make-up: Increase the gain so that it is 1–3 dB above pre-compression level

BAND 4: TO INCREASE THE CLARITY OF MID- TO HIGH-FREQUENCY INSTRUMENTS:

- Frequency: 1.5–8 KHz
- Ratio: 2:1 or 3:1
- Threshold: Just above the quietest notes so that it is triggered often but not continually
- Attack: 5–10 ms
- Release: 100–140 ms
- Gain make-up: Increase the gain so that it is 1–4 dB above pre-compression level

BAND 5: TO REDUCE UNWANTED ARTEFACTS SUCH AS A 'ROUGH' OR 'HARSH' TOP END

- Frequency: 8–16 KHz
- Ratio: 2:1 or 3:1
- Threshold: Set so that it is triggered occasionally on the highest peaks
- Attack: 5 ms
- Release: 100–140 ms
- Gain make-up: Increase the gain so that it is 1 dB above pre-compression level

As always, these are not precise settings and are only listed as a general guide to begin with. With these settings dialled into the compressor, you will need to experiment by increasing and lowering the make-up gain, the threshold, the attack, the release and the ratio while listening for the difference each has on the mix. Also, remember to bypass it often and compare it with the uncompressed signal, ensuring that you're not destroying the dynamics of the mix.

HARMONIC EXCITERS

Following compression, some engineers may send the mix out to a harmonic exciter to add some extra sparkle to the mix. This is a moot point among many engineers, as some regard any harmonic excitement as sounding too artificial. Nevertheless a proportionate amount of music released today will have been 'cooked' with some form of enhancement to increase the overall resolution of the music.

The main reason behind applying any form of psychoacoustic enhancement to a mix stems from the introduction of modern recording techniques. As we now have the ability to record and re-record music, the clarity and detail of music is usually compromised, particularly in the higher frequencies.

This results in the music's frequency content seeming dreary, dull and lifeless. When an enhance is used, these dulled or missing frequencies can be restored or replaced, in effect making the music appear clearer, brighter and crisper. While this seems relatively simple in principle, in application it is much more complicated as there are various enhancers available and each uses different methods to brighten the mix.

The first exciters were developed by the American company Aphex in 1975 and were, according to the story, discovered by accident. A stereo amplifier kit was put together incorrectly ending up with only one channel working, the other giving a thin distorted signal. When the channels were mixed together the resulting signal came out sounding cleaner, brighter and generally more enhanced.

This principle was researched and developed before a conclusion was reached that adding controlled amounts of specific harmonic distortion to an audio signal can actually enhance it. Originally, the first Aphex exciters were only available for rent to studios at around £20 per minute of music, but due to the overwhelming demand they soon became available for purchase. These commercial units work by allowing you to select low-, mid- or high-range frequencies, which are then subsequently treated to second- and third-order harmonic distortion. This distortion is then added back to the original audio before the whole signal is treated to small amounts of phase shifting.

This form of synthesizing additional harmonics from an original signal is only used by the Aphex exciters. Many engineers prefer this but some don't like the idea of applying any distortion to a mix and so use enhancers instead. Aphex registered the name Aural Exciter to its family of exciters; all other manufacturers refer to their exciters as *psychoacoustic enhancers*. These are available from Behringer, SPL and BBE, to name a few.

Each of these utilizes slightly different forms of enhancement, many of which are secrets closely guarded by the manufacturers. Most of this is probably because even they can't explain why what sounds normal to our ears informs our brains that something extra special is happening. And although I would love to explain the details of why psychoacoustic enhancements sound so special, (a) it would take up most of this book to explain and (b) it would also involve me knowing why. Nevertheless, I can say that they are all based around some form of dynamic equalization and phase alignment.

Much of what we determine from a sound is derived from the initial transient, but during the recording process and subsequent processing the phase can be shifted, which results in the transient becoming less noticeable. Realigning the phase using an enhancer makes these transients more defined which results in extra clarity and definition.

Most enhancers and exciters are single-band, either allowing you to specify the frequencies you wish to enhance or simply enhancing all of the frequencies at once. However, some enhancers, such as Izotope's Ozone, also offer individual control over numerous bands of 'enhancement', allowing you to process different frequency bands separately. Generally speaking these types of enhancers introduce second-order harmonic distortion into the original signal which makes them suitable for adding the typical tube emulation throughout the frequency range.

Ultimately, although applying an extra sonic sheen to music will make pretty much anything sound better, enhancers should be used conservatively. The results they produce can be extremely addictive and our ears can grow accustomed to the processing quickly. This can result in overprocessing (often referred to as 'overcooking' or 'frying') which results in a sound that can be very fatiguing to the ear. Thus, like all other effects they should be used sparingly.

STEREO WIDTH

Following the excitement effect, a proportionate amount of mixes will also be sent through a stereo widening effect to increase the overall image of the mix. This is a very simple effect and requires little explanation, as usually if will only offer a percentage parameter. When this is increased, the width of the stereo image will be perceived to be wider than it actually is.

This works on the principle that sounds that are shared in both left and right channels can appear to be in the middle of the mix rather than panned to either side, even if they are. However, if you were to subtract one of these channels from the other then the phase would be adjusted, resulting in a wider stereo effect.

Like any other mastering process, overuse of stereo widening effects can have a detrimental effect on the music, so they must be applied carefully. This is because as you widen the spectrum further left and right the phase adjustments can create a 'hole' in the middle of the mix and as the bass, kick drum and vocals are usually located here the result can be a distinct lack of body, or in extreme cases a total removal of body. To avoid this, some mastering modules will allow you to adjust frequency bands individually. In this case, it's worthwhile widening the high frequencies but applying less to the mid-range and the bass consecutively.

LOUDNESS MAXIMIZERS

The final link in the chain is the loudness maximizer. The basic premise of this is to reduce the volume level of any peaks so you can increase the relative volume of the rest of the mix without fear of overloading the signal. This is similar to the operation of a limiter, in that loudness maximizers establish a strict dynamic maximum but unlike typical limiters, loudness maximizers are designed to more natural sound by 'rounding off' any peak signals rather than cutting them dead. This approach to increasing the relative volume of a mix is preferred to normalizing on an audio editor as this can introduce unwanted artefacts.

As previously discussed, when audio is normalized, the file is first scrutinized to locate the loudest waveform peaks. Then the volume of the entire section of audio is increased until these peaks reach the digital limit. Theoretically, this should raise the volume without introducing any unwanted artefacts, but any digital recording will sound much better if the waveform was captured as a loud signal in the first place. If you artificially increase the gain using normalization, it will often reveal the limitations of digital recording. What's more, as it raises the volume in direct relation to the peaks, while it may appear louder we do not perceive volume by the peaks in a signal but through the average volume.

Loudness maximizers limit the distance between the peak and main body of the music, which increases the volume more substantially. This does, however, mean that the more maximization that is applied to a mix, the more the dynamics will be reduced, so like any other processor loudness maximizers must be used cautiously. Ideally you want to keep a peak-to-average ratio of at least 4 or 5 dB, or perhaps more if you want the kick drum to really stand out (keep a large excursion).

Due to their simple nature, most maximizers will consist of only a few controls: a threshold, a make-up gain (often known as a margin), a release and the option to use either brickwall or soft limiting. These functions are described in the following sections.

THRESHOLD

Similar to a limiter and compressor, the threshold on a loudness maximizer is used to set the level where the limiting will begin. When the loud is reduced, more of the signal will be limited, whereas increasing it will reduce the amount of the signal being limited. Where this should be set is determined by the mix in question but as a general rule of thumb, it should be adjusted so that only the peaks of the waveform (usually the kick drum) are limited by a couple of decibels while ensuring that the excursion between the kick and the main mix is not compromised to less than 4 dB. For most competent mixes (i.e. those that have a good signal level) the threshold will usually be set around −1 to −4 dB.

MAKE-UP GAIN (AKA MARGIN)

This is similar to the make-up gain on a typical compressor and is used to set the overall output level of the mix after 'limiting' has taken place. Broadly speaking, it isn't advisable to increase this so that the output level is at unity gain. Rather, it is more prudent to set this 1 dB lower just in case any further processing has to take place.

RELEASE

Again, the release parameter is similar to that on a compressor and determines how quickly the limiter will recover after processing the signal. Ideally, this should be set as short as possible to increase the volume of the overall signal, but if it is set too short the signal will distort. The best way to set the release parameter is to adjust it to its longest setting and gradually reduce it while listening to the mix. As soon as the mix begins to distort or slice, increase the time so that it's set just above these blemishes. You'll usually find that the more limiting that is applied to an audio file, the longer this release setting will need to be.

BRICKWALL/SOFT LIMITING

Most mastering limiters or loudness maximizers are described as being brickwall or soft, and some will offer a choice between the two modes. Fundamentally, if the maximizer is particularly good then these will both produce similar natural results, but soft limiting may appear more transparent than brickwall, depending on how hard the limiter is being pushed.

When you use a soft limiter, the overall level can exceed unity gain if it is pushed too hard but if not it will produce results that are generally seen as more natural and transparent than with brickwall. Alternatively, when you use a brickwall setting, no matter how hard the limiter is pushed it will not exceed unity gain, but if it is pushed too hard the limiter may have a difficult time rounding off the peaks of the signal resulting in a sound that's best described as 'crunchy' or 'digital'. The choice on which to use is dependant on the mix, but a soft setting will often produce better results on a mix.

GENERAL MASTERING TIPS

- Avoid the temptation to master your own music

All mastering engineers are impartial to the music and they will have a subjective viewpoint on the music. You will not be aware of your bias due to your involvement in the project. Even if you are not willing to employ a professional mastering engineer, ask a knowledgeable friend or fellow musician to master your music for you. Another person's interpretation can produce results that would never have crossed your mind.

- If at all possible use full-range, flat monitors

Both vinyl and CD are capable of producing a frequency range that is beyond the capabilities of all nearfield monitors and if you can't hear the signal you can't judge it accurately. Mastering engineers will use monitors that are incredibly flat but which, although they sound a little bland, guarantee that the music will play properly on all sound systems.

- Monitor at a fixed volume

Adjust the monitoring volume so it is at a fixed comfortable volume. Loud monitoring volumes will tire your ears more quickly and affect the frequencies of the mix due to the Fletcher Munson contour control. Once the volume is set, listen to some professionally mastered CDs and compare them with your own mix.

- Always reference your track with a professionally mastered track

This is possibly the best way to train your ears. By continually switching between your own music and a professionally mastered track you will be able to keep some perspective on your current work.

- Always perform A/B comparisons

Our ears become accustomed to tonal changes relatively quickly, so after applying any processing, perform an A/B comparison with the unaffected version.

- If it isn't broke, don't try to fix it

One of the biggest mistakes made by many musicians is to immediately jump in and start adjusting the EQ and/or multi-band compression without actually listening to the mix first. Itchy fingers will inevitably produce a poor master, so listen to the mix thoroughly before touching anything. The less processing that is involved during the mastering stage, the better the overall results will be.

- Avoid using presets or wizards

All music is different, so always use your ears rather than someone else's idea of what the EQ or compression should be. Also, bear in mind that the approach is different from one track to the next, so don't think that mastering settings from a previous track will work on the next.

- Do not finalize anything as you're recording to external hardware such as DAT

Unlike computers, DAT machines do not have an undo function and if you apply the wrong settings at this point there is no going back.

- Do not normalize a mix

All normalizing algorithms look for the highest peaks and then increase the volume of the surrounding audio, but the ear doesn't respond in the same way. Rather, it judges the loudness of music by the average level.

- Do not apply any processing that is not vital to improving the sound

Simply because some may mention that a certain process is used in mastering do not take it for granted on your own mix. Every DSP process comes at the cost of sound quality. The less processing you apply to a mix during the mastering stage, the better it will eventually sound.

- If possible, use a VU meter to measure the levels of the mix

Although VU meters are considered old hat compared to the latest digital peak meters, VU meters are much more suitable for mastering. A digital peak meter only measures the loudest peak of a signal; thus, if you have a drum loop measured by one it will seem as though the level is particularly high even though it may be quiet. Conversely, a VU meter generally measures the overall loudness (RMS).

- Keep the bit rate as high as possible and only dither (if required) before it is recorded

Use as many bits as possible in the mix until it is ready to be copied to the final media. If you continually work with 16-bit throughout the recording and mixing, further processing such as raising the gain will truncate the overall sound. Always bear in mind that the more bits that are contained in a signal, the better the signal-to-noise ratio with less chance of distortion being introduced.

- Use a top-quality soundcard with the best software you can afford

Poor-quality soundcards will reduce the overall quality of the end results, and poor dithering algorithms will reduce the quality further.

- Make sure that the headroom is optimal

Ensure that the faders on the mixing desk are not too far below their optimum 0 dB and that the main mix output signal is around 0 dB (or −15 dbFS for digital signals). Keeping them too far below this can often result in compensating for it by adding more volume at the amplifier which can result in a loss of transients at the recording stage.

- Learn your mastering processors

Learn the effect that each processor has on a mix, and unless you're experienced, create a number of takes using a mix of gentle and aggressive settings. Write these settings down, copy the audio to CD and then play it on as many systems as possible. All engineers have learnt the most from previous mistakes so don't be afraid of making them. It's much better to make a mistake in private than in public.

- Avoid using a noise gate

Although noise gates are useful for removing any potential noise during silent passages, they can also result in clipping as the sound stops or starts. It is more prudent to leave any potential noise in. You can remove ir by applying the noise reduction software over just the noisy part and you can use it to create a noise profile if it is present throughout the track.

- Avoid using rapid fades

If there is noise present at the end of a recording, don't try to avoid it by rapidly fading the music down to nothing as this may result in cutting off the end of an instrument. A better approach is to leave the recording running for a few seconds longer and then use noise reduction software to remove the problem. As with the above tip, any noise present here can also be used as a noise profile.

- Always check your mastered mix in mono

If the track is destined for radio play, it is imperative that it sound well in mono as well as stereo. Although a majority of radio stations broadcast in stereo, slight signal discrepancies can result in the mix being received in mono. If the mix has not been checked for mono compatibility it could sound drastically different.

- Consider decreasing the tempo of the master

It isn't unusual for some radio stations to increase the tempo of a track so they can cram in more commercials throughout the day. This is only by a few beats per minute but nevertheless, if a song is destined for radio airplay it is sometimes worth time-stretching the track to a slower tempo. This way, when the radio station speeds the track up it will be back at its original tempo.

- Use the best CD burner that you can afford

If the final media for a mastered mix is CD, then it is vital to use a high quality CD burner. These are not only more reliable but will also reduce the number of errors that can be introduced onto the disc. Additionally, although many CD burners can record at speeds of 40× and above, it is advisable not to record at speeds greater than 4×. Apart from preventing the chance of buffer underrun (CDs are recorded in one continuous pass and if the computer cannot deliver data fast enough an interruption will occur and destroy the recording), if the burn rate is too fast crackling and distortion can appear on quieter passages of music.

- Use quality blank CDs

This may seem like common sense, but many musicians use cheap quality CDs in an effort to save some money. Silver and gold dye on CDs will invariable perform better than dark green, and if you value your music it is worthwhile spending a little extra for a better recording quality.

- Do not use CD labels

The typical sticky paper labels should be avoided because they can often increase the CDs error rate. This is not usually perceptible on many CD players, but it will degrade with age and low error rates are required by mastering plants. Ideally, you should use a specifically desinged for CDR pen and try to write outside the area where the data is recorded.

■ Avoid using 80-min CDs

Despite the manufacturer's claim that 80-min discs are as reliable as 74-min CDs, they have a tendency to introduce distortion, so stay with 74-min CDs.

■ Use a professional mastering plant

If the record is for the public market (hi-fi players, vinyl etc.) you should very seriously consider taking the recording to a professional – the difference can be remarkable and worth the additional outlay. Mastering is a precise art and you really do need golden ears to produce an accurate master copy.

CHAPTER 21
Publishing and Promotion

'I signed away Voodoo Ray for a hundred pounds so that I could buy a new drum machine. I had no idea that the record would be so big...'

A Guy Called Gerald

On completion of a dance track (or album), you may want to release it to the general public but you can't simply burn it to CD and send it out to everyone. Indeed, you will first need to copyright your music and also register your band name.

Registering your band name isn't absolutely vital but it can be worth the effort, if only to ensure that no one else in your country releases a record using your band name. By registering, the Worldwide Registry notifies artists and labels if there is a potential name conflict and also registers your claim to the band name. The most recognized way to register your band name is at http://www.bandname.com.

Once you have your name registered, you'll also need to copyright your music. This is to ensure that wherever the music is played, you'll be paid or credited for it and is defined by UK law as *"the right granted to the creators of original literary, dramatic, artistic and musical works to ensure that copyright owners are rewarded for the exploitation of their works."* Basically, this means that as long as you taken the appropriate steps to copyright your creation, you have the right to determine what people can do with what is essentially 'your property' and allows you to prevent anyone from:

- Broadcasting the work through radio or television
- Copying the work onto any medium
- Distributing the work to the public
- Renting the work out to anyone
- Adapting any part of it, for instance remixing it
- Performing in a public place
- Using it as 'background' music on films or corporate videos

Although the current UK law states that as soon as you produce an original piece, it is automatically copyrighted to you, it's vital that you take steps so that you can offer physical proof that it *is* your work before you sign it over to a record company or release it commercially.

The most common perception of how you should keep this evidence is to write a copyright statement on the media, seal this into an envelope and post it to yourself. As long as you don't open it, you have proof from the date stamped on the postage.

However, while this approach is still commonplace amongst many musicians, it is sometimes *not* considered suitable evidence of copyright ownership. Therefore, a much more sensible approach is to write the copyright statement on the media and mail it to a solicitor or bank and ask them to keep it (unopened) for you as these prove to be much more reliable should the copyright ownership be questioned. A typical copyright statement simply consists of:

©Artists names, Year, All rights reserved.

Also you can only copyright the recording of the music (which lasts for 50 years after first publication) and the melodies and words used in the music (which lasts for 70 years after the death of the last surviving composer). This means that you cannot claim copyright to the musical arrangement or the song title!

Once your material is copyrighted, you can then look to releasing it in the general market. Generally speaking, this is possible in one of the following three ways: you can sign the record over to a record label, press and release it yourself as a white label or promote and sell it yourself over the internet. Each of these requires a different approach and so we'll look at each of these options in detail starting with the most obvious route, signing the recording over to a record label.

RECORD COMPANIES AND CONTRACTS

Sending a 'demo tape' to a record company has to be the most infamous way of acquiring a record deal. Even though this approach is looked upon with a certain amount of scepticism today, it cannot be denied that many artists still use this route and, as a result, some of them have been signed on the strength of the demo. Before we go any further into this, though, there are some home truths that you should be aware of if you have big plans to sign to a large label and become an overnight 'mega star':

1. Signing to a large label is *not* a guarantee that you will ever reach superstar status.
2. The record company will *expect* the rights to everything you ever plan to do.
3. The record company will *expect* you to become who *they* want.
4. The record company will expect you not to change your musical style midway through the contract but reserve the right to turn away any music you produce and *tell* you to make it more 'commercially viable'.

5. The record company will *expect* you to do exactly what they say, when *they* say.
6. The record company will *expect* you to appear on humiliating music programmes.
7. They will invest money in what *they* want, not what you want and will *expect you* to pay for these expenditures out of your royalties.
8. You will be expected not to make any demands or requests to the record company whatsoever.
9. They reserve the right to drop you if they see fit, but you do not have the option to drop them if their performance is terrible.
10. It costs between 1.5 and 2 million to launch an artist into the 'public' eye and the record company will *expect* you to do *everything* to earn this money back, and then some, before you receive any royalties.

If you're still interested in being signed to a large label for superstar status, my advice would be to seek urgent medical attention or psychiatric help. Of course, when presented with a contract and a cheque for a substantial amount of money, it is a challenge to say no but bear in mind you could be signing away your life for the next 5 years.

Rather, if you want to take the record label route, it's much more advisable to forget stardom and aim for the smaller labels that specialize in the genre of music you write. These are much more reliable, make fewer demands and will appreciate you for who you are.

Even taking this route, however, the days of sending a 'demo-quality' CDs have long since died; they will expect the music to be professionally mixed or, at the very least, have some good production values behind it. More demos have been thrown in the bin from crap production values than anything else and literally hundreds of musicians, many of who are talented, have destroyed their careers because they didn't have a good mix to back their ideas up. Therefore, if you're serious about sending a demo to a label and are not fully confident with mixing, then it should be put in the hands of a qualified engineer.

A typical well-kitted studio will charge around £40–60/h and to mix a track containing vocals, harmonies and overdubs should take around 20 h. Before visiting a studio you should also ensure that the arrangement is complete; attempting to make it up on the spot while the studio's clock is ticking can be incredibly stressful and the attitude that a further instrument could be added during studio time can prove to be an expensive endeavour. Bear in mind that a typical session musician will charge anything from £50 to 100 an hour! Note that even employing a session musician to play a tambourine can cost in excess of £50/h. Don't think that you can play one yourself either; as simple as it looks it is difficult.

Most studios will now burn the record direct to CD, but if they offer DAT, there is no harm in accepting that *as well as* the CD. If they only offer DAT, you need to seriously consider whether this is appropriate. DAT, while a popular recording medium for studios in the 1990s, is now slowly becoming defunct and therefore you should request the mix in 24-bit data form on CD and also 16-bit audio. After all, demos that are delivered to a label on CD are more likely to be heard than the one that arrives on DAT as not many offices or cars have access to a DAT machine.

When it comes to posting the mix, ensure that you put your name, address and phone number on absolutely everything you send, along with a publicity shot of yourself which is in keeping with your musical direction. This way, the label can immediately tell what music you write and where you get your influences. The addresses of most record companies can be found printed somewhere on their latest release, or alternatively in a publication known as the *White Book*, which contains the names and addresses of most reputable labels.

It is also worth including a single sheet of A4 paper containing everything that you want to say about yourself and your music but this should be kept short and to the point – the same applies with your demo CD. The basic premise behind sending in a demo is to give the company an idea of your talent, not your entire song-writing history. Therefore, you should start by sending only two or three songs, ensuring that your best music is first. This way – assuming that they do actually listen to your demo – the first 20 or 30 s are the ones that matter the most. If that doesn't grab them, they may listen to 20 s of the second, but if that doesn't work then the game is over.

This means that the musical arrangement will also need to be closely analysed, especially for club-based dance music. As it isn't unusual for club music to consist of the drum track alone for the first 16–32 bars, if you're submitting to an A&R department, this should be shortened so that the main elements of the music, which would commonly start around 2–3 min into the track, start within the first 10–20 s. Ideally, the intro should be kept particularly short and new melodies should be introduced as soon as possible to maintain attention. That said, care does need to be taken that you don't introduce too much at once; it's a fine line that has to be trodden carefully.

You should also prepare yourself for a negative response, if you receive one that is. Larger labels very rarely respond unless they're interested in signing you, but smaller companies are more likely to either phone or write to you informing you of why they've rejected your music.

Rejection is something that you will become accustomed to, unless you happen to be a musical genius, but it's important to bear in mind that it's nothing personal, and you should spare your ego the devastation of their laughter by phoning or contacting them asking why. Just keep in mind that they're simply business judgements that are made according to the current market status.

If you are accepted, then you may be notified by post but more usually by phone. Naturally, you will be tempted to say yes to anything they have to offer, but only fools rush in. With smaller labels, this call will probably be from the director/owner of the label, but with larger companies it will almost certainly be the head of the A&R department. At this stage, do not agree to anything contractual over the phone and insist that you meet them before you agree to anything. This is important as some companies will send the contract by post and expect you to sign and return it, but without meeting the people in charge you don't have the opportunity to negotiate a better deal.

When the time comes to meet the director or head of A&R, make an effort to look presentable, reputable and trustworthy. Though you may have been offered a contract on the strength of the music alone, you haven't signed at this point and first impressions count. It's vital that you adopt a professional and business-like attitude. These companies exist to make money and you need to prove that you're dedicated enough to be both reliable and sensible, which also means turning up on time. The excuse of 'I missed the bus' isn't particularly concerting. They will have literally hundreds of artists lining up waiting to get on their books, and if you present a poor or unreliable image, they'll simply go for someone else.

While it's doubtful that you'll be asked to sign any contracts at the first meeting, it is sensible to contact a solicitor who has dealt with the music industry before negotiating the terms of a future contract. This not only helps to ensure that you turn up on time but if the label does present a contract, the solicitor can look through it for you. As tempting as it may be to save money at this point by reading up on the law yourself, I *strongly* recommend that you employ one to check that the terms are acceptable. A typical record contract is crammed full of numerous clauses, and if any escape your attention, you could end up losing not only money but also the rights to your current music and any future releases.

ROYALTIES

When presented with a contract, the part you're most likely to be interested in are the royalty rates as these dictate how much money you will earn from each sale. Most labels will either offer a publishing deal or simply choose to license the music, and the royalties you receive will depend on whether you're being offered a licensing deal or a publishing deal.

Publishing deals generally pay a much higher percentage for royalties but in return you're giving them the rights to release any future material you produce over a specific number of years that is stipulated in the contract. This is commonly set over 5 years, but some labels are known to sign for a period of 10 years depending on the past sales performance of the artist. As these types of contract are long term, they should be scrutinized very carefully by a qualified music savvy solicitor, as a missed clause on your behalf can result in a long, drawn-out legal battle against the label if they're not performing up to your expectations.

Despite the temptation of being on a label's book for a number of years, from personal experience it is not worth signing these types of deals because you will be tied to the label. Instead, if it is offered, the label involved is clearly confident that your music will sell and it will be open to negotiation and could be persuaded to sign a licensing deal instead. This is a more secure route to take, not only allowing you to assess the label's competence but also giving you the freedom to walk away and find another label for your next release if at all it goes horribly wrong.

Licensing deals are less complicated and are more common on the dance music scene. Essentially, by signing such a contract your music is licensed to the record company for a specific period of time in which they can do with it however they see fit. These are most usually signed for a period of 1 year, although it can sometimes stretch to 2 or 3 years depending on how popular the label thinks the track may become. It can therefore be generally accepted that the longer the time period you're expected to sign, the more potential the record company believes the track will have, and this can open up some bargaining power. In fact, it is often worthwhile asking for a few changes in the contract if there is something that you're not happy with.

From past experience, I've learned that many labels don't even understand their *own* contracts and therefore occasionally are only too happy to change some aspects as long as it isn't changing the percentage of the royalties you or they receive. Even some of the larger labels didn't understand their contracts when they were dissected!

Perhaps the most important aspect to look out for here is a clause stipulating that they have the right of first refusal on any other tracks you produce. This means that you could be forced by a legally binding contract to hand over any subsequent releases to the same label regardless of their previous performance.

Alongside the royalties received from the record company, you may also receive payments from the Music Publisher, Performing Rights Society (PRS), Mechanical Copyright Protection Society (MCPS), Association of United Recording Artists (AURA), Performing Artists' Media Rights Association (PAMRA) and Phonographic Performance Limited (PPL). Some of these payments will be made directly to the record company who will pass these, along with your royalty payments, onto you or your manager while others will pay them directly to you, depending on the circumstances.

To better understand the roles that each of these companies play and how it all ties together in the real world, we'll use an example by envisaging two artists, James and Jon. They have just finished engineering a track and featured an artist (Emma) to perform the vocals on the record. So the artists can continue to make music rather than deal with the recording industry; they've also employed a manager who we'll call Steve.

Unless there is a specific written contract between the band members, the manager or the record company stipulating that they are all entitled to an equal share of *all* proceeds, the members may be able to claim different sums of money from different organizations, as well as the record company, depending on the circumstances. Figure 21.1 outlines how this works.

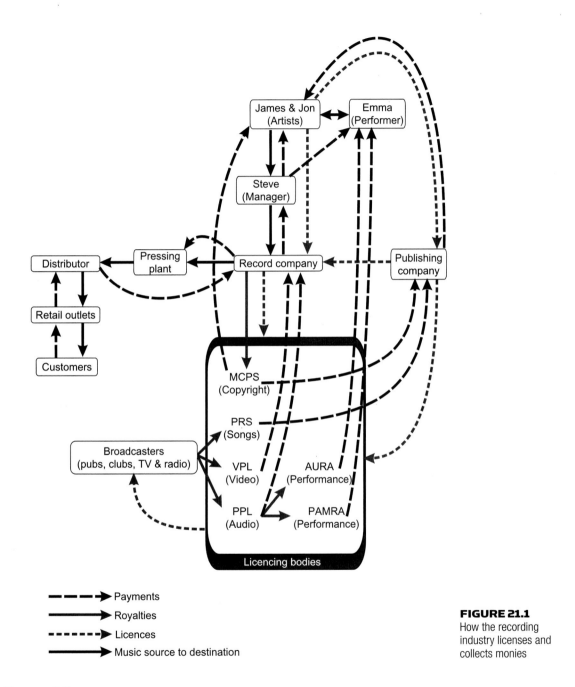

Payments
Royalties
Licences
Music source to destination

FIGURE 21.1
How the recording industry licenses and collects monies

Record Company

The record company will pay any royalties due to you from record sales or licensing deals. If you have a manager, he will receive the payments and then distribute them to you and any other group members while keeping an agreed percentage for his managing duties.

Publishing Company

Publishing companies work very closely with record companies and deal with releasing the recording and tracking the national royalty systems around the world on behalf of the original author of the work. For this they take a percentage of any of the royalties, paying the rest to the original author.

Phonographic Performance Limited

The PPL licenses records to radio stations, clubs and pubs. This is because every time the record is played to a collective public audience, whoever is playing the music must pay for the privilege of playing your music. The PPL then collects these royalties and pays it directly to the record company involved and to any UK featured artists. For this service, the PPL keeps a percentage of the royalties collected.

Video Performance Limited (VPL)

Similar to the aforementioned PPL, the VPL licenses the music video to be played by broadcasters (such as MTV), pubs and clubs. These royalties are paid directly to the record company involved, and for supplying this service the VPL keeps a percentage of the royalties collected.

Association of United Recording Artists

If a piece of music has been engineered by a professional producer or it features an artist (session musicians etc.) who is not a member of your group, then AURA may become involved if the featured artist is not from the UK. These collect royalties from the PPL and pay them directly to the producer or featured artist. It is up to the producer of the featured artist to join this association, and although membership is free, AURA keeps a percentage of the royalties that are collected.

Performing Artists' Media Rights Association

PAMRA is similar to AURA and collects any royalties that are due to the performers. Like AURA, it is up to the performing artist to join, but while membership is free, they keep a percentage of royalties for offering this service.

Performing Rights Society

If the song is going to be performed in public by a third party, then royalties are due to the original writer of the song every time it is performed. These royalties are collected by the PRS and paid directly to the author of the song. Like AURA and PAMRA, it is up to you to join if you wish to collect. For the rights to 15 songs, membership costs £400 with the PRS also taking a small percentage of the royalties. Joining the PRS is rare for dance musicians since it is rarely performed by another party.

Mechanical Copyright Protection Society

Once signed to a label, the record company will make a number of copies of the record for distribution and sale. These are referred to as 'mechanical copies' and all the artists involved are due 'recording royalties' based on the value of the number of copies made. The MCPS collects these royalties, which are currently 8.5% of the wholesale value, and pay them directly to you. Again, it's up to you to join this organization that costs £50 for membership along with a percentage of the royalties collected.

Naturally, there is no universally accepted amount that you should receive from any of these companies but especially from the record company as it depends entirely on the size of the label, your past performance and the way the label works its accounts. Most labels will operate on an 'all-in' basis which is derived from the suggested retail or the wholesale price. An all-in royalty rate means that if a third-party producer, musician, instrumentalist or song writer was employed during the recording process, their payment is derived from your own royalties rather than the record companies.

Whether the royalties are based around the suggested or wholesale price is also an important consideration. The suggested retail price (SRP) is usually double the wholesale price, so if you have a 40% wholesale-based royalty rate then you'll get paid the same as a 20% retail-based royalty. This percentage is quite normal for a royalty rate, and if it's any less you should try to negotiate a better deal.

PACKAGING

One of the biggest mistakes any artist can make is asking whether their single or album is released in commemorative or novelty packaging. While it is true that leather cases or steel boxes sell more copies due to their novelty value, they are incredibly expensive to manufacture and this additional cost will come out of *your* royalties.

In fact, though jewel cases and booklets aren't actually part of your music, they're essential for delivering it to the audience and *you* pay for their manufacture, not the record label. This is usually described by many labels as a 'sales allowance' and, depending on the label involved, the charge can be anywhere from 80 to 20% of your net royalties, although this can quickly rise to above 50% if you're adamant that your recording be released in novelty packaging.

This allowance is also used to cover any damage that the packaging might suffer from on its way to the shops. Along with this sales allowance, most labels will also take a percentage of royalties for 'recoupables'. These are one-off expenses that are incurred by the label for promotional purposes. This includes tour support, personal advances, photo sessions, artwork, publicists, copies given away for promotion and music videos.

Today, more than ever, videos provide a huge promotional tool and the current MTV generation needs videos to associate with the music. The price of making a music video can vary tremendously from as little as 10 000 to over 2 million and you will be expected to pay a percentage towards making it which will be subtracted from your royalties. That said, it is very doubtful that any record label would allow you to make a video for more than 10 000 if you're a relatively new artist and, in some cases, they may insist on a particularly small video budget – so forget about that video featuring you in the Bahamas surrounded by scantily clad women or men. It's far more likely you'll find yourself in a disused warehouse in a particularly dubious part of town, trying to mime to a track that's occasionally drowned out by police sirens.

Again, the percentage you pay towards a video will vary from company to company but you should insist/negotiate on not paying more than 40% towards it. What's more, you should insist/negotiate that no more than 10% of your recordings are given away as promotional items. By providing promotional copies to record stores, a sales team can purchase both shelf and window space, but if you let them, they'll give away a huge amount of records to secure a good space in the window and/or shelves which ultimately results in loss of income for you.

As a word of warning, all major record labels will not allow you to find your own supplier of CD cases, artwork or pressing plants even if they are cheaper than the amount being paid by the record company (and, of course, ultimately you). Whatever the label decides to do, you do not have the opportunity to audit, and no matter how much the label decides to spend, you'll have to pay for it from your own royalties!

More importantly, ensure and then double (and triple) check that some of the promotional copies are sent to some related magazines *before* they arrive in the shops rather than after. Most magazine reviewers will not review a record that is freely available in the shops, so if they receive them after they are available they will usually not be reviewed, reducing your potential sales. This is a surprisingly common occurrence and an apology from the label for overlooking this 'small problem' will not help flagging sales.

To prevent this from happening, many artists take this into their own hands by requesting 20–30 copies of the record before they go into the final manufacturing stage. Often referred to as test pressings (TPs), you can use the excuse that it allows you to listen to what the mastering engineer has done to your record and so you can hand out copies to friends and relatives. In truth, you're sending them out to magazine reviewers yourself to ensure that they receive copies to review before they're released in the general market.

One final issue that is vital to fully comprehend is the advance payment. All record companies, no matter how big or small, should offer an advance on signing of the contract. This is the label's proof that they are committed to promoting your record but also that you're also paid for the work.

You shouldn't expect a sum amounting to millions, similar to those often reported by the press, apart from often being blown totally out of proportion by the media – you need to have a long history of best sellers and a remarkable reputation with the public. For a first release, a label may realistically offer between £200 and 5000, depending on the size of the label and how popular they believe your music is going to be.

Advances are refundable from the profits, though, which means that you will not receive any royalties until the sum total of your royalties from sales have exceeded the advance. It is worth ensuring that there are no 'hidden' clauses in respect of this advance too, by making sure that it is only recoupable from the profits. If they are not, the label may have the right to recoup the money if they don't release the track or if the record does not sell as well as they initially expected it to.

PROMOTION

Once the final details have been accepted by both the company and the artist, the promotional campaign will start and the recording will hit the shops. The first 3 months are the most crucial stage; many record companies and shops will judge the selling potential of the record within this time period. If only a few copies are sold, it's very likely that the shop will return them to the record company and ask for a percentage of what they paid for the record back.

Most of the high-street stores settle for a 25–50% return rate but many of the smaller independent shops will demand a 100% return if they can't sell it. Consequently, whenever a record is released, all record companies set up what's known as a reserve royalty account. This allows the label to withhold anywhere from 40 to 80% of your net earnings and hand them out over a 2-year period, preventing them from making a loss by paying you too much.

This can cause some substantial problems if you've signed more than one record or a publishing deal since they will view it as one financial deal rather than a number of individual ones. Thus, if your first release is a failure but the second or third is a hit, they are within their rights to recoup any losses from the failures by taking a larger percentage from the hit. In a worst case scenario this could result in your earning nothing whatsoever.

Ultimately, here we've only looked at the most common pitfalls and recording contracts can contain many more clauses. They are a tedious read but it is vital that you go over every aspect with a good solicitor because if you make a mistake now it will stay with you for the rest of your recording career. Also, as you become more successful, it is advisable to hire a good manager to take care of everything for you, so you can continue making music rather than dealing with all the paperwork. Until you have a manager, it's almost certainly worth joining the Musicians' Union as they can offer solid advice relating to both contracts and publishing deals. They can be contacted at: www.musiciansunion.org.uk.

The MCPS can be contacted at: www.mcps.co.uk.

The PRS can be contacted at: www.prs.co.uk.

The PPL can be contacted at: www.ppluk.com.

The PAMRA can be contacted at: www.pamra.org.uk.

The AURA can be contacted at: www.aurauk.com.

INDEPENDENT RELEASE

While sending off a demo to a record label is still viewed by many as the best way to get your recording released in the general market, actually being accepted by a label is becoming increasingly difficult. Indeed, many labels will not entertain the idea of signing an artist with no previous history of releases, and even if they are willing to sign you, they pretty much have all the bargaining power giving them the freedom to dictate much of the contract you'll be expected to sign.

Consequently, an increasing number of musicians are releasing their singles independently. This does mean that it's you who has to do all the legwork involved, from sorting out the artwork and distribution along with incurring all the costs, but if you expect a record label to have the faith in you and put its time, money and effort into it then why shouldn't you be confident enough to do it for yourself?

Releasing your own music isn't as complicated or difficult as many are led to believe, and although it does take some financial capital to get it off the ground, any earnings are yours to keep. Additionally, if it does well, it also means that you have a history of a release giving you some bargaining power if you ever decide to sign to a record label.

Over recent years CDs have invariably become the ultimate format on which to record and distribute music. They're cheap and easily available, and while many DJs are now turning to this format, alongside MP3 – to allow them to remix, and remix using software such as Ableton Live – to the same amount of DJs, vinyl turntables are the definitive system to play dance music on. So much so that many DJs will not take a dance mix that arrives on a CD seriously. Apart from the fact that it doesn't particularly show any 'real commitment to the scene', vinyl reproduces the lower frequencies much better than a CD. This obviously means that if the music is bass driven, as dance music is, vinyl is always going to be the number one choice for dance.

This does, however, pose some problems if you've mixed down with CD in mind because there's a tendency to overcompensate to create a deep bass. If this is then pressed to vinyl, the bass can be so loud that the stylus can skip along the record. What's more, the louder the bass the wider the groove needs to be on the record, which can force you to reduce the overall length of the track.

Broadly speaking, a typical dance record with a heavy bass should be no longer than 10 min at 33 RPM or 8 min at 45 RPM. It is possible to go a minute or so longer than this at both speeds but it may result in a drop of volume or, in some cases, distortion in the high-frequency elements.

At the beginning of a record the outer circumference of the groove is around 36″, but as it continues towards the centre it continually reduces in size until it reaches around 13″. This means that as the record progresses there is less and less space to hold all the frequencies, so the audio quality will reduce commonly introducing distortion in the higher frequencies. Consequently, it's sensible to avoid boosting any frequencies above 8 KHz while also rolling off everything above 16 KHz. It's also worth running any high-frequency sounds through a good de-esser while mixing for extra security. Both these techniques have to be applied cautiously, though, as once the mix has been committed to vinyl there's no going back if it's wrong. Therefore, I strongly advise that if you want to submit a mix to vinyl, you send it to a professional mastering house.

Although we've already covered the general principles behind mastering in the previous chapter, without a doubt, there is absolutely no substitute for employing a professional mastering engineer. Even though there are numerous programs available that promise to produce 'professional' results, which some can, none of them replace a mastering engineer's experience and ears.

These guys have years of experience behind them and will not only give a different perspective on your music which you may not have considered but also make sure it's mixed correctly for vinyl. On top of this, there may be no need to adjust a mix that's been prepared for a CD, it can just be sent to the mastering house as it is. That said, it is worth checking with the mastering house before you send any material as they may need the CD prepared in a particular way. Also, you should include all important details such as track numbers, the length of each song in minutes and the total length of a side including the pauses between the tracks, provided there is more than one.

Most mastering houses today will accept CD in data form often called pre master compact disc (PMCD) but they will also accept Exabyte, U-matic 1630 and perhaps DAT. This latter medium is becoming less popular as a recording standard as many reputable mastering engineers do not consider this to be of a high-enough standard. This is because DAT can suffer from drop-outs and other tape-related problems that can make the mastering engineer's job even more difficult which can result in one of the following two:

1. They'll charge you a hefty sum of money to 'repair' the problem on site.
2. They'll charge you for the time they've spent listening to the DAT before deciding that they have to return it to you.

The prices mastering houses charge vary from plant to plant but at the time of writing a typical price for mastering and preparing two tracks (remember there's usually a B side!) is around £250–400, depending on the reputation of the mastering engineer and plant involved. On the subject of a B side to the record, it's a wise idea to use this for the instrumental version and include plenty of exposed segments consisting of just the drum loops, along with instrumental riffs and vocals. This way you open up more options for the record:

- If the A side does incredibly well, it will not only increase the demand for it but also allow any producers to remix the record to promote it further.
- If the A side does poorly, a producer or well-known DJ may decide to remix it using the elements included on the B side, increasing its chances of becoming a hit record. If it does well, a label will look to sign it and you'll receive royalties for every sale – a very profitable venture.

- A producer may drop different vocals over the top of the B side which could promote the record further.

Once the track has been mastered, the plant will often return a copy of the master back to you in a format that you request, allowing you to verify the quality but this copy should not be sent to the pressing plant. The mastering engineer will, or at least should, have recorded the mastered version onto Exabyte or U-matic 1630 which is then sent directly from the mastering house to the pressing plant to produce the record. This is because the quality required to produce the 'glass master' must be of an excellent quality and CD or DAT are not of a high-enough standard. This does, of course, depend on the pressing plant involved and some will accept these lesser quality formats simply to get their hands on your money.

Glass masters are so called because they are destroyed during the production process. After the recording has been copied onto the glass master it can be transferred onto a stamper, a round metal form that is then used to process the lacquer to produce the final record. This glass/stamper/lacquer process generally costs around £95 per side but, as always, the prices will vary from company to company.

Once this is completed, the plant will press a small amount of records known as TPs to give you the idea of how the finished product will sound. Again, these are not free and you'll be expected to pay anything from £2 to 4 per TP. Many plants will refuse to make just one TPS, insisting that you have a minimum of 10. Nevertheless, with these TPs, you can then begin to look towards distribution deals, publishing deals, and more importantly, registering yourself with the MCPS.

As you're planning to release records, you're essentially becoming a record label, so you should register with the MCPS for a number of reasons. Firstly, you'll need to check that no one else is using the name you have decided upon (you can also search bandname.com). This mistake has been made many times with the most notable example being made by Chemical Brothers when they decided to change their name to Dust Brothers. Rather than checking with the MCPS first, they simply used the name which understandably resulted in the 'original' Dust Brothers being a little more than upset.

Secondly, the MCPS will ensure that your record is copyright protected and that you receive any mechanical royalties that are due to you. For providing this service, they are usually paid 8.5% of the total income from every unit sold. You do not have to pay this if you press and release less than 500 records (provided that you reach an agreement with them first) or for records that are used for promotion. These promotional copies must, however, be prominently marked with a non-removable, non-erasable notice carrying the words 'PROMOTIONAL COPY – NOT FOR SALE' on the sleeve *and* packaging *and* the record itself – you cannot simply just attach this label to the dust jacket. Plus, of course, they must be supplied free to the DJ, reviewer or club!

After registering, you can begin to look towards getting a distribution deal using the 10 or so TPs you've received. The reason behind looking for a distributor now rather than when all the records have been pressed is that you may be able to secure a pressing and distribution deal. This will depend on the quality of the record, and if your chosen distributor believes that it has a good sales potential they may offer to pay 50% towards the price of pressing. This investment will obviously be recouped by taking a percentage of profits from every sale, along with their distribution percentage and further percentage for organizing the manufacturing. Though this probably sounds like a poor deal for you, it is the best deal to make as the distribution company will be heavily motivated to sell your records so that they can recoup their expenditure. Additionally, they will also be obliged to sort out the bar codes that help record shops detail their sales, something that is very difficult to organize yourself.

Many distributors may also want you to sign a contract, giving them exclusive rights to distribute your record for a minimum of 1 year. Although a relatively small press of 2000 records may not sell over this time period, but if they do, you will be tied to the same distributor until the end of the contract. This means that you should take your time to find a good one.

Finding a good distributor can make the difference between your record reaching the best independent DJ and high-street shops or ending up with a backstreet dealer frequented by 'wedding DJs'. Consequently, it is worth shopping around, phoning the companies and asking with whom exactly they deal with and if they have any previous history of distributing 'hit' dance records. Typical questions to ask are:

- Whom do they distribute to?
- Have they ever distributed a hit record?
- Do they have plenty of contacts?
- Are they in contact with any radio stations, producers or DJs?
- Are the contactable 24 h a day?
- What publicity plans, if any, do they have?
- If you let them distribute the record, when would they plan to release it?

The last question is important in establishing whether they have their finger on the pulse of the music industry. Obviously, you'll want to release your record as soon as possible, but if they plan to release it during Christmas or at the same time as one of their other, more popular, artists then it will be lost in the flood of releases or not receive their utmost attention.

Enthusiasm, faith and an understanding of the music you produce are the most important qualities of any good distributor, and this is something you should be able to ascertain after speaking with them for about 10 min.

Remember that it's your future being placed in their hands, so ask them to inform you of their release schedules and, assuming that they have heard your recording, ask what they think of it. A good distributor will not be obsequious but offer solid advice on the selling points of your music and let you know of any ideas they have that may increase its sales potential. The price for all this commitment is usually around 20–30% of your earnings but don't underestimate the power of negotiation. If you adopt a diplomatic attitude, you may be able to reduce this to 15–25%. Whatever happens, though, it's unwise to sign any contracts allowing them to take over 30% of the income.

If you do not want to become involved with a distributor, preferring to handle it all yourself, then you can use a vinyl broker instead. These companies are solely concerned with just pressing the vinyl and configuring the bar codes and, as they have a good relationship with pressing plants, they can usually get records pressed cheaper than if you approached a plant yourself.

Assuming that you have a distribution deal or have decided to approach it all yourself and have approved the quality of the TPs, the plant will continue to process the amount of records that have been ordered. Generally speaking, 130 g lacquer is the most commonly used for 12″ records but some plants may offer 180 g. From experience, this grade produces a heavier bass, a sharper high end and a clearer stereo image, and suffers from less distortion, but it costs approximately twice as much.

It is possible to get very good quality from 130 g but you stand a better chance of quality if the vinyl is virgin and not recycled. A proportionate amount of pressing plants will use recycled lacquer unless you state otherwise, so it's always advisable to ask before you commit yourself.

A typical price for pressing a single 12″ 130 g record is around 60 pence but very few plants will press less than 500 and, on average, many will not press less than 1000 copies at a time. Although many unestablished artists would prefer to press an initial run of 500 records, the less that are pressed, the more expensive it is and, ideally, you should look to earn enough from pressing records to finance pressing your next release. This means that you're generally better off making a minimum press run of 1000 or, if you can afford it, a run of 2000.

Another option that may be open to you is the choice of coloured vinyl, rather than black. It's a fact that coloured vinyl sells more copies but it is considerably more expensive. Blue, red and yellow vinyl can cost as much as £1.50 per unit while white vinyl or a customized colour can cost in excess of £3 per unit. Bear in mind, though, that the choice of colour can govern how many records you should press. Most pressing plants will reduce the price of each unit with the more records you order, so while it is recommended that you press over 1000 black records so that you can recoup the money you're investing, if a different colour is used you should look at pressing more than 2000 to make them cheaper.

ARTWORK

The very name white label should give some clue as to how much artwork is involved on the record's sleeve. A majority of white labels have little more than the name of the artist and a contact number or, more commonly for a record that contains an illegal sample, an e-mail address from an 'untraceable' supplier such as googlemail or yahoo.

For obvious reasons, I cannot publically condone the use of illegal samples in any record but I can say that a proportionate amount of white labels that have made their way into clubs and onto the charts have contained illegal samples and by only including an e-mail address you have the option of not replying if a record company gets in touch to press charges against you.

That said, in a majority of cases if the recording is doing exceptionally well and the label who owns the original sample believes that it has a big sales potential, they may wish to release it commercially, but be warned, it gives them the upper hand when it comes to negotiating a deal.

Many pressing plants will offer to design and print the artwork on your behalf but this obviously costs extra and it's usually a better idea to actually supply the plant with your own artwork. As long as it's quite simple, the plant will only charge you a small 'cover preparation' or 'film charge' which is used to prepare the printers to create the labels, and then charge 3 or 4 pence per label, provided there are no colours involved. Alternatively, if money is tight, a cheaper option is to leave the labels from the plant blank, design your own, print them out and then stick them on the sleeve yourself – if you have the patience to stick labels on over 1000 records – and it does become incredibly tedious. Remember that time is money to everyone, including you, so calculate the time it will take to prepare the labels and balance it against what the pressing plant may charge – plus the pressing plant will put the labels on 'straight', there is nothing more unnerving than seeing a record spin with an off-centre label.

While a majority of white labels contain little more than the name of the track and some contact details, a trick employed by some musicians is to design a label that not only contains these details but also gives the impression that it's a special edition import. The ego of many DJs should not be underestimated and the elitism of owning an 'import' can immediately raise the hype so that DJs are more interested in playing it on the dance floor.

Most plants will also give you the option between an inner or an outer sleeve. An inner sleeve is the thin 'tracing paper' type cover you often find inside normal record sleeves to protect the vinyl from being scratched by the outer sleeve and these cost around 6 pence each. The outer sleeve, often referred to as a 'dust jacket' or 'disco bag', is constructed from cardboard; thus, it is more substantial and normally comes in black only. Obviously, this means that they are more expensive, costing between 15 and 20 pence each, but they do offer more significant protection for the vinyl and are preferred by many DJs for transporting their records.

Thus, assuming that you decide to press 2000 copies of your recording and you have no manufacturing deals, the price so far would be:

- Mastering: £200 × 2 = £400
- Lacquer: £95 × 2 = £190

- TPS: £4 × 10 = £40
- Film Charge: £8
- Labels: 4 pence × 2000 = £80
- Dust Jacket: 20 pence × 2000 = £400
- Pressing: 60 pence × 2000 = £1200
- Total: £2318

Once you've received the completed order of vinyl, you'll need to look into promoting the single before it hits the record shops. Simply releasing it without any publicity to back it up isn't going to help it sell as few people will walk into a record shop and buy a track without knowing anything about it. There are two options available for this: you can either employ a press company to deal with it all or attempt to create some press yourself.

In an ideal situation, you would employ a press company. These people, who often work in league with distributors, have all the contacts needed to publicize your record and will get your name in every magazine or paper that really matters. To accomplish this, they will ask for a minimum of 100 free copies of the record so that they can send them out to the people who are important, i.e. editors and journalists, who will, hopefully, give you rave reviews, increasing your sales potential when the record finally hits the shops. If you do decide to take this route, then finding a good publisher is as vital as finding a good distributor and it is worth shopping around to find a company that deals in the genre of music you create.

In finding a good publishing company, typical questions to ask are:

- Whom have they dealt with in the past?
- Which magazines or papers are they in contact with?
- What are they prepared to do to publicize your record?
- Will they take out adverts in music papers?

The latter question should result in a resounding no if the press agency is reputable as unless you're an established artist adverts are an absolute waste of time and money. Employing a publishing company obviously isn't free and, generally speaking, it can cost between £500 and 1000.

Whether you decide to use a press company or not, you should *seriously* consider employing a DJ promotion agency. These form an essential aspect of dance music promotion and without them it is doubtful that any record will ever get off the ground. Essentially, they have various lists of clubs and well-known DJs suited towards the genre of your record – and will mail it directly to the ones that matter.

Both clubs and DJs get the records for nothing, which means that you have to supply them for nothing too, but in return they fill out reaction sheets that report how each record is going down on the dance floor. This gives you a good idea of how well the track is doing on the circuit, allowing you to estimate its popularity.

Obviously, this means that you should go with established agencies that have the addresses of the best clubs and DJs for the type of music you produce; you don't want your records ending up in the hands of wedding and party DJs. Agencies such as *Waxworks* or *Rush* are both well respected (and whom I fully recommend), as both these have large lists and will advise you on how many records you should supply them. Obviously, this service is not free and they charge 2–3 pounds per record and have a minimum mail-out of a hundred. They will also expect payment upfront rather than after they've been mailed – just in case you go 'out of business'.

It certainly isn't advisable to miss this stage out in the hope of saving a little money. Hanging around in clubs in the hope of catching DJs and giving them a copy is a time-consuming affair, time that could be put to better use writing another record. Let the professionals handle the promotional aspects.

This brings the price so far to:

- 2000 130 g black vinyl records, with sleeve and labels: £2258
- DJ promo agency: £300 and 100 records
- Total: £2558 and 1900 records

On average, most record shops will pay approximately £2 for a 12″ single, so provided that you could sell them all, it would equal a profit of £1442. Alternatively, if you only pressed 1000 records, the profit would be considerably lower, equalling around £382. However, while the latter still provides a profit, it isn't much to put towards a second release and also remember that many record shops operate on a no sale return policy. On top of this, you may also have to pay song-writing royalties to the MCPS.

As an extra precaution against losing money, a number of artists also send their labels to overseas distributors, effectively licensing the record to them. The benefit of this approach is twofold: it provides a larger potential audience and they'll press any additional recordings themselves. On some occasions they may also be willing to pay you a licence fee as an advance for the records, but if the record doesn't sell many may well expect this back.

In an ideal world the record would fly out of the shops and the money you make can go towards releasing the next single until ultimately you have a number one hit. You start your own record company and start signing other artists or another label offers you a sub-licensing label deal. This has happened with many small independent labels, where a larger record company such as Sony or Virgin help to not only finance your work but also use their marketing influence to reach a much wider audience, in return for a slice of your profits, of course. This kind of affiliation is more common that many realize and a proportionate amount of 'independent' labels depend heavily on the backing from large companies. So, if the premise that the end of large labels is on the horizon from the evolution of self-promotion on the internet and illegal MP3 downloads holds true, the smaller labels will fold first!

Some of the main distributors in the UK for dance are:

- Intergroove: 020 8838 2000
- 3MV: 020 7378 8866
- Kudos: 020 7482 4555
- Pinnacle: 01689 870622

DISTRIBUTING VIA THE INTERNET

While white labels invariably provide the most promising way of getting your track noticed, especially if you produce dance, if money is tight then internet promotion may prove a more viable option. Indeed, the lines between web designer and musician are drawing closer as more and more musicians are developing their own websites and nearly all have a website even if its only purpose is to keep fans informed of when and where the next gigs are.

Designing a website is a professional business, and as it's essentially acting as the face of your music, it must be well developed, easy to navigate and able to exude a professional feel. This latter aspect is especially vital if you plan to charge users to download the music – if the site looks cheap and tacky, it isn't going to instil a lot of confidence in the buyer.

Once a website has been created, it has to be promoted. The first step towards this is to register a domain name, and this should be kept short, sweet and memorable. An IP address that uses long drawn-out page names or squiggles is not only difficult to remember but a pain to type in, and is considered tawdry by many web users. More importantly, do not use banners or pop-ups on your own site. These are often voted as the most annoying part of the web and many users avoid any sites that feature them like the plague. Similarly, it is not worthwhile paying for a website to advertise yours with a pop-up window as a proportionate amount of knowledgeable internet users will employ software that prevents them from appearing.

Once you have a domain name, it can be submitted to a search engine. Additionally, a number of search engines will allow you to add your URL to their engine by clicking on the appropriate link but it is also advisable to purchase software that is specifically designed to submit URLs to most search engines. Wolf Submit Pro is considered by many as the best software for this but there are various other recommended software packages such as Addme.

Of course, there's much more to registering a domain name and simply placing it in a search engine because unless you already have a fan base no one is going to be able to locate the site. Simply typing in MP3, independent bands or dance into a search engine isn't going to result in just your site being displayed, there will be thousands to choose from. To prevent this, it's a much better idea to create a site dedicated to a subject that your potential fans might be interested in, or something a little more fun such as flash-script-based games with your music playing in the background. Alongside this, you could place the band history and full MP3 downloads on a separate page with a link to them from the main website. This way, if anyone finds the music interesting they can jump to the 'real' page.

If you take this approach, remember that not everyone has access to a broad-band connection, so don't attempt to place a full track in the background, just choose the best snippets and loop them. As a general rule of thumb, it's advisable that any web page should load faster than you can hold your breath, using a typical V90 56 K modem as a reference speed.

Furthermore, try typing in some keywords that relate to your genre of music and find out who the top bands are. Once you have their addresses, contact them and ask if they would be willing to exchange links with you. Similarly, join or start your own webring. This is a group of sites dedicated to the same subject that are interconnected together with links. Essentially, this forms a virtual chain allowing a visitor to go forwards or backwards through it, visiting each site in turn. You can find out which webrings deal with your genre of music, or start your own by visiting www.webring.org.

Notably, it isn't advisable to set up a new webring if there is one already covering your interests. Most webrings will insist that a number of factors are met before you can join in. Consider translating the website and the vocals in the music, if there are any, into different languages such as French and German too. This opens up the site to more visitors which will hopefully lead to more clicks on your site, and by translating the songs to different languages there is a better chance of them being played on a foreign radio station. For instance, many stations in France and Germany insist that a certain percentage of the music they play is in the native tongue.

Although all these methods will go towards promoting your site and music, it isn't a good idea to charge visitors to download any music unless you are well established. Although offering small samples of the music before charging to download the full track may seem sensible, it will easily distract potential customers. A much better approach is to allow users to download all of your music for free, but ask them for contributions if they do.

Most people will take anything if it's offered free, and not only does it allow them to evaluate all of your music, if they like it they'll tell friends about it which helps spread the music. As an additional incentive, you could create a members-only area for those making contributions, giving them inside information on the band, access to 'limited edition' tracks or the chance to remix one of your tracks. This latter approach can obviously be beneficial to both parties. Above all, remember that when starting out promotion is substantially more important than royalties from sales; the more promotion you receive, the bigger the chance that a label may snap you up.

To further promote your music, it is also worth seriously considering signing to an internet label (iLabel). Not only they have much more financial clout and promotional influence but some also pay royalties based around how many times your music is downloaded or listened to. This, however, can be fraught with complications similar to when signing with a normal record label, so it is sensible to shop around. What to look for depends on what you want to gain from the signing but there are some things you should look out for.

First and foremost, find out how long the iLabel has been online and how popular it is. The label needs to be both big and professional enough to attract visitors for you to sell your music. Naturally, registering with any iLabel is useless if no one can find your music, so it's a good idea to look for one that features plenty of genre categories. A website that encapsulates every genre of dance music under one name in the search isn't going to help those looking for a specific style of music. Using narrower categories will help to funnel potential listeners towards your song. Also ensure that the site categorizes these genres as accurately as possible. After sitting through a download, you expect it to at least be the right genre.

Another important consideration is what, if any, software do you need to access and download from the site. A few iLabels use their own proprietary download system, resulting in a visitor having to download and install software before they can even access the music. Others will not let you download a small snippet of music without requiring you to fill in various forms first, while some put stringent requirements on the version of browser that's needed to access the site and others are so laden with graphics that those with a slow internet connection have to wait 5 min before they can even begin to browse the site. Any of these can distract potential customers. Many visitors will simply be browsing and only those who are incredibly patient or enthusiastic will put up with missing around to access the music, and even their patience can wear thin if it's all too convoluted.

On this same subject, the overall design of the site is important. It's a safe bet that they have all been programmed by professional designers but some are complex to navigate and finding a specific genre of music can be a drawn-out experience, almost bordering on torturous. Ideally, they should be quick to navigate and you should be able to find the genre within a minute of entering the site. More crucially, the site should also reflect the style of your music. If you're producing music that's considered 'cool', you need to register with a site that projects a cool image. Moreover, ensure that the website vets all music before it can be posted to the site. All successful record companies only sign and release music that they think will sell and make an effort to ensure that they are professionally recorded and mixed. The internet doesn't have any of these filters and there can be nothing more annoying than spending time downloading an MP3 that promised to be good but turned out to be self-indulgent rubbish. To avoid this and provide some security of quality, iLabels such as Peoplesound.com vet the recordings first and hence stand a larger likelihood of being visited.

If you already own a website, then it's worth checking that the iLabel will allow you to place a link to your own website. Some iLabel sites only allocate a single page to each artist and it is difficult, if not impossible, to fit all the information about yourself on one page. By including a link to your own page, potential customers can jump to it to find out more about you. If you don't already own a site and have no plans to start one, then look for an iLabel that offers promotional pages and permits you to advertise upcoming gigs, new album releases or merchandise. What's more, some iLabels will even offer web space with an address that you can use and these may be the best option if you have no home page.

While in a few cases all this is supplied free, many iLabels will charge you for the services they provide by taking a percentage of any royalties you are due. The amount varies widely from site to site, with large sites that have hundreds of visitors every day taking a higher percentage than the one that only attracts a couple of hundreds a week. This begs the question as to whether you will make more money signing to a large site that pays fewer royalties, which is frequented by lots of visitors, against a site that is not as popular but pays a higher royalty rate. Obviously, this is something you'll have to consider carefully but a few factors that could help you decide are the terms of the iLabels contract.

Most reputable sites will not ask for exclusive rights but signing to those that do means that you can't post the music to any other sites, thus limiting the potential market. On top of this, if a major record label discovers you, the iLabel may not let you terminate the contract with them until it has run its course. Few sites impose these kinds of restrictions but it is worth examining all the contractual clauses before you agree. Finally, consider that some sites will not send you a cheque for royalties until they have crossed above a certain amount while others will send them religiously every month. Both these methods are potential double-edged swords: if you're paid only when the royalties have crossed over a specific total, it could be a while before you see any return, but if they're sent religiously every month, you may be cashing cheques for only a few pounds.

Above all, although promotion forms a vital part in getting your music heard, a much more important factor is the quality of the music you produce. If it's good enough, you can guarantee that people will hear about it and a number of artists have been signed to labels through an MP3 demo alone.

MUSIC CHARTS AND PLAY LISTS (UK ONLY)

The ultimate promotional tool for any artist is being placed on BBC Radio One's play list and this is the holy grail of most record labels. In fact, labels want their artists on this list so much that the producers of Radio One are wined and dined continually.

These producers meet at the beginning of every week and listen to all the latest releases to decide which make the A, B and C play lists for that week. If a single makes it onto the A list, then it's played three of four times every day. If it makes it to the B list, then they recommend that it should be played once a day and if it's on the C list it's entirely up to the DJ hosting the show. Once you're on any of these lists, you stand a much better chance of making it to the shops that go towards making up the official UK music charts.

The official music chart is compiled by a market research group known as GALLUP. The figures they use to compile these charts are obtained from the sales records from just under a thousand shops spread across the country. When anyone buys a record from one of these shops, the sale is registered and stored on a computer. Every Saturday, at the close of business, GALLUP randomly chooses 300 of these chart return shops and from them compiles the charts for radio play on a Sunday evening.

CHAPTER 22

Remixing and Sample Clearance

'Everybody is influenced by someone else.
Thus, while complete originality is impossible
individual style is not...'

Leonard Bernstein

There can be little doubt that remixing plays a major role in much of today's music industry. It is now regularly used as a viable form of promotion even if the original record isn't dance, yet while some artists still scorn it as *untalented theft* or the *plague of today's music scene*, remixing has played a vital role throughout music history since we've all been inspired by previous releases. Even classical musicians such as Mozart, Bach and Beethoven could be seen as remixers since it wasn't considered unusual for classical composers to borrow entire sections of someone else's composition and rework it into their own style.

Although some may view this as stealing, it's a natural progression that has formed the basis of all music today. Artists have always derived the foundation of their own music from what has gone before; disco formed the foundations of house music, rap formed the basis of lo-fi and trance borrowed heavily from techno and house. Even pop music has encapsulated the most fashionable pieces of every genre, creating a hybrid of all previous musical styles resulting in what it is today.

The roots of remixing in the context of dance music, however, can be traced back to 1976 when a New York DJ, Walter Gibbons, altered Double Exposure's single 'Ten Percent'. He blended the track into a disco fusion which went on to become the first ever commercial 12-inch vinyl single, something that to many was seen as the first ever modern remix. Building on this foundation, when the first electronic drum machines and samplers appeared on the scene, DJs such as Kevin Saunderson and Ron Hardy began to chop up vocal and instrumental passages mixing them with four-to-the-floor loops, manipulating them into something more danceable.

From these humble beginnings remixing has since become an industry all unto itself with DJs such as Armand Van Helden, Joey Negro, David Morales, Timo

Maas and Grammy winners (who established a remixer category in 1997) Deep Dish now worshipped as club legends. Remixing is also beneficial to everyone involved, as both the author and the remixer profit from it because while the author still retains copyright and thus is entitled to some of the proceeds, the remixer gets to reap some of the benefits of (hopefully) having a hit without the hassle of doing everything from scratch.

While the basic premise behind remixing any track is simple enough – you take someone else's record, chop it up and then stick it all back together in your own style – it's much more complicated than it first appears and taking any song and transforming it into the next club 'hit' is far more difficult than it first appears. Since it commonly involves making the music suitable for the dance floor it requires plenty of familiarity with the dance music movement. Plus, since you're essentially turning a 4 min pop mix into a massive 8 min club hit, it also requires its own style of production and arrangement.

SESSION TAPES

The beginning of any remix starting with the source material, which in the majority of cases is resourced from copies of original session tapes recorded by the studio involved in producing the original track. For many, this is the favoured approach because having the individual tracks allows to contort the parts without the need to worry about finding exposed parts of the original that you can 'borrow' and utilize in your own mix.

Today, nearly all producers keep the original samples. Provided you have some charm and a little luck you may be able to get your hands on the samples from the studios along with written permission to remix the material from the record company involved. Even with this written permission, however, the studios may charge you to rent the session tapes from them, or simply refuse to let you have them at all.

This isn't as unlikely as it probably sounds, since they may not trust you with the original tapes. The original artist may implicitly refuse to release them, the producer may not want to let them go, if it's an old recording they may have been lost, or they could have been destroyed in a studio fire. Although this latter occurrence is rare, with all the electrical equipment contained within a professional studio it's certainly not exceptional. If, however, the session tapes are available then you can expect to receive the original recording on one of the five formats.

DASH

It is acronym for digital audio stationary head. These use two formats: either quarter-inch tape that uses 8 digital tracks to record a stereo signal or half-inch tape capable of recording up to 48 digital tracks. Price wise they can cost as much as a three-bedroom house in the most expensive area of your town, and therefore it's unlikely that you would receive a recorded session on this format. That's not to say it doesn't happen, though, and some remixers (including

myself) have received this format and usually end up taking them to a professional studio to have them copied onto a more acceptable media.

PCM

PCM derives its name from the method it uses to record: pulse code modulation. This form of recording is also employed in the A/D converters in most
of today's computer soundcards. PCM machines use ¾-inch video cassettes to
record but as the machines themselves are exceptionally expensive to purchase,
few studios use them for recording and so it is unlikely (although not improbable) that you'll receive the track on this format.

DAT

DAT recorders are not as popular as they once were but if the track for remixing
is quite old you may receive it on this format. They record on a very narrow,
slow-moving tape and achieve a 48 kHz bandwidth by using two heads on a
rotating cylinder. The tape, when 'loaded', is always contacting the head even
while using fast forward or rewind and so allows you to 'scrub' (forward and
rewind the audio while hearing it) to locate parts that you want to use quickly.
This is almost certainly the format you'll receive the remix parts on.

ADAT

Alesis digital audio tape works on a principle similar to that for DAT machines,
although they can record eight multi-track channels using SVHS video cassettes.
Although, like DAT, it isn't as popular today, it was a popular format for studio
recording as it not only offers digital quality but also costs significantly less
than an equivalent analogue machine and allows individual instruments to be
stored in separate tracks onto a single tape for mix-down. Moreover, ADAT can
also be connected digitally to a computer for editing using the ADAT optical
interface, occasionally referred to as Lightpipe. This is another format you may
receive as a remixer. Some labels will transfer from ADAT onto CD for those
who do not own an ADAT machine, but they may charge for this service.

Pro Tools

This is perhaps the most common format you'll receive, since most producers
today work on a Pro Tools rig. Generally, you'll receive the Pro Tools session
files which consist of a *.ptf* file with two folders 'audio files' and 'fade files'. If
you own a Pro Tools rig, you can just open the *.ptf* file direct into Pro Tools; if
you do not, the audio can be found in the audio files folder and imported into
any sequencer. However, be forewarned that not having access to a Pro Tools
rig will mean that you're missing out on any effects, editing or processing that
have been applied by the engineer or producer and you may have to replicate
these again in your sequencer. One session file I received had a note attached
saying *Don't solo the lead vocal! It's been auto-tuned to shit!* The standard vocal
audio file contained in the audio files folder was totally unusable and would

have taken a good few valuable days of work to correct if I didn't have access to a rig.

BOOTLEGS AND INTERNET REMIXING

If you're just starting out then permission is going to be incredibly difficult to obtain as most large labels are not willing to let little known artists remix their artists; for that, they want big-name remixers. Consequently, many aspiring remixers choose to illegally remix a track in the hope that upon a small unofficial vinyl press and release to club DJs, it will become a massive club hit and a record company will search them out. These illegal remixes, known by the industry as 'dubs' or 'bootlegs', are often created from creatively sampling the original stereo recording.

A proportionate amount of remixes that have done particularly well during the past few years have been dubs since the general agreement is that it's better to 'ask for forgiveness than beg for consent'. If permission is refused before the track is even constructed, it could result in a potential floor-filler disappearing forever.

For example, *Richard 'X'* created an instant hit when he mixed a certain old soul tune with Gary Numan's *Are Friends Electric* to create *Freak Like Me* for the lovely Sugarbabes and *Eric Prydz* certainly did well with his remix of *Valerie*. This type of remixing is becoming increasingly popular since it doesn't require you to be a musician; you just need a keen ear and a little luck.

In fact, many top DJs are renowned for spending weeks slicing and dicing yesteryears popular tracks, removing the drums and other elements and replacing them with more 'club' style timbres to create special versions of tracks for their sets. These are often created with just the original track and a laptop running a sequencer – usually Ableton Live.

While it would, of course, be irresponsible to encourage these illegal activities, some larger labels will request a dub remix to gain an idea of what you intend to do before they will release the session tapes, so I feel it is only practical to discuss the techniques used.

At the time of writing there are no processors, hardware or software that can reliably remove a single instrument or vocal from a complete mix, so many dub remixes are constructed from a mixture of exposed parts (commonly at the intro and middle eight) and hardline EQ'ing to home in, and hopefully, expose specific instruments.

The familiar parametric EQ found on most mixing desks isn't suitable for this form of aggressive tonal cutting, though, requiring the use of specialist processors such as the *Electrix* EQ *Killer*. This is an extremely aggressive three-band EQ unit, allowing you to pinpoint and extract specific instruments. It doesn't provide the perfect solution however, since its performance at stripping frequencies from a mix depends on the type of music you're working with and isn't suitable for extracting vocals.

Rather, for these, it's more prudent to search record stores or the internet and file sharing programmes such as Torrents for 'accapellas'. These sometimes appear on the 'B' side of a record and are a vocal only mix which can be gratuitously sampled, contorted and rearranged to sit on a new arrangement alongside the exposed instruments from the original mix. Of course, downloading or copying accapellas is illegal, and something I cannot condone – and I urge you to purchase the original!

If this sounds too much like hard work, or you prefer to stay on the right side of the law, the internet provides plenty of downloadable session tapes from smaller labels and lesser known artists simply for the sake of promotion. Often referred to as 'sample packs' and usually in MP3 format they allow you to download and remix their work which they can later judge and often release if it's first-rate.

However, if you choose to acquire the original parts, the most important aspect of starting any remix project is deciding which parts should remain and which should be discarded. Naturally, this judgement is most likely already made for you if you're creating a dub remix as there will only be certain elements exposed in the mix that you can use.

Alternatively, if you're approaching a remix using the original multi-track recordings you have to make a conscious decision. Nevertheless, both will require you to leave just enough of the original to deem it a remix while at the same time not including too much of the original so that it simply becomes a trivial variation.

Fundamentally, a good remix should feature all the good things that the original does, and more. Therefore, it's important to know what the good features of the original are. This means that you need to listen to the original as often as you can until you know the track by heart, so much so, that you not only hear but actually feel where the focal points are.

Usually, this is down to the main hook of the record: the instrumental melody or vocal lines that you just can't get out of your head. However, in many instances there is more than just one hook line in any popular record. In fact there are usually two, three or sometimes four, all of which manage to indent their way into our subconscious minds but in such a way that we only perceive that there is one. As a result, while you may feel that the main instrumental hook or vocal is the one of importance to keep while remixing, our minds work on a much higher level and simply hearing one of the smaller, less frequent hooks can often remind us of the original record too. To gain a better understanding of this, we need to look at the common tricks employed by producers and songwriters when producing a 'pop' record.

Although dance music is often scorned as being far too repetitive, all pop records use the same amount of repetition to drive the message of the track home, only in a more devious way. One of the most classic techniques employed is to subliminally reinforce the message of the music by replacing the vocals with instrumental copies. For instance, many tracks will open with the vocal melody being played on an instrument; this will then be replaced with the vocals during the verse and 'hook' chorus, before being replaced again with an instrumental version of the 'hook' vocals during the breakdown. In some cases this instrumental hook then appears again at the end of the track to finally drive the message home.

These instrumental 'clones' are not always played on the same instruments, though, so as not to appear to repetitive; rather they may be played on piano, followed by a wind section, then a string section to finish. This is where we, the listeners, derive the main melody or hook from a record, but there are also other hooks that although not driven home as much can still form a proportional part of the music we hear.

Alongside the main hook of the record there will be any number of secondary hooks that are used to maintain listener's interest. These are commonly short riffs, motifs or small vocal ad-libs that appear between the lines of a song and are created to fill in the gaps. These ad-lib hooks and vocals are often ignored because we're concentrating on the main hook, but they are still subconsciously imprinted on our minds as part of the record. Just using these small ad-libs and synthesizing the remainder of a track can produce a remix that still remains immediately recognizable without having to worry whether the main hook will suit the genre you have in mind.

At this stage, it's also worthwhile envisaging the final arrangement as this will often determine which parts of the original are to be kept aside. There's usually little need to worry too much about how the finished product will sound, but rather have a good idea of how the arrangement will fit together using some of the parts from the original recording. It's also at this song mapping stage that you can determine what genre of music the remix is aimed at: is it going to be trance, house, garage etc.

As with all forms of music this isn't a fixed method and every artist will use their own methods and ideals; however, it's only on very rare occurrences, or bootlegs, that you will be given total freedom from musical style. If a record company requests a remix from an artist, they have been carefully chosen by the A&R department because of the style of music they have produced in the past.

It's not contemptuous to believe that some big-name remixers are approached simply because of the name that would be attached to a recording for the sole purpose of profit margins but, optimistically, they are solicited simply through their previous stylistic endeavours. By requesting Oakenfold to produce a remix, they wouldn't expect a hip-hop recital; instead, they would expect a club-style hit typical of his previous releases. After all, there is little point in taking a record that is already a hit in R 'n' B circles and reworking it in the same style – the public market is the same and, in many instances, only the diehard fans of the original will want to purchase the remix. The result is that it hasn't been promoted, but is merely an over (or under) produced original aimed at the same listeners.

Another benefit from using this form of 'song mapping' is that it can often help to envisage the remix in more ways than one. Along with helping with figuring out which of the original samples are going to be used, it can also offer an idea for the general tempo of the piece. As already discussed in earlier chapters the tempo can divulge a great deal about any piece of music, especially in dance music circles. As the original will most probably require some time stretching to suit the new tempo, it's an idea to stick to your chosen tempo to avoid introducing unwanted digital distortion from continually stretching a piece of audio each time you change your mind.

These days it can be difficult to encapsulate a tempo for any remix as they are constantly being reinvented but, roughly speaking, most tend to fall between 120 and 145 BPM. Very few remixers with a tempo over this have ever reached mainstream clubs and even fewer will be accepted by a record company. This is because the vast majority of regular club goers are not able to dance to a rhythm at this speed while still managing to look cool, and if nobody can dance to it then the DJ isn't going to play it.

A 180 BPM drum 'n' bass remix of Madonna's latest hit may capture the imagination of some but this is only a small percentage when compared to the millions of clubbers who go dancing every weekend. On this same note, creating a remix that constantly switches between, say, 4/4 and 3/4 signatures or has a constantly fluctuating tempo will only serve to confuse and alienate listeners further. While it may seem inequitable to suggest that the public is fickle, unimaginative and short of attention span, a great deal of today's popular artists have released experimental work that has failed despondently because they ignored the elements that their fans have become comfortable with.

I have no doubt that many unsuccessful musicians and remixers will undoubtedly argue this point, insisting that they create mixes to please only themselves, yet the most successful are the ones who know what the clubbing generation want and how to deliver it. Morales, Oakenfold, Deep Dish, BT, Joey Negro, Tiesto and Timo Maas have continually proven these points and, as a result, are currently considered some of the best in club land.

Of course, there is certainly nothing wrong with innovation but unless you're already a well-known remixer with a fan base it certainly doesn't have the same promotional power as the more conventional four-to-the-floor club mixes. It is important to have a distinctive style but it's equally as important not to get too carried away. Dance music is an exact science, and close scrutiny of the current market is crucial. This is one of the key reasons why DJs usually produce the best club remixes.

Another important factor as far as the tempo is concerned lies with that of the original recording. If you're remixing on behalf of a record company then the chances are that you'll be given the original working tempo of the track, or in some rare events, you may even receive the original track sheets. These are written records usually taken by the tape ops at the time of the original recording and note the start times, EQ settings, channel assignments and general working tempo. Having these to hand obviously makes remixing much easier, and if you're not automatically supplied with one, it is worth asking if there is one available. There are no guarantees that there will be, though, as the filing system in most studios usually consists of throwing them into the nearest cardboard box and some don't even bother compiling track sheets at all. Notably of every remix I've been requested to do, I have never received a working tempo sheet; original track sheets and the audio parts have been somewhat lacking too.

If you are provided with a tempo of the original, it still shouldn't be taken for granted that it will be this quoted tempo when you begin working with it. A mistake often made is not taking into account the clock in the sampler or computer being used for the remix which can often lead to small discrepancies in tempo as the track progresses. While it would seem natural that something recorded at, say, 120 BPM would remain the same on everything, all computers and samplers utilize a crystal clock that controls the rate they execute their instructions and not all crystals operate at exactly the same speed.

Additionally, the operating system being used can also affect this clock resulting in a stated tempo of 120 BPM being anything between 1 and 2 BPM faster or slower. This is significant if you decide to approach a remix using the cut, drag and drop method – a system mainly used as an idea generator consisting of importing the original track into a sequencer, cutting it up into bars and then moving the segments around.

Consequently, it's worth calculating the tempo yourself, even if you've been supplied with the original tempo. Most of today's sequencers offer various methods for calculating the tempo of any piece of music. Generally speaking you can import a segment of the original track; select 4 or 8 bars of the sound file and then tell the sequencer how many bars are present. The sequencer will then determine the BPM and adjust the tempo to suit. This method isn't applicable to all audio sequencers and some, such as Steinberg's Cubase SX, require you to tap in the tempo of a piece of music using the keyboard before it calculates it. Using this method can be a little hit and miss, so a preferred method is to calculate the tempo manually.

To begin with you need to place the original track into a wave editor and cut out a 4-bar loop for importing into the sequencer. It's usually best if you choose a segment where there is a drum solo featuring just the kick, but any part where the kick is present will be suitable as this allows you to see exactly where the beats occur allowing you to determine a 4-bar loop. It's vital that the loop starts with a kick and ends just before the final kick so that it can loop over and over with no glitches. Once achieved, this loop can then be imported into a sequencer set to constantly cycle over these 4 bars whereby you can manually adjust the sequencer's tempo until the audio file loops clearly. Alternatively, if you know the length of the loop in seconds and the number of beats in a measure then it's possible to use mathematics to calculate the tempo:

60 divided by loop length in seconds = N

$N \times 8$ (assuming you have 2 bars) = BPM

(The 8 is for 2, 4 beat measures.)

While these are the widely accepted for determining the tempo of any piece of music, if the original was recorded before the 1980s the process can be a little more difficult. Before this and the intervention of step-time drum machines, tracks often speeded up naturally as they progressed and frequently the producer would also slow down the tape machine by 8.5% on the final mix.

The reason behind this is that when the track is played back at normal speed the mix becomes a semitone higher in pitch, in effect tightening up the playing and adding an additional polish. This clearly makes calculating the tempo throughout the entire song a complex procedure as it can fluctuate by as much as 10–15 BPM throughout the track. If this is the case, then it's advisable to import each loop from the original that you plan to use in the remix and time stretch each constituent loop to the tempo that the remix will be.

Time stretching is available on most audio sequencers, audio editors and samplers but the resulting quality relies entirely on the algorithms that are used and the material being stretched. Time stretching adjusts the length of a sample (effectively changing its BPM) while also leaving the pitch unchanged. It does this by cutting or adding samples at various intervals during the course of the sample so that it reaches the desired length, while, to a certain extent, smoothing out the side-effects of this process on the quality and timbre of the sound. This is quite obviously a complex, processor-intensive process and is not suitable for extreme stretching. For instance, stretching an 87 BPM loop into 137 will undoubtedly introduce unpleasantness to the rhythmic flow of a loop, producing a side-effect often referred to as digital clutter.

Occasionally, this digital clutter can be used to good effect, but on riffs and more especially vocals, it's unadvisable to stretch them by any more 25 BPM. In fact vocals, more than any other instrument, can be adversely affected by the process, even if it's only stretched by a small amount. The problem with stretching vocals derives from when vibrato is used while singing. We've become accustomed to the human voice so even small time-stretching adjustments can become immediately obvious, more so if there is an element of vibrato involved. With the advancements in technology, there are various methods to avoid vocals sounding deliberately affected, often involving spending days cutting vocals into individual syllables to make them fit a new groove.

We've already seen the uses of sample splicing software in earlier chapters as a form of reconfiguring and rewriting sampled loops, but it can be equally useful as a form of stretching audio. By placing cuts manually at specific points throughout a riff or vocal line it's possible to change the tempo without introducing any unwanted digital artefacts. The basic principle is that if you need to stretch a vocal from 80 to 140 BPM, you can thinly slice the audio and after importing them into a sampler or audio sequencer at this new tempo, the decay of each sliced sample will be stretched or reduced to prevent gaps or any overlapping.

Yet another advantage of taking this approach is that when a loop is sliced, each slice is allocated a different MIDI note gradually ascending in semitone steps.

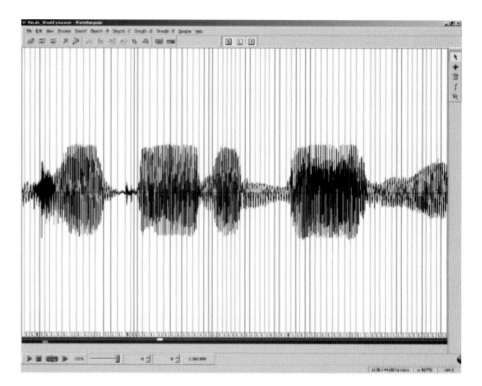

FIGURE 22.1
Slicing vocals into smaller segments for stretching

These MIDI notes, or the audio slices if you prefer, can then be rearranged to produce alternative vocal lines or riffs from the original (Figure 22.1).

Clearly, this type of editing can be an incredibly time-consuming affair and, in the majority of cases, doesn't work too well if there are vibrato or legato phrases in a vocal performance. To get around this, a common practice is to 'comp' the stretched vocals with the original vibrato or legato parts. Comping plays an important part of any production process and consists of mixing numerous vocal takes together to produce the finished article. In the context of remixing, this often involves cutting each word of a vocal phrase into separate segments and stretching each to suit the mix. Using this method if there are any vibrato, legato or held notes they can be paired up with the stretched parts and cross-faded together to produce a more natural sound.

Failing this, another popular method involves pitch-shifting, rather than time stretching the audio to get it in time. By importing the vocals into a sampler, you can use the pitch-bend wheel of an attached controller keyboard to find the best speed. Clearly, this affects not only the speed but also the pitch; however, by using a sequencer's controller editor you can try various pitch-bend values until you find the right speed, and then use a good-quality pitch-shifter to move the vocals back to the correct pitch. This often removes the additional digital 'clutter' introduced by more conventional time stretching.

On some occasions, while most popular remixes will retain the vocals from the original there isn't the need to use all of them. In most forms of music they commonly form the most influential and dynamic part and many producers will treat them with the utmost care but in dance music and remixing the part they play is negligible. Pure dance music commonly has very little lyrical content, dealing in just one thing – feel. Vocals in dance are simply a form of distraction from the main event: the underpinning groove, the rhythm that keeps everyone up on the floor, the DJ in work and the record companies in profits. Subsequently, when remixing popular music that consists of the typical vocal verse/chorus for the dance floor, there is little need to keep all the vocals intact and running throughout the remix. In reality, it's common practice to lift just the most prominent parts of the vocal performance, that is those that make the track instantly identifiable, and disregard the rest.

Eric Prydz's number one remix of *Valerie* is an excellent example of this technique. He ignored a large proportion of the vocals choosing to keep just a few small unaltered vocal snippets, practising the 'minimal vocals for minimal attention spans' technique.

Having said that, even if you cut up vocals or use the various methods to transform and stretch them, there is a limit as to how much of a tempo adjustment you can make from the original. This is not solely down to the vocals as stretching riffs can also introduce unwanted artefacts, so after auditioning the original parts the song's tempo must be determined in order to have a starting point from which to derive the remix's tempo. If a song is already 120 BPM a house remix will be a pretty easy stretch assuming you actually want to make the record faster, but if the song is 90 BPM, reaching this tempo is going to be much more difficult. It's significant, therefore, to consider the remixes tempo carefully because the less you time-stretch (or time-compress) a track, the better it's going to sound.

As an example, suppose that you've been given an original vocal recording with a tempo of 90 BPM and that you want to produce a remix of 130 BPM – a tempo that's typical for most club records. You have two options: either stretch up to 130 or go down to 65, half the tempo of 130. Considering that 90–130 is a stretch of 40 BPM, it would be prudent to compress down to 65 as this is only a difference of 25 BPM and in the mix have the vocals doing half-time to the 130 beat.

That said, keep in mind that not all songs have been written close to a dance tempo and some hooks that make the original what it is will not suit any other tempo. As we've seen, using the aforementioned techniques it may be possible to make anything fit any tempo if you're willing to put the additional time into it, but stretching some hooks can ruin the integrity of the original song. It isn't worth ripping the foundations of a track apart and sacrificing artistry just to pump out a mix at 130 BPM.

Of course, not all remixes use unaltered vocals and there are many popular techniques for twisting vocals, and riffs, in remixes, although in many cases the techniques employed will depend entirely on the artist that's being remixed. What follows is by no means an exhaustive list of possibilities as the market is constantly changing but currently these techniques are proving to be the most popular.

- Import large sections of vocals into sample splicing software such as Wavesurgeon and use this to slice the vocals/melodies into individual phrases. Once completed, use either MIDI note commands to a sampler or the actual audio and rephrase the line. This technique is more popular with melodies as the timbre remains the same as the original, yet the melody is different.

- On this same note, assign each vocal/melodic phrase to a specific key in a sampler and then use a keyboard to play different pitches, layering them over one another to create chords.

- Vocals and melodies can also benefit from low-pass, high-pass and band-pass filters. By assigning a controller keyboards mod wheel to control the filter it allows you to filter specific phrases live. A variation on this technique that's also popular is to use different filters every few phrases.

- Although the stutter effect is becoming rather cliché it is helpful in bringing attention to the vocals or riff. These can be created by setting a sampler to play a sample only while the key is held and then sending 1/32nd or 16th MIDI note on commands to the sampler.

- An alternative to the previous method is to import a riff or vocal into an audio sequencer and then place cuts every 1/16th before deleting every other 1/16th. This recreates the typical gating effect.

- To bring more attention to the vocals, an often practiced method is to create a vocal phrase that is physically impossible to sing. The methods used to create this are varied but often include the use of autotuners set to adjust the key of some phrases by a large amount so that the side-effects are noticeable.

- Time stretching can also help towards creating phrases that, in the real world, would be impossible. By stretching some vocal phrases by a large amount and leaving others unaffected, the vocal line can take on a whole new meaning. Similarly, using cross-fading techniques try to loop part of a vocal phrase for longer than would be physically possible so that it still sounds natural.

- Reverse parts of a vocal phrase while leaving others playing normally.

- Effects play a large role in creating interesting melodies and vocals. Simply double tracking the vocals and offsetting one from the other by just a few clicks can create an interesting effect.

- Reverse reverb can also work particularly well, especially if you do not want to affect the vocals too much. This can be accomplished by reversing the vocal phrase and then applying reverb with a large hall setting, or similar, applying it to the waveform and then re-reversing the vocals so that they're the correct way again. Using this method the reverb will build up slowly to the vocal line creating a sweeping effect.

- Specialist effects, such as Steinberg's GRM Tool's version one and two, Voice Machine or Spektral Designs Ultra Voice, can be used to create strange results. The pitch accumulation from GRM Tools v1 is a popular effect in many remixes, creating a culmination of pitching and filtering.

- EQ should also not be underestimated; rather than save it solely for mixing. Use it to remove the bottom end from vocals to create a 'telephone' or 'megaphone' vocal by rolling off all frequencies above 3 kHz and below 1 kHz.
- If you have access to a hardware reverb unit, use a long decay and apply it to the vocal line. While playing back the vocal, record only the reverb's return signal and then offset this against the vocal phrase.

These are only suggestions and, as always, experimentation and close scrutiny of the current market is essential for creating great effects. A visit to a local record store frequented by club DJs with £30 in your pocket will help you follow the latest manipulation trends with remixing.

Before you begin to splice, dice or mangle vocals in any way it is worth checking with the record company first, or if it's an illegal dub remix, avoiding overprocessing altogether. Whereas some styles of music such as happy hardcore almost invariably use 'chipmunk' style vocals and the current trend is leaning towards additional processing with effects such as Antares Auto-Tune, a proportionate amount of labels do not approve of their 'selling-point vocals' being mutated into something totally incomprehensible. Getting too carried away with vocal manipulation may not only annoy the label, who coincidentally will probably demand that they remain unaltered anyway, but also upset the original artist and alienate the audience.

ARRANGING

The method for arranging a remix varies from producer to producer, but while it's quite probable that the familiar verse/chorus will be used in the original mix, this does not mean to say that the remix should follow suit. Indeed, for clubs it's quite rare for any track to follow this structure and most will follow the atypical 'club' structure that, incidentally, is very similar to classical music. In other words, you begin with a theme followed by a variation, before eventually returning back to the original theme. Although how this style of arrangement is approached varies widely, a typical example is given in Table 22.1.

This type of structure often forms a vital part of any remix since most need to be at least 8 min long and the original track will more than likely be half of this. Additionally, even though the given table is an example and isn't a strict rule it is considered a rule to keep both the intro and outro as drums only and leave them 32 bars in length. Most DJs prefer to mix beats than mix tuned instruments from one track to another (known as harmonic mixing) since they have to adjust the pitch accurately on the fly, plus it can also make some tuned instruments sound utterly terrible. By keeping with a percussive intro and outro, they only have to mix the beats, so they are more likely to play the mix.

More importantly, to create an 8 min mix from only 4 min of music, many of the bars are continually repeated, and to maintain interest and keep the mix

Table 22.1		Example of a Club Arrangement
Part of Arrangement	Length in Bars	Analysis
Intro	32	Commonly consists of nothing more than a building of the drum track as to allow the DJ to mix the track in with another record.
Body 1	32	A melodic instrument is introduced. Most usually the bass to establish the groove of the record which may be followed 16 bars later with another melodic instrument. This latter instrument is rarely a part of the original track but a motif programmed to sit around the original parts.
Drop 1	16 or 32	Most percussive elements bar the kick are commonly dropped from the mix followed by the introduction of a cut-up main hook line or vocal to establish (or give some clue) that it *is* a remix to the clubbers.
Body 2	32 or 64	All the previous elements may return, playing alongside the cut-up hook. Small ad-libs either from the original or programmed are sometimes introduced after 16 bars, mixed among snare rolls, cymbal crashes and sound effects. Vocals from the original are sometimes also introduced here but more often than not they have been cut-up, so the entire vocals are not revealed, adding to the tension of the music.
Drop 2	16 or 32	The main drop of the track is often similar to the first but occasionally the kick is removed and the main hook may be removed or filtered down. This could be followed with a 16-bar snare roll with the filter cut-off on the main hook opening, creating a building sensation.
Body 3	32 to 64	All instruments return to the mix, alongside the full main hook or vocal line. Further ad-libs may be added after 16 bars, alongside skips in the rhythm, small breaks, rolls and cymbal crashes.
Breakdown	32	Typically, all the instruments are dropped gradually from the mix signified by drum rolls or cymbal crashes so that only the drum track remains.
Outro	32 bars	Similar to the intro this commonly consists of nothing more than the drum track to allow the DJ to mix the track in with another record.

exciting, production techniques play a vital role. This includes using effects to create more interesting timbres, cross-fading between samples, creating stutter effects, altering transients (with a transient designer) as the track plays and using filter sweeps to add movement to static areas of the arrangement. Sudden breaks in the rhythm, snare rolls, small breaks, skips, scratches, sound effects and slight changes in the drum pattern, melodies and motifs are all used to add tension and a dynamic edge to the music. As ever, experimentation often produces the best results.

CONTRACTS

If you've been commissioned to remix a tune, you're obviously going to need some kind of agreement with the label involved to ensure that you both fully agree what you're being paid for. Although remix contracts differ from label to label, they will all consist of a number of clauses that must be fulfilled if your remix is to be commercially released.

Firstly, as a remixer you will need to agree on the type of remix that the label requires, this means that you both need to concur on the genre of music and a suitable length for the track. Hopefully, the label will have asked you to perform the remix based on the same genre of music as your previous releases, if you've had any, but a suitable length of track can vary immensely.

The widely accepted length for a typical club remix is around 8 min, but it is important that you know exactly how long the label want it to be because if it's over or under the specified length they may refuse to pay you. Additionally, you will need to confirm with the label how many versions of the remix they want from you since it's not uncommon to be asked to supply not only the club mix with vocals but also an instrumental version.

You also need to agree on the terms of delivery to the label. Most will only accept your completed mix on DAT; however, CD is becoming a more widely accepted format. Also there will be a rather annoying clause that states 'the mix must be of a commercially acceptable standard'. All record companies adopt this and it's nigh on impossible to persuade them to remove it from the contract. It is particularly vague but essentially it gives them the right to refuse your remix if they don't believe there is an acceptable market for it, or it's been poorly mixed, produced or mastered.

As previously touched upon, it's also essential to check what media the label will use to deliver the original track to you. Although larger labels will only supply a remix in ADAT format, smaller labels may put it to CD, but more often, they commonly supply it on DAT with the master tape recorded track by track with the time code. The benefit of this is that it's then possible to accurately recreate the track in a hard disk editor or software sequencer, but of very little use if you're working with just a sampler.

Finally, you will also need to be in agreement on when the master tapes will be delivered to you and, of course, the deadline that they want your completed remix back. Deadlines can vary enormously with some labels expecting you to turn around a professional remix in a matter of weeks while others may offer a few months. It's rare that any remixer is given longer than 4 months to complete a mix, so before accepting a project you should ensure that you can perform it fast enough so as not to annoy the label and ruin your chances of being offered any further projects.

Assuming that you have come to a mutual accord on the terms and conditions of completing and delivering a remix, you need to reach an agreement on how much you should be paid. It's rare for any label to offer royalty-based payments and, more often than not, a label will pay you a one-off fee – usually half upfront and then the other half if the remix is accepted.

If it isn't accepted then you could be expected to pay back the previous payment, so this is something you should check before taking on the venture. The amount you can expect to be paid is dependent on the label, the artist you're remixing and your past performance. Little known remixers can be paid as little as £500, and in some cases, a label may ask you to remix for free, only agreeing to pay you when the remix is returned and if they want to release it. This may not seem so great, but as you produce more and more remixes, your name gets around and you can begin to charge thousands. It's not uncommon for a well-known remixer to charge in excess of £5000 to remix a successful artist who's signed to a big label.

No matter how large, or small, the payment it isn't advisable to try and barter for more money unless your remix happens to be selling incredibly well. If it is, try contacting the label asking for royalties, but depending on the remix, this can be a bit of a grey area. Many remixes usually find their way onto the B side of a record and you could have a hard time trying to ascertain that it's your remix that's selling the record and not the artist who's featured on the A side.

CLEARING SAMPLES

If the remix has been created without the authorization from a record company and it is doing particularly well in a club then you may need to look at releasing it commercially. Ideally, the record company that released the original track should be your first port of call because these will often charge less for the use of their samples than if you signed to a rival label.

The subject of clearing samples has often been overly exaggerated as being a long-winded, drawn–out, expensive process but this isn't necessarily the case. Since the first case of illegal sampling in 1991 when Gilbert O'Sullivan successfully sued Biz Markie, sampling other artist's records has become an art form in itself and numerous companies now exist to clear samples on your behalf.

Generally, from experience, most record companies and artists are only too happy to work out a deal and allow you to use their samples simply because they will receive an income for doing absolutely nothing. This is especially the case if you can make an old, out of date record hip again. However, be warned that if the record that has been sampled was originally a hit, then the record company and artist will take a bigger slice of any royalties. Similarly, the record company holds all the cards when it comes to negotiations because you broke the law in the first place by illegally sampling and releasing a track featuring their artist or music.

This same principle applies even if you haven't created a remix but have sampled an artist to create your track. In these latter cases you have to be diplomatic as the record company will quite obviously try to take all the proceeds from the record by arguing that the samples form a major part of the record. For instance, say that you've constructed a track but only relied on a few small guitar hits in the record. The record company will begin by arguing that they contribute, say 90% towards the track, whereby you can argue that they only contribute, say 40% to the music. They will then return by saying that it contributes 70% to the music, where you can reply that they only contribute 50% and so forth. Notably, you will have to be amenable since if the record company is not happy with your dealings they can simply refuse to release the track while you watch an instant club hit disappear into obscurity.

A much better way is to avoid this type of negotiation by writing the track but to clear the samples before pressing to vinyl and releasing it. This has the immediate benefit that the record company doesn't know if it will be a hit or not, so you stand a better chance of getting a good deal. If the record is already released and doing well in the clubs, then the sample's owners know it's a hit and will try and squeeze you for as much money as possible.

If you go for sample clearance before release, 80% of the time the record company will offer clearance so long as their terms are accepted, but it's unwise to attempt to clear this yourself and you should look towards using a sample clearance company. Typically, these firms charge around £300–£400 for all the legwork involved but they stand a much better chance of getting you a good deal when it comes to paying for the sample's clearance. This is especially the case if you've sampled from a record that's originally signed to a label from France, Germany or America. From some very personal experience, these guys are notoriously difficult to deal with and the cigar-chewing fat cats are equally adept at chewing you up and spitting you out again unless you know exactly what you're doing.

Of course, it is perfectly possible to clear samples on your own so long as you adopt a professional nature but you will have to clear two copyrights: the mechanical rights and the publishing rights. The mechanical rights, sometimes referred to as the 'recording rights' or 'master rights', concern the physical recording of the music which, in many cases, belongs to the record company since they have put up the money for the studio time.

The performance rights belong to the composer of the music but this is usually handled by a specific music publisher who deals with the performance of the recording on behalf of the original artist. The first step to finding out exactly who these are is to speak to the Mechanical Copyright Protection Society (www.mcps.co.uk). This organization has a list of every UK record label and music publisher and will have all the information on who owns the mechanical and performance copyrights. Alongside this, they will give you the contact details and offer advice on the best ways to deal with the situation.

If the record has been released through a large label, then clearing a sample is quite clear-cut as most companies have adopted EMI's online sample clearance contracts. Using these, you simply fill in an online form and wait for a response from the clearance team. On the other hand, if the record is particularly old (but still in copyright) or was released by a smaller label then you may find yourself having to go to the middle of nowhere to ask for permission.

This can be particularly complicated if more than one artist was involved in the original project as you'll have to obtain permission from each one of them, or their closest relatives if they've passed away. Typically, they will give permission since it will earn them money, but it's important to note that they do not have to give approval or offer any reasons why they won't. If they do consent then you'll have to come to a written agreement on how much they receive in royalties and, in many cases, how much they want as an up-front payment.

The amount of money asked for varies and depends mostly on the popularity of the artist. Famous artists such as The Beach Boys, Beatles and Michael Jackson expect at least 80% of royalties and demand an incredibly substantial cash sum up-front. As a guideline, when I went for sample clearance on the mechanical rights to use a lick of Sting's guitar it cost an up-front advance of £10 000 along with 70% of any royalties while Barry White asked me for an almost unbelievable £60 000 per *word* and half of any and all royalties. Needless to say, his vocals never touched my sequencer…

On the other side of the coin, little known disco records from yesteryear usually demand £300–800 along with 20–30% of all royalties, depending on how much of the original is used in your music. On top of this, you also have to pay for the publishing rights of the artist and again this varies depending on the artist. Commonly, they will only ask for 20–30% of all publishing royalties.

In some cases you'll be confronted with the dilemma of the record company saying yes and the artist saying no, and if this is the case, you're pretty much stuck. By law, you cannot include a sample in your music if the original artist has refused permission, but if the artist has given consent and the record company has refused then there is work around. Since the record company only owns the mechanical rights, you're free to recreate the sample yourself or alternatively employ a company or studio to recreate the sample for you.

Above all, if you do plan to sample another artist's record for use in a commercial track then it's prudent to join the Musicians Union. This costs around £80 a year but they will offer to look over any contracts or agreements that you have and ensure that legally you are protected just in case the record company or original sampled artist change their minds. This has been known to happen and it's always better to be safe than sorry.

CHAPTER 23

A DJ's Perspective

It's fair to assume that dance music would be nowhere if it wasn't for the innumerable DJs spinning the vinyl and keeping us all up on the floor. With this in mind, it seems that there is no better way to close the book than to interview a DJ and get his perspective on dance music and DJ'ing in general. For this, I decided to interview DJ Cristo, a regular figure on the international circuit, contributor to DJ magazine (amongst many others), record reviewer and regular host on a Miami radio show to offer his comments. However, after discovering just how much he had to say on the subject, it seemed more prudent to simply hand over a few questions and leave the entire chapter to him.

I'd like to begin my chapter by saying that it is a great privilege to be asked to contribute my words of wisdom to this phenomenal book by Rick – dance music stalwart and luminary who is one of the most knowledgeable, astute guys I know on dance music production scene. When he talks shop, I listen. He has a constant string of quality productions and remixes under his belt – longer than my arm and all under various pseudonyms – even I can't keep up and I'm a reviewer! His dance music and remix seminars are also incredibly insightful and it's great to at last see words written about such a specialist area as dance music by someone who's actually actively involved in it for a change.

WHAT IS A PROMO AND HOW DO REACTION SHEETS WORK?

A promo is a record or CD that record companies release way before it's due to be released to the general public, normally to gauge how well a track is going to do commercially when it eventually goes on general release. For this, they use promotional companies who push their tracks and ensure that they land in the right hands – that is, from the top DJs right the way down the scale. These companies have a massive list of nationwide club, bar and radio DJs of all levels, scenes and genres, most of whom are also members of the mailing lists of *Power* in London, which is one of the biggest and most powerful in the industry.

Getting on these lists isn't a walk in the park. All DJs must first apply to an organization such as *Music Power* and pay an annual subscription to them before they can apply to the promotional companies. In order to be accepted on these you don't necessarily have to be Carl Cox but you do need to be a proficient DJ playing regularly enough to reasonable-size crowds. Even then, the selection process can be rigorous because they need to know they can depend on their DJs to support, play and chart the tracks that they want to promote. Notably, most jocks (*acronym of Disc Jockey*) that are on these lists are professional or semi-professional and are on the club or radio circuit.

Incidentally, the funding from the DJs/members goes towards keeping the whole ship afloat as it were. If they didn't pay to subscribe, there would be a massive reduction on the number of 'free' records sent out and fewer records being promoted in far smaller quantities. This would have a major impact on the vast majority of DJs, reducing the number of weekly chart returns to hugely influential magazines like *Music Week*. If this ceased, so would one of the central strands of generating interest in music both through club and radio play, and through club chart success. The charts have a fundamental role in gauging records and it would have a devastating effect on promotional companies and smaller labels alike.

Nevertheless, if you are selected to be onto these lists, you have to maintain your position and abide by certain rules in order to remain on them. It's naive to think that you can just sit back and receive numerous free records because when you receive a promo copy, you also receive a reaction form that has to be filled in and returned. This is basically a sheet that details the artists, title, label and some history about the track, including where it originated from, who's currently spinning it, which big DJs are supporting it and which radio stations are playing it. At the bottom of the form are a number of boxes for your input which need to be completed. These are for your opinions on how well the track went down at the club/bar/venue (i.e. your dance floor) along with marks out of 10 for how well it was received by the crowds. It also asks whether the track was enquired about (by Joe Public), which position it occupies in your charts out of 20 other promo records and what clubs you're currently playing at.

When you have gathered all this information and completed the forms, you then need to return them. However, not only do you have to return them to the original promo company but also to all the relevant magazines, via post, fax and e-mail, because this goes towards compiling charts magazines such as *Record Mirror*, *DJ*, *Music Week*, *Seven* and *DMC Update* to just name a few, along with all the E-groups and internet charts such as the influential DMC DJ chart (www.dmcworld.com). What's more, every DJ signed to the promo company lists is closely monitored from their chart performance response and staying on the list can depend on how often and how high you place their records. Of course, you must always stay true to yourself because you are what you play, but you do have to be willing to meet the promo company halfway, otherwise they have the power to remove you if they feel that you're unacceptable. This is a very competitive business and nothing comes easy. That said, one of the many good points (apart from obviously receiving free vinyl!) is being able to see the changes in trends. I'm lucky enough to be sent all the big tracks along with having the privilege of receiving all the peculiar tracks that are just about to break.

The record companies certainly have it sussed and it's a very finely tuned hype machine, but you must understand that while you'll receive plenty of vinyl way before the general public, the big boys such as Judge Jules and Pete Tong will receive more exclusive records than you will. These guys are very highly targeted because of their heavy influence on the radio and club circuits. Average-level DJs will receive the same records eventually, but it will be much later than the bigger players. It's a well-oiled machine and it all adds to the allure of the exclusivity of promos.

As many music enthusiasts will already be aware of, it's also easy enough to buy promos from various record shops and websites as I used to for many years. By doing so you're onto a winner because you get your hands on a top tune weeks before it's released, and if you're especially lucky it might be a TP for a bargain price. This, however, can have a polarity shift effect because when a tune (typically on vinyl) becomes in demand a buzz will circulate around it and it will increase exponentially in value. This is especially the case if it's on an independent label or a TP/white label and subsequently becomes signed to a big UK record company. With internet 'shops' such as eBay, previously unsigned material becomes responsible for extreme bidding wars, where DJs and collectors want the promo before the official release date. This happened just recently with a double pack of the artist Jurgen Vries and 'The opera song' featuring Charlotte Church. When I last looked, it was being bid at well over $100 and that's just for a piece of vinyl! It also largely depends on the current trends, and a simple supply and demand structure is in place that determines the price of the records. For instance, coloured vinyl always seems to be in vogue, and despite the fact that some vinyl connoisseurs will argue that the sound quality suffers, it's nevertheless highly sought after and some people will pay ridiculous prices for it.

IN YOUR OPINION WHAT 'WORKS' ON THE DANCE FLOOR?

I feel that for a track to work on the dance floor all the basic elements have to be there. Firstly, for a DJ to be able to mix in and out of tracks, I find that it's generally best to have a steady build-up of beats and drum loops, along with a solid, crisp pattern. It's a very loud environment in clubs, so it helps if there is a minimum of 16 bars of crisp percussion to begin with. In fact, 32 bars are even better but this obviously depends on how long the track is and on the style of music. Even with shorter tracks of any genre, though, 16 bars of percussion are necessary for the intro and outro, as you can build and structure a set using these, slowly progressing as the night goes on. Also, most jocks will thank you for making tracks 'DJ friendly' by not starting the track with loads of percussion and some melodies too. These are incredibly difficult to mix and make your life that little bit harder.

I spend a lot of time playing house and trance, and like most genres of music it's true to say that it is somewhat formulated. The drum patterns and rolls are very similar and many of the melodies share plenty of similarities in the sounds that are used. That said, people who really love dance music understand that it is a vibe as well – a feel-good factor – and this is what really separates the music. Dance music has to have soul and it has to connect with you on the right level. When I buy new tunes (yes, even with promos we still buy lots of vinyl, especially imports) I can hear immediately if a track has the right elements. It has to stand out from the rest with an indefinable groove, that extra-special something that makes you want to move your feet. I also often find that sometimes the simplest riffs create the best reaction on the dance floor. Tracks like Layo and Bushwacka's *Love Story* with the acappella vocal of *Finally* over the top is a massive tune in clubs at the moment, reaching anthem status of late. Also, Jam X and De Leon's *Can U Dig It?*, which sampled a famous 1980s riff in a trance style, is doing incredibly well. Both these use very simple melodies but to a devastating effect that sends the crowds nuts.

I would also say that familiarity in dance records is a good thing. Take Room 5's *Make Luv and Listen to the Music*, I, along with a few other DJs, managed to get hold of a copy approximately a year before the UK label Positiva snapped it up. I used to play it out and get an average reaction from the crowd, but, when it became more popular and featured on a certain deodorant advert the reaction when played out was outstanding. People would cheer and sing along to it as it became the favourite tune of the moment. Familiarity plays a large role in the scene. Ultimately, though, a track has to have a thumping, pounding four-to-the-floor-groove that just makes you want to shake your butt. And from the many years I've been spinning vinyl and playing everything from house to progressive to techno to trance, I have to say that the hands in the air, spine tingling euphoria (trance) always seems to get the best reactions from the crowd.

COULD YOU DESCRIBE SOME DJ TECHNIQUES?

This is possibly the most difficult question to answer because there are so many genres of dance music and each has a slightly different method of mixing. The basic principles are all the same but some require more skill than others, along with some dexterity and flair thrown in too. DJ'ing and mixing has come a long way since the early 1980s. It originally started when New York DJs like Cool Herc (one of the founders of hip-hop and DJ'ing) used to use two rickety old turntables with rotary pitch control, a stereophonic mixer (with no cross-fader) to cut, scratch and mix two copies of the same record together. The breaks from old soul, jazz and funk records were mixed so that the music played continuously while MCs (acronym for Master of Ceremony) such as KRS 1, MC Shan, Sugar Hill gang, Steady B and Afrika Bambaata would rap fresh rhymes over the top. This was the birth of DJ'ing as we know it today and many hip-hop artists are still cutting up old soul and funk tracks to make fresh beats. Early idols of mine such as the legendary Jazzy Jeff, who invented the transformer cut/scratch, with partner in crime, quirky rapper Will Smith led the way for a plethora of DJs to experiment, create and innovate. In England, back in the late 1970s, a little known fact, believe it or not, is that one of the first non-dance music DJs (as we know it) to 'mix' two records together was none other than

Sir Jimmy Savvile – how's about that then! In fact, years ago he was one of the first guys on air at the BBC to fashion this technique.

Nowadays, hip-hop DJ'ing aside, this same mixing principle still applies and the typical DJ technique is to mix and 'beat match' two songs together. This is to physically blend and craft the sounds together to make one continuous sound and to do this, and do it well, is an art form in itself that is dependent on the genre. For example, a techno DJ will mix differently than a drum 'n' bass DJ would, while a house DJ will spin much differently than would a hip-hop DJ. There are lots of different techniques and to describe them all would take me a long, long time, so I'll describe them briefly. Firstly, there's the technique known as the blend. This is used mainly in house-style mixing but is also common with trance and consists of merging two tracks together, keeping them in sync so you can fade from one to the other. Personally, I get both tracks in sync and also in key, with room to build both to a crescendo, as they tend to be quite epic.

Hip-hop DJs use 'cutting' techniques where you use the cross-fader in conjunction with the turntable, bringing in parts of the track as you manipulate the faders and 'cut' parts of the track, beat or vocal up. A classic set-up of the decks for hip-hop DJs is to turn them around vertically so that the pitch and tone arm is at the top (as the pitch control is rarely used much with this style), giving more room to manoeuvre for techniques like scratching – which is cueing up a beat, noise or vocal on the up beat, and physically moving the record back and forth quickly to create a 'scratch' (Figure 23.1).

FIGURE 23.1
The hip-hop scratching technique

There's also the rougher cutting and chopping style used in drum 'n' bass and breakbeat (the music's a lot faster, as is the style of mixing), which consists of using the percussion from one record and mixing it with different elements from another. These are all advance techniques that are difficult to encapsulate in words and the only real way to learn them is to watch the 'masters' at work and pick up on their tactics. What's more, a lot of these techniques can be applied to various styles, it depends on how dextrous and creative you are.

WHAT GEAR WOULD YOU RECOMMEND TO THOSE WANTING TO DJ'ING?

The classic set-up for a DJ is two decks and a mixer (even though you probably want to start mixing on six decks, it's best to start out with the classic set-up for now). To those just starting out, I would recommend investing in a pair of second-hand belt drive decks. However, there are plenty of arguments when it comes to this kind of advice, since some will recommend that you should go straight to the professional decks immediately. This is because when you get to a club they will almost definitely have the 'industry standard' Technics (which they invariably will the world over). I disagree.

I learnt the hard way by using cheap gear and building my way up and I still believe that this is definitely the way to go. By taking your apprenticeship on less than pro equipment, it puts you in good stead when you get to the pro gear, and keeps your skills sharp. If you can mix on wobbly old belt drive decks near perfectly, then you can mix on anything in any condition – once you have learnt to adjust the differences of the two. As a result, I recommend the Gemini PT2000s or the Numark range, but there are literally hundreds of starter decks and packages available. Alongside these you'll also need a cheap two-channel mixer to start mixing and matching beats (if they're not included in the starter pack) and I'd recommend the Kam Made To Fade or the Gemini 626. These can be hooked directly into your hi-fi system through the AUX-IN and it saves you additional expenditure having to purchase an amp and speakers. Conversely, if you want to mix with CD rather than vinyl, then, if you're just starting out, get a second-hand pair of Pioneer CDJ100s, or 500s (the old models), and learn the basics on these. When you feel you are ready, you can then step up to the intermediate or pro equipment, depending on what level you are at and what your budget is.

PROFESSIONAL TURNTABLES

There was a time when the mighty Technics deck reigned supreme and remained unchallenged and, although they still are, there are some serious contenders in the turntable ring. There are now three viable and dependable main makes of pro turntables:

- Technics
- Vestax
- Stanton

FIGURE 23.2
The requisite Technics
SL-1210MK5.
Photograph printed
with permission.
© Technics 2003

The Technics 1200 and 1210Mk2 'wheels of steel' professional turntables have been knocking around for well over 20 years and are still the industry-standard turntables in serious clubs the world over. They have heavyweight clout and are the firm favourites of DJs; they were the DMC world mixing championship's only turntable for well over a decade. That said, they have released only two variations on the theme since the initial production of the Mk2. The limited edition 1210Mk3 was released a few years ago (and is still selling well), and now they are set to release their newest models, the 1210Mk5 and 5G. These all retain the classic sexy design with slight improvements on performance, yet keeping the old familiar feeling with the torque strength. This has been a problem with other decks from different manufacturers as DJs are only used to the Technics' design and feel. Any professional relying on a set standard of equipment for a performance wants to know exactly how it works, where everything is and how it reacts. It's dreadful arriving at a gig, especially abroad, only to find a deck you're not familiar with. In fact, because of this I now specify what equipment I will use in my contract, and I only stipulate Technics or Stanton turntables (Figure 23.2).

Vestax has, over the years, come on leaps and bounds in the turntable market, having been a manufacturer of quality mixers for many years which are favourite installations in pro DJ booths worldwide. Of late, they have turned their intricate knowledge to decks with their quality PDX range, including the PDX D3s, and PDX2000/2300. Although the features of these decks baffled many DJs, as they started to include features like ultra pitch, instant reverse, digital pitch display, joystick controls, etc., they proved their worth, though, especially with younger clients and scratch DJs alike, as the straight tone arm skips less when performing in DJ battles. Vestax keeps updating its range regularly, with new and innovative models coming out all the time, so it's difficult to keep up and recommend a specific turntable.

Stanton, a name associated and regarded highly for cartridges and stylus' (the classic 500MkII has been an industry standard for many years), have also turned their knowledge to turntables, with ground-breaking effect. A little over a year and a half ago, they released the 'ST' range to critical acclaim. Not only does it come in two different models, STR8 100, with more traditional S-shaped arm, and ST150 (their latest model), with straight (str8) tone arm, but also it had some very interesting features. They are as rock solid as a Technics with a 10% greater motor torque, they have an instant reverse (I've never quite figured out the reason for this function yet!), a pitch bend of up to ±25% and a line directly through the deck for connecting portable media such as Minidisk, DAT or CD players. This allows you to play them directly through the mixer, so there is no need to plug anything into the sound system halfway through your set! What's more, they also have a master tempo function with key *aDJust* which allows you to speed up the track but keep the vocals at the same pitch, so not to wind up with Mickey Mouse vocals! If you have any doubts, Ministry of Sound were so impressed that they have just installed six pairs of them in the main and box room and done away with Technics completely!

CD DECKS

Over recent years, CD mixing has come to the fore. What were once received as rather shoddy twin-slot CD decks with a basic mixer rack at the bottom and a few buttons and lights have now come into their own. Many DJs used to be totally 100% vinyl purists (some still are) including myself until about 2 years ago, as we love our back-breaking heavy black blobs of plastic with a passion! But with huge advancements in the technology market from the likes of Pioneer and Denon, CD mixing and stand-alone decks have really come up to speed with the twenty-first century. In fact, you'll find that most pro DJ booths will now have quality CD decks or dual players nestling alongside the turntables and mixer. These offer numerous benefits over vinyl, since if you also produce your own music, CD media is incredibly cheap as you can copy your music and road test it at your next gig. What's more, boxes containing a hundred CDs will weigh considerably less than a hundred vinyl records, and as some professional decks now play the MP3 format (for instance, the Pioneer's DMP-555 MP3), it's possible to fit over 30 tracks on one CD that you can spin like vinyl.

Some of the most prominent CD decks are from Pioneer with CDJ1000 and CDJ800. These both use jog-wheel technology that cleverly emulates the practicalities of mixing with vinyl – you can spin the wheel the same as you would spin the vinyl to create scratching effects. Pioneer were also the first to introduce smart card slots in their CD players that enable DJs to turn up at any club that are using them, insert their smart card, and the decks automatically configure the

FIGURE 23.3
The Pioneer CDJ 1000.
Photograph printed
with permission.
© Pioneer 2003

players for that particular DJ, recalling all the cue points from the DJ's personal collection. These decks also feature a plethora of effects and useful real-time functions, such as instant reverse, internal cue/loop memory, real-time seamless looping, anti-shock and much more. It's a serious deck with heavyweight clout from a well-respected maker of quality DJ gear (Figure 23.3).

Slightly more expensive than the CDJ1000 is the Pioneer DMP-555 MP3 CD player, which is the world's first digital media player. It offers all the effects that are available for the CDJ1000, plus it's also capable of playing MP3 directly from a CD. Again, you can use Smart Media cards to store all your local settings and it also features full EQ trickery. What's more, you can hook it up to your PC via digital output or USB, and hit comes with a DJ booth CD software that allows you to control virtual decks and make loops and samples (Figure 23.4).

While these decks are considered to be the best by many DJs, if you require more effects and scratch capabilities for DJ'ing hip-hop, then American Audio's Pro Scratch 2 deck, or the PSX, could be the answer. It offers features similar to the Pioneer decks along with digital scratching, loads of effects, flash cue, hyper pitch and a host of other facilities that are suited towards 'cutting' beats. Notably, there are also many other fine makes and models of dual CD decks in the market such as Denon, Vestax, Numark, Tascam and Gemini. Some of these have built-in mixer and effects, all of which are good, and they all have ranges to suit different budgets and needs. If you want an option easier on the pocket, then there are many makes out there that will do the job.

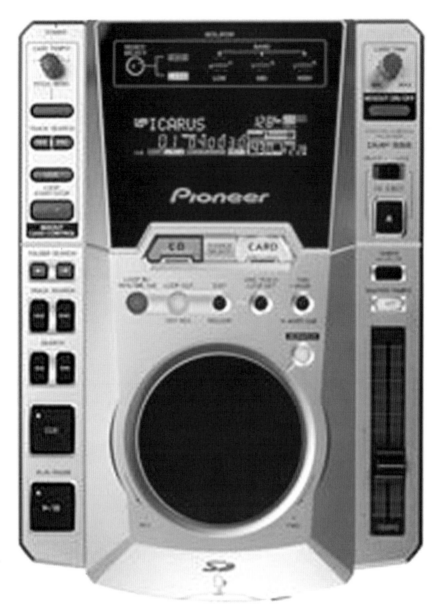

FIGURE 23.4
The Pioneer DMP-555.
Photograph printed
with permission.
© Pioneer 2003

MIXERS

Having two decks or two DJ CD players isn't much use unless you have a mixer to join the two together and though many starter kits include these, some of the more professional gear will not. Nevertheless, I'm going to be brief here, as the market is absolutely flooded with DJ mixers, some with fancy flashing lights, and knobs, while others have a host of various bells and whistles (that can be confusing if you're just learning). Again, which to choose would depend on what you want and how much gear you have to plug into the mixer. For instance, if you have only two decks (CD or vinyl, it doesn't really matter), then you will probably only need a two-channel, four-input mixer, i.e. 2 × phono and 2 × line. On the other hand, if you have untold number, then you will need to invest in more channels.

Some mixers have built-in beat counters, and effects, but these won't necessarily help you mix any better. Personally, I've found that the more the flashy parameters, the more it tends to put you off the job at hand since they tend to be a hindrance more than anything else. It is worth trying them out, though, and seeing what you think – it's all about what suits you and your budget. If I were to recommend one, however, then it would be the Ecler Smac 32. I had one of these for ages and it was a brilliant mixer with a lovely warm sound and very slim line to fit between the decks. Ecler makes some fine stuff, with an expanding range like the impressive HAK 310 and 360 series.

Alternatively, Vestax has a staggering range of quality mixers such as the classic PMC-05 Pro, or the PCV 275. Equally, Stanton also has a good line with the SMX401 and 501 and SA8, all of which with are two channels, and Pioneer has also made some DJ mixers that are worth checking. The most notable is the DJM300, 500 and 600 series with two and four channels, respectively. The 500/600s have a built-in FX and sampler that are very user-friendly. All these makes are some of the best I have ever used, but the true crème de la crème has to be Soundcraft range of DJ mixers, but these are very expensive and so are usually limited to the DJs who earn substantial amounts of money.

MP3

I mentioned MP3 earlier when talking about the Pioneer CD DJ decks, but while some say the difference in sound quality is noticeable when compared to CD or Minidisk, there really isn't a lot in it for the average listener. If it's played on a quality system, it's still going to sound good, but watch it if you're mixing vinyl into it! The sound difference between MP3, and CD for that matter, is quite substantial because vinyl wins every time for warmth and depth of sound. Nonetheless, the technology around the MP3 format has, as I've mentioned, evolved into the DJ market, and programmes have started to appear that are specifically for DJ/performance use, whether live at the club/festival or in the studio or at home. There are fantastic programmes like Native Instruments' (NI) Traktor DJ studio and Soundgraph's D-Vinyl and PC-DJ, but the real daddy of them all is Stanton's Final Scratch system. This is certainly leading the way for this technology while very cleverly keeping exactly the same techniques DJs use.

Tracktor, D-Vinyl and PC-DJ are merely software-based in that they allow MP3 and wav files to be mixed together via a computer, and are extremely good. The look is similar to a dual CD analogue desk and the layout is pretty much the same, with similar buttons, pitch bend/aDJust, etc., with NI's Traktor being one of the leaders. With Stanton's (FS1and2) Final Scratch, however, they have gone one step further, and then some!

They have cleverly devised a system that comprises of a central hub, to which you hook your turntables up before connecting to your mixer and then into your PC. By then using two pieces of time-coded 'vinyl' you can 'virtually' play, scratch, mix and blend the wav and MP3 files together, using a clever piece of software, but exactly the same as mixing two records together on your turntables. There isn't actually any sound on these special 'vinyl' records; they just transmit special signals to the software. Essentially, this means that you can record your music collection into your PC, plus new tracks, and literally have hundreds and thousands of tunes in your 'virtual' record box (hard drive) that you can then mix in the same physical manner as normal vinyl! Plus, this system is truly portable, so all you need to take to a gig is a hi-spec laptop, the FS hub and your two pieces of special vinyl. We are seeing increased popularity at gigs now that include this set-up, alongside the traditional gear, and it seems to be gaining momentum and respect from a quality name in the industry at the forefront of technology.

This obviously makes MP3s a viable DJ format and although most jocks still have a passion for vinyl – I know I do – this technology has embraced the past as well as the future, so there's no need to give your priceless collection of vinyl to Oxfam just yet! You can indeed use both, to a staggering effect. If you want to embrace new technology, yet still use all the essential skills of mixing, then this is certainly a way to go. It's an amazing, forward-thinking system, and is rock solid, as it has been road tested by Ritchie Hawtin for many years.

COULD YOU OFFER ANY TIPS FOR ASPIRING DJs?

First and most important, you need a passion for music. Getting into the scene just for fame, sex and money will not get you very far; it's a hard slog being a DJ and the only way you'll be able to stick at it is if you love what you do. The obvious place to start is by learning the basic skills of DJ'ing first. This means learning the fundamentals such as cueing up a record (assuming your chosen medium will be vinyl), becoming accustomed to the weight of the vinyl (heavier vinyl has a better sound quality) along with figuring out how to drop a record on the first beat and where the beats are by simply looking at the vinyl. When you are comfortable with this, try the basics of beat mixing by syncing two records together and mixing between them. Of course, at the moment these may seem incredibly simple and you may be twitching to get to the more advanced techniques, but trying to learn the advance techniques before you can even beat match will inevitably end in disaster. If you take it easy, it will come and even though you may get the hang of the basics quickly it doesn't mean that you can DJ.

Once you feel confident enough, start doing small parties, playing on small sound systems and in front of small crowds. It's a whole different ball game than playing at home in your bedroom, you'll be amazed at the difference in sound, plus you have many other distractions in a club, the lights, the smoke, rude people pestering you in the mix! But just do it for a laugh at first and see how you go, if you want to be the next Cutmaster Swift or Judge Jules, you need to learn the basics of mixing first. DJ'ing is in some ways an art, but hip-hop scratching/cutting, etc. is an art form in itself, and like advance DJ'ing skills it literally takes years to learn and perfect. The only real way is to start from the very bottom and work your way up. This means you will need to do gigs for free until you're at a proficient level and have proved yourself to be worthy of rubbing shoulders with the major players. It certainly won't happen overnight, but be persistent, believe in what you do and try to be individual – it's always a mistake to play all the same tracks as your peers.

When you feel confident enough, it's prudent to start recording your sets and then listen back to them with a critical ear. Ask yourself questions such as is the mixing tight and are the songs well synced. More importantly, be honest with yourself because even the DJs at the very top will admit that they do not 'know it all' and are constantly learning new techniques.

If at first you don't succeed, try again and then again, and then again until you do. As boring as this will probably sound, if your 'practiced' demo sounds dodgy then you won't get a look in, so keep practising until it is right, and it will gradually become easier. You have to be dedicated, as there are plenty of quality DJs out there and you have to outshine them to stand a chance.

It's also worth putting your best polished demo onto CD or minidisk to use as an advertisement for your skills (more on promoting yourself in a moment). The best demo in the world won't get played if it looks pants; a crappy dusty old tape with a name and number scribbled in dodgy biro isn't going to do you any favours. A large number of demos I get look like dog's dinner! If that's the amount of trouble they've gone on the outside, what's the recording and mixing going to be like? Naturally, I'm not saying you have to get a gold-plated CD and cover, but a little presentation doesn't go amiss. Don't forget the bigger promoters receive hundreds of demos all the time, so you need to make an effort for them to notice you. Get together a DJ biography, what you've been up to, what clubs and radio gigs you've done, etc. and print it on a nice paper. Also, learn how to network, pester promoters and be knowledgeable. The internet is a good way to promote yourself, so construct a site or get a few pages on various free sites and upload your MP3 mixes to it. I do lots of radio work, both internet and FM, and that's another angle to get you out to a wider audience; it boosts your confidence and is another string to your musical bow.

HOW DO DJs PROMOTE THEMSELVES?

There are various ways you can do this, some of which I've just touched upon, but it depends on what level you are at and how much experience you have. As I've mentioned, the internet is probably one of the best ways and its worthwhile building a website and placing a biography and an MP3 mix on there. When you're at a reasonable level, consider joining an agency, as they will profile you, but obviously you need to have a certain amount of experience and achievement behind you.

I still spend many, many hours tirelessly sending hundreds and hundreds of e-mails out and spending a lot of time on courtesy and follow-up calls, chasing promoters, making new contacts, following up contacts and networking all the time. You have to try to have something going on all the time and in the future, even several months away. The busier you become, the further down the line you have to plan, and it helps if you have a few fingers in a few musical pies so to speak, be determined, have drive, as it's a very cut-throat business. I review my website regularly and always make an effort to keep abreast of any new music that comes out.

More importantly, you need to be adventurous. I've found myself accepting bookings in many weird and wonderful places resulting in me DJ'ing up mountainsides, at the edge of lakes, in Russia during wintertime, at festivals in Armenia in front of 40 000 people, in swanky clubs of Bulgaria and even in disused military arms bases in Germany! If you have an audacious streak, you'll have the chance to play at some amazing places and meet some wonderful people and make new friends. What's more, they will value you much, much more than your local club in whichever major city you live in. This will allow you to build up a massive fan base, a lot of whom are budding DJs themselves, and will strive to get you more work in these places.

Another good tip to promote yourself is to get a portfolio together, achievements, flyers, discography if that applies, DJ biography, etc., but make it ostentatious so that you will have something presentable to show to the promoters and the like. Also, you have to start building up a reputation for yourself, and it's hard, you have to take the rough with the smooth. If you're on the lower level, you'll often find yourself playing the warm ups and earlier times, and you'll probably be playing to the floorboards for quite some time. Most of this will be for gratis, but it will give you the vital experience you need. When I started DJ'ing, it was almost a dirty word and I had to collect glasses in between playing. What's more, when the club shut I had to clear up and sweep the floor before I could get paid. That's what a resident meant back in the early 1980s, there certainly wasn't such a thing as the 'superstar' DJ nonsense that there is today. You learnt the hard way, but gained invaluable experience along the way, playing to tough demanding crowds, learning essential skills of dance floor dynamics. It's the only way to get the experience you need to be a DJ. All too often I see the sorry sight of 'DJs' with 40 records and 12 months of experience under their belt with a pair of dodgy Sound Labs who think they are something special. Remember who you are and be true to yourself, that's what counts, and know where you're at and where you would like to be, because if you disappear up your own backside, you'll quickly lose respect from everyone.

ANY OTHER ADVICE YOU CAN OFFER?

Possibly the best advice I can offer is to look after yourself. I'm going to go into some aspects quite deeply here but it's not to put you off; on the contrary, it is an insight into the potential hazards of being a DJ.

Firstly, I'll start with what is the most important aspect of the job – your hearing! It's all too easy to have the opinion that it *'won't bother me'*, but if you spend a lot of time in clubs subjecting yourself to very high levels of sound as DJs, rock musicians, lighting and sound engineers do, then I would fully recommend investing in some professional musician's ear protectors. These work by using a diaphragm that vibrates at a lower level than your eardrum, therefore reducing the noise by 15–25 dB. In layman's terms this means that in a loud environment such as a club you can still hear everyone very clearly and the plugs don't spoil any of the enjoyment of the music; it just feels as if someone has turned down the volume control at the back of your head. These are very different from the traditional foam plugs that just block out noise and make everything appear muffled. Without protecting your ears you can suffer from tinnitus. This is the medical term for a ringing in the ears, which is often the case after a night out working/clubbing. The ringing is our body's natural defence mechanism when we have been subjected to too much noise; for most of us this will pass, but the damage that the ringing has signalled won't. Many old skool DJs, including myself, used to play and listen to massive amounts of noise on monstrous sound systems which admittedly were great at the time, but we were unaware of the damage we were doing to our hearing. We never had this kind of advice back then, so I'm glad to be able to share some of my knowledge with you now.

TIPS FOR YOU EARS

- Wear specialist earplugs if you're working in or frequently attending loud clubs/concerts, etc.
- Take regular breaks from the dance floor if you're a clubber or pro dancer.
- Be careful on drink or drugs, as your perception of what is too loud will be impaired.
- Don't get dehydrated, as it can increase the risk of tinnitus.
- Don't push the levels to the max when DJ'ing (as is too easy to do on a big system).
- Avoid shouting conversations into each other's ears, remember those earplugs!?
- Don't stand next to big speaker stacks at big clubs and festivals.

Another hazard for the DJ (especially those that play mostly vinyl) is back pain. I'm a towering 6'2" and occasionally get it so chronically that I have to cancel gigs! Carrying heavy record boxes and bags about is fundamental for DJs because you have to get your precious records or CDs to the gig. Over the years of doing this, I now have one shoulder higher than the other, as well as a trapped nerve in my neck (as does Tall Paul), which is worsened with stress, and being cramped in airplane seats and cars for long journeys does not help things. This is unfortunately one of the inherent and unglamorous elements of the job.

Apart from these problems, the most damage is done is when you're actually standing, stooped over the decks for hours. This can cause a lot of damage and is hazardous for your spine, especially over a number of years. Mobile jocks are even more susceptible as they will be carrying even more cumbersome heavy equipment around. Also, another problem is hunching one shoulder, pressing the headphones to one ear, which is not good for you neck. It isn't just me, a lot of DJs, especially tall ones, have the same problem; Judge Jules and Tiesto are two high-profile jocks who are equally affected. Obviously, any stance that stops you bending over the decks unnecessarily is a good thing. As mentioned, I'm particularly tall, so sometimes have to part my legs when the decks are a bit low, as this eases the stress on your lumber region. Try and find a stance that is good for you, as it's in your interest to adopt this pose when you're playing regularly; try and make it a habit.

TIPS FOR YOUR BACK

- Invest in a pair of decent headphones you can use on one ear without hunching your shoulders too much; it will make a lot of difference.
- Change position and stretch regularly when you're playing.
- Be careful on drinks so as not to strain while dancing/throwing shapes!
- Carry equipment/boxes, etc. close to your body, with the heaviest side closest to you. Share heavy loads between friends where possible.
- Get some strong painkillers that will ease the pain if it gets chronic.
- Back pain can become serious, get it checked out and see a specialist if necessary.
- Carry a muscle relief spray or gel with you when you travel.
- Physical exercise plays a big part in keeping your back healthy, so try some stretching exercises, especially after you land from a flight, and after you're set.
- If playing gigs internationally, have a massage when you get to your destination, at a reputable place, as this helps unwind the stress and tension that causes back and neck pain.

Also tied in with the lifestyle of the DJ is the stress and fatigue factor. On the surface of what appears to be a very glamorous, never-ending party image of the busy international DJ, there are gruelling physical and mental demands. With hectic schedules and busy flight patterns, jocks can burn out if not too careful. Many of the top DJs are now coming clean about how demanding this actually is, with people like Sasha admitting that he has suffered years of serious panic attacks, and many other top DJs collapsing from exhaustion. Some not so wise DJs started to abuse heavy amounts of narcotics to deal with the near superhuman aspect of the job and ended up getting in a spectacular mess as a result. Both Brandon Block and Nicky Holloway have been into rehab through this. It's vital that you try and look after yourself; it is still possible to enjoy the highlights of the lifestyle if you take care of yourself and keep your mind and body in good shape.

TIPS FOR FATIGUE

- Allow yourself to fully recover after a long period awake (sleep and rest well).

- Carry effervescent multi-vitamin energy tablets with you and take them after a long journey to give you a boost and combat exhaustion.
- Avoid dehydration by staying off booze and caffeine during and after flights, even though it's tempting with complimentary sauce; drink lot of mineral water instead.
- Get a good, undisturbed sleep, using eye mask and earplugs if necessary but not tranquilisers, as they can mess up sleep patterns (if necessary, use herbal remedies).
- Exhaustion affects the immune system, leading to illness, so try and maintain a healthy diet. Eat lots of fruits and take daily vitamin supplements.
- Take melatonin tablets, a natural remedy thought to help promote sleep and reduce the effects of jet lag (unfortunately not available in the UK, so stock them up in places like America).

USEFUL LINKS

- www.energyinternetradio.com (Miami's best radio station)
- www.grooveexponents.co.uk (DJ agency)
- www.elacin.nl or call 0207 323 2076 (musicians ear protectors)
- www.hearnet.com (useful tips on your hearing gear)
- www.backcare.org.uk (Look after that back of yours)
- http://www.pioneer-eur.com/eur (forward thinking DJ gear)
- www.technics.co.uk (home of the world leaders and the mighty Technics deck)
- www.stantonDJ.com (home to the finest cartridges, and the world famous Final Scratch)
- www.vestax.co.uk (purveyors of quality DJ gear)
- www.dmcworld.com

Binary and Hex

All hardware circuits, even computers, are based on a series of inter-connected switches that can be in one of two states: either on or off. By continually switching these on and off in different configurations it's possible for the hardware to count and perform complex calculations. This switching is accomplished by sending a series of bytes down an electrical cable. If it is equal to 1 then the switch is turned on while if it's a 0 then the switch is turned off. Because of this, all computers must count using base 2, or 'binary' rather than our usual method of counting which is base 10.

COUNTING IN BINARY

Our normal method of counting was developed because we have ten figures. Using this system, any one digit in a number has a value that is based upon the position of the digit in the number. Thus, when we are working to the power of 10, every time a digit moves one to the left its result is to the power of 10, then by 100, then by 1000 and so forth. Consequently, the number 17 593 could be seen as follows.

10000000	1000000	100000	10000	1000	100	10	1
0	0	0	1	7	5	9	3

Or $(1 \times 10000) + (7 \times 1000) + (5 \times 100) + (9 \times 10) + (3 \times 1) = 17360$

From this above example we can see that each position in a number essentially adds a 0 to the meaning of a digit. Or, the value of each position is ten times the value of the previous position (moving from right to left). This method of counting is also implemented in binary, but rather than use a base 10 system, we use a base 2 system. So, rather than saying that 10 to the power of $0 = 1$, 10 to the power of $1 = 10$, 10 to the power of $2 = 100$, 10 to the power of $3 = 1000$, it's 2 to the power of 0, 2 to the power of 1, 2 to the power of 2, 2 to the power of 4 and so forth.

128	64	32	16	8	4	2	1
0	0	0	0	0	0	0	0

Looking at the above table we are assuming that there are eight hardware switches grouped together producing what is essentially an 8-bit byte. Also, as they are all switched off, the sum produced will be zero. However, if we were to introduce some positive bits we could sum them together to produce a decimal number.

128	64	32	16	8	4	2	1
0	1	1	0	1	1	0	1

$(0 \times 128) + (1 \times 64) + (1 \times 32) + (0 \times 16) + (1 \times 8) + (1 \times 4) + (0 \times 2) + (1 \times 1) = 109$

Thus the decimal equivalent of the binary 01101101 would be 109. Additionally, we could also determine that the maximum number that could be calculated via binary would be 255 (11111111).

As MIDI uses this same 8-bit system when communicating with any devices, it would seem sensible to assume that it should be able to offer this same maximum parameter (of 255) but this isn't the case. Similar to CC messages, two forms of information need to be transmitted; a status bit and a data byte (which is composed of 7 bits). The status bit informs the synth of an incoming message that is arriving, while the following data byte informs it by how much the parameter should be adjusted. Because of this initial status bit, only 7 other bits are left to provide the information, resulting in a maximum decimal value of 127; hence, the maximum number of any CC message can only be 127. To transmit numbers larger than this, an 8-bit byte has to be split into two halves and then converted into another numeration format, hexadecimal.

When we split a byte into two halves, both halves are commonly referred to as 'nibbles'. If the previous example were to be broken down into two nibbles, it would become 0110 and 1101. These could then be individually summed together as 0110 = 96 and 1101 = 13 (96 + 13 = 109) to produce the result again. However, by splitting a byte into two and then converting it into a hexadecimal value, it's possible to produce much higher values. The reason behind this is that hexadecimal works to base-16, meaning that it is possible to access up to 16 383 parameters in a synth, much more than the standard 127 offered through CC messages.

COUNTING IN HEXADECIMAL

Hex uses a base-16 numbering system, but as there are not enough symbols to represent 16 different digits, as soon as the number 10 is reached it has to be converted into letters. Thus to represent number 10–15 the letters A–F are used.

Recall how we count in decimal. Counting upwards, we count from 1 through 9 and then we place a 1 to the left of this and go back to 0, making the number 10 (which actually means 1 to the power of 10 plus 0 to the power of 1). With hexadecimal we count beyond 9, but because our entire numeric system is only based on 9 digits, with hexadecimal, letters are used instead as below so that we can count up to 15.

F	E	D	C	B	A	9	8	7	6	5	4	3	2	1	0
15	14	13	12	11	10	9	8	7	6	5	4	3	2	1	0

When we count the zero, just as we do with the decimal counting system, we have used a total of 16 numbers. You can see from the decimal to hexadecimal conversion table that whenever we count up to F, increase the number to the left by 1. This is also the same as the more common decimal counting system, for example, if you count from 11 through 19, there are no more digits to use to you change the 9 to 0 and increase the number to the left by 1, giving us 20.

You can see that hexadecimal counting is exactly the same as decimal, except that in hexadecimal we count up to F before increasing the number to the left by a factor of 1.

It follows that in decimal, the number 24 actually means '2 × 10 to the power of 1 + 4 × 10 to the power of 0 = 24', which is what you or I already understand it to be because we use this every day like counting on our fingers.

In hexadecimal, the number 24 actually means '2 × 16 to the power of 1 + 4 × 16 to the power of 0 = 36'.

Logically it follows then that the number CE in hexadecimal means '12 × 16 to the power of 1 + 14 × 16 to the power of 0 = 206' in decimal.

Using this principle, it's easy to convert hexadecimal into decimal.

Using this method, it's possible to count until we reach FF, which essentially means 15 × 16 + 15 units or 255 in decimal. Creating a new position gives 100 or 16 to the power of 2 which is 256 in decimal. Continuing to count using this base system it would be possible to produce two nibbles, both with a value of 7F resulting in a total of 16 383. (127 × 128 + 127 = 16383!).

For instance, suppose we wished to convert the decimal number **12720** to hexadecimal:

- The largest power of 16 that fits into 12720 is $16^3 = 4096$
  It fits **3** times and gives a remainder of 432. This number is derived from the fact that

  (3 × 4096) = 12288
  (12720 − 12288) = 432

  The first hex number is **3**

- The largest power of 16 that fits into 432 is $16^2 = 256$
  It fits 1 time and gives a remainder of 176. This number is derived from the fact that

  $(1 \times 256) = 256$
  $(432 - 256) = 176$

  The second hex number is **1**
- The largest power of 16 that fits into 176 is $16^1 = 16$
  It fits 11 times with no remainder (this remainder equates to 1 to the power of nothing which is nothing and will make 0 the last digit in the hex number). The third hex number is therefore B
- The hex equivalent of the decimal number 12720 is therefore: 31B0

To clarify, to convert 31B0 hexadecimal into decimal the maths is as follows:

$0 \times$ (16 to the power of nothing) = 0
B (which is equivalent to 11 in decimal) $\times$ (16 to the power of 1) = 176
$1 \times$ (16 to the power of 2) = 256
$3 \times$ (16 to the power of 3) = 12288

$0 + 176 + 256 + 12288 =$ **12720** in decimal

As another example, suppose we wanted to convert the decimal number **14683**

- The largest power of 16 that fits into 14683 is $16^3 = 4096$
  It fits 3 times and gives a remainder of 2395
- This remainder is derived from the fact that

  $(3 \times 4096) = 12288$
  $(14683 - 12288) = 2395$

  The first hex number is **3**
- The largest power of 16 that fits into 2395 is $16^2 = 256$
  It fits 9 times and gives a remainder of 91
- This remainder is derived from the fact that

  $(9 \times 256) = 2304$
  $(2395 - 2304) = 91$

- The second hex number is **9**
- The largest power of 16 that fits into 91 is $16^1 = 1$
  It fits 5 times and gives the remainder of 11
- This remainder is derived from the fact that

  $(5 \times 16) = 80$
  $(91 - 80) = 11$

- The third hex number is **5**

■ The remainder of 11 has no power of 16 that will divide into it as a whole number so the last digit of the hex will be the hex equivalent of 11 which is B. This gives the final hex number which is therefore 395B.

Again we can check this by converting 295B hexadecimal into decimal. The math is as follows:

B × (16 to the power of nothing) = 11
5 × (16 to the power of 1) = 80
9 × (16 to the power of 2) = 2304
3 × (16 to the power of 3) = 12288
11 + 80 + 2304 + 12288 = **14683** in decimal

Decimal to Hexadecimal Conversion Table

Dec	Hex	Dec	Hex	Dec	Hex	Dec	Hex	Dec	Hex	Dec	Hex
0	0	19	13	38	26	57	39	76	4C	95	5F
1	1	20	14	39	27	58	3A	77	4D	96	60
2	2	21	15	40	28	59	3B	78	4E	97	61
3	3	22	16	41	29	60	3C	79	4F	98	62
4	4	23	17	42	2A	61	3D	80	50	99	63
5	5	24	18	43	2B	62	3E	81	51	100	64
6	6	25	19	44	2C	63	3F	82	52	101	65
7	7	26	1A	45	2D	64	40	83	53	102	66
8	8	27	1B	46	2E	65	41	84	54	103	67
9	9	28	1C	47	2F	66	42	85	55	104	68
10	A	29	1D	48	30	67	43	86	56	105	69
11	B	30	1E	49	31	68	44	87	57	106	6A
12	C	31	1F	50	32	69	45	88	58	107	6B
13	D	32	20	51	33	70	46	89	59	108	6C
14	E	33	21	52	34	71	47	90	5A	109	6D
15	F	34	22	53	35	72	48	91	5B	110	6E
16	10	35	23	54	36	73	49	92	5C	111	6F
17	11	36	24	55	37	74	4A	93	5D	112	70
18	12	37	25	56	38	75	4B	94	5E	113	71

Dec	Hex	Dec	Hex	Dec	Hex	Dec	Hex	Dec	Hex	Dec	Hex
114	72	139	8B	164	A4	189	BD	214	D6	239	EF
115	73	140	8C	165	A5	190	BE	215	D7	240	F0
116	74	141	8D	166	A6	191	BF	216	D8	241	F1
117	75	142	8E	167	A7	192	C0	217	D9	242	F2
118	76	143	8F	168	A8	193	C1	218	DA	243	F3
119	77	144	90	169	A9	194	C2	219	DB	244	F4
120	78	145	91	170	AA	195	C3	220	DC	245	F5
121	79	146	92	171	AB	196	C4	221	DD	246	F6
122	7A	147	93	172	AC	197	C5	222	DE	247	F7
123	7B	148	94	173	AD	198	C6	223	DF	248	F8
124	7C	149	95	174	AE	199	C7	224	E0	249	F9
125	7D	150	96	175	AF	200	C8	225	E1	250	FA
126	7E	151	97	176	B0	201	C9	226	E2	251	FB
127	7F	152	98	177	B1	202	CA	227	E3	252	FC
128	80	153	99	178	B2	203	CB	228	E4	253	FD
129	81	154	9A	179	B3	204	CC	229	E5	254	FE
130	82	155	9B	180	B4	205	CD	230	E6	255	FF
131	83	156	9C	181	B5	206	CE	231	E7		
132	84	157	9D	182	B6	207	CF	232	E8		
133	85	158	9E	183	B7	208	D0	233	E9		
134	86	159	9F	184	B8	209	D1	234	EA		
135	87	160	A0	185	B9	210	D2	235	EB		
136	88	161	A1	186	BA	211	D3	236	EC		
137	89	162	A2	187	BB	212	D4	237	ED		
138	8A	163	A3	188	BC	213	D5	238	EE		

General MIDI Instrument Patch Maps

Programme Nos.	Instrument Set (Piano)	Programme Nos.	Instrument (Chromatic Perc)
1	Acoustic Grand	9	Celesta
2	Bright Acoustic	10	Glockenspiel
3	Electric Grand	11	Music Box
4	Honky Tonk	12	Vibraphone
5	Electric Piano 1	13	Marimba
6	Electric Piano 2	14	Xylophone
7	Harpsichord	15	Tubular Bells
8	Clav	16	Dulcimer

Programme Nos.	Instrument (Organ)	Programme Nos.	Instrument (Guitar)
17	Drawbar Organ	25	Nylon Acoustic Guitar
18	Percussive Organ	26	Steel Acoustic Guitar
19	Rock Organ	27	Jazz Electric Guitar
20	Church Organ	28	Clean Electric Guitar
21	Reed Organ	29	Muted Electric Guitar
22	Accordian	30	Overdrive Guitar
23	Harmonica	31	Distortion Guitar
24	Tango Accordian	32	Guitar Harmonics

Programme Nos.	Instrument (Bass)	Programme Nos.	Instrument (Strings)
33	Acoustic Bass	41	Violin
34	Finger Bass	42	Viola
35	Pick Bass	43	Cello
36	Fretless Bass	44	Contrabass
37	Slap Bass 1	45	Tremolo Strings
38	Slap Bass 2	46	Pizzicato
39	Synth Bass 1	47	Orchestral
40	Synth Bass 2	48	Timpani

Programme Nos.	Instrument (Ensemble)	Programme Nos.	Instrument (Brass)
49	String Ensemble 1	57	Trumpet
50	String Ensemble 2	58	Trombone
51	Synth Strings 1	59	Tuba
52	Synth Strings 2	60	Muted Trumpet
53	Choir Aahs	61	French Horn
54	Choir Oohs	62	Brass Section
55	Synth Voice	63	Synth Brass 2
56	Orchestral Hit	64	Synth Brass 2

Programme Nos.	Instrument (Reed)	Programme Nos.	Instrument (Pipe)
65	Soprano Sax	73	Piccolo
66	Alto Sax	74	Flute
67	Tenor Sax	75	Recorder
68	Baritone Sax	76	Pan Flute
69	Oboe	77	Blown Bottle
70	English Horn	78	Skakuhachi
71	Bassoon	79	Whistle
72	Clarinet	80	Ocarina

Programme Nos.	Instrument (Synth Lead)	Programme Nos.	Instrument (Synth Pad)
81	Square Lead	89	New Age Pad
82	Sawtooth Lead	90	Warm Pad
83	Calliope Lead	91	Polysynth Pad
84	Chiff Lead	92	Choir Pad
85	Charang Lead	93	Bowed Pad
86	Voice Lead	94	Metallic Pad
87	Fifths Lead	95	Halo Pad
88	Bass and Lead	96	Sweep Pad

Programme Nos.	Instrument (Synth Effects)	Programme Nos.	Instrument (Ethnic)
97	Rain FX	105	Sitar
98	Soundtrack FX	106	Banjo
99	Crystal FX	107	Shamisen
100	Atmosphere FX	108	Koto
101	Brightness FX	109	Kalimba
102	Goblins FX	110	Bagpipe
103	Echoes FX	111	Fiddle
104	Sci-Fi FX	112	Shanai

Programme Nos.	Instrument (Percussive)	Programme Nos.	Instrument (Sound FX)
113	Tinkle Bell	121	Guitar Fret Noise
114	Agogo	122	Breath Noise
115	Steel Drums	123	Seashore
116	Woodblock	124	Bird Tweet
117	Taiko Drums	125	Telephone Ring
118	Melodic Toms	126	Helicopter
119	Synth Drum	127	Applause
120	Reverse Cymbal	128	Gunshot

GENERAL MIDI PERCUSSION SET

MIDI Key	Drum Sound	MIDI Key	Drum Sound
35	Acoustic Bass Drum	59	Ride Cymbal
36	Bass Drum	60	Hi Bongo
37	Side Stick	61	Low Bongo
38	Acoustic Snare	62	Mute Hi Bongo
39	Hand Clap	63	Open Hi Bongo
40	Electric Snare	64	Low Conga
41	Low Floor Tom	65	High Timbale
42	Closed Hi-Hats	66	Low Timbale
43	High Floor Tom	67	High Agogo
44	Pedal Hi-Hat	68	Low Agogo
45	Low Tom	69	Cabasa
46	Open Hi-Hat	70	Maracas
47	Low Mid Tom	71	Short Whistle
48	High Mid Tom	72	Long Whistle
49	Crash Cymbal	73	Short Guiro
50	High Tom	74	Long Guiro
51	Ride Cymbal	75	Claves
52	Chinese Cymbal	76	Hi Wood Block
53	Ride Bell	77	Low Wood Block
54	Tambourine	78	Mute Cuica
55	Splash Cymbal	79	Open Cuica
56	Cowbell	80	Mute Triangle
57	Crash Cymbal	81	Open Triangle
58	Vibraslap		

General MIDI CC List

CC	Function	CC	Function	CC	Function
0	Bank Select	19	General Control 4	49	General Purpose Controller 2
1	Mod Wheel	20–31	Undefined	50	General Purpose Controller 3
2	Breath Controller	32	Bank Select	51	General Purpose Controller 4
3	Undefined	33	Mod Wheel	52–63	Undefined
4	Foot Controller	34	Breath Control	64	Damper Pedal (on/off)
5	Portamento Time	35	Undefined	65	Portamento (on/off)
6	Data Entry	36	Foot Control	66	Sustenuto (on/off)
7	Channel Volume	37	Portamento Time	67	Soft Pedal (on/off)
8	Balance	38	Data Entry	68	Legato Footswitch
9	Undefined	39	Channel Volume	69	Hold 2
10	Pan	40	Balance	70	Sound Controller 1 (Sound Variation)
11	Expression	41	Undefined	71	Sound Controller 2 (Timbre)
12	Effect Control 1	42	Pan	72	Sound Controller 3 (Release Time)
13	Effect Control 2	43	Expression Controller	73	Sound Controller 4 (Attack Time)
14–15	Undefined	44	Effect Control 1	74	Sound Controller 5 (Brightness)
16	General Control 1	45	Effect Control 2	75	Sound Controller 6
17	General Control 2	46–47	Undefined	76	Sound Controller 7
18	General Control 3	48	General Purpose Controller 1	77	Sound Controller 8

CC	Function	CC	Function	CC	Function
78	Sound Controller 9	93	Effects 3 (Chorus) Depth	120	All Sound off
79	Sound Controller 10	94	Effects 4 (Detune) Depth	121	Reset All Controllers
80	General Purpose Controller 5	95	Effects 5 (Phaser) Depth	122	Local Control on/off
81	General Purpose Controller 5	96	Data entry +1	123	All Notes off
82	General Purpose Controller 5	97	Data entry −1	124	Omni Mode off (+ all notes off)
83	General Purpose Controller 5	98	Non-Registered Parameter Number LSB[1]	125	Omni Mode on (+ all notes off)
84	Portamento Control	99	Non-Registered Parameter Number MSB	126	Poly Mode on/off (+ all notes off)
85–90	Undefined	100	Registered Parameter Number LSB	127	Poly Mode on
91	Effects 1 (Reverb) Depth	101	Registered Parameter Number MSB[2]		
92	Effects 2 (Tremolo) Depth	102–119	Undefined		

[1]Least significant bit.
[2]Most significant bit.

Sequencer Note Divisions

The following charts display the number of clock pulses for each note value for the four most popular PPQN resolutions – 96, 192, 240 and 384.

Note Value	PPQN	Note Value	PPQN	Note Value	PPQN
96 PPQN Whole	384	Dotted whole	576	Triplet whole	256
Half	192	Dotted half	288	Triplet half	128
Quarter	96	Dotted quarter	144	Triplet quarter	64
Eighth	48	Dotted eighth	72	Triplet eighth	32
1/16th	24	Dotted 1/16th	36	Triplet 1/16th	16
32nd	12	Dotted 32nd	18	Triplet 32nd	8
64th	6	Dotted 64th	9	Triplet 64th	4
128th	3	Dotted 128th	N/A	Triplet 128th	2
192 PPQN Whole	768	Dotted whole	1152	Triplet whole	512
Half	384	Dotted half	576	Triplet half	256
Quarter	192	Dotted quarter	288	Triplet quarter	128
Eighth	96	Dotted eighth	144	Triplet eighth	64
1/16th	48	Dotted 1/16th	73	Triplet 1/16th	32
32nd	24	Dotted 32nd	36	Triplet 32nd	16
64th	12	Dotted 64th	18	Triplet 64th	8
128th	6	Dotted 128th	9	Triplet 128th	4

Note Value	PPQN	Note Value	PPQN	Note Value	PPQN
240 PPQN					
Whole	960	Dotted whole	1440	Triplet whole	640
Half	480	Dotted half	720	Triplet half	320
Quarter	240	Dotted quarter	360	Triplet quarter	160
Eighth	120	Dotted eighth	180	Triplet eighth	80
1/16th	60	Dotted 1/16th	90	Triplet 1/16th	40
32nd	30	Dotted 32nd	45	Triplet 32nd	20
64th	15	Dotted 64th	N/A	Triplet 64th	10
128th	N/A	Dotted 128th	N/A	Triplet 128th	5
384 PPQN					
Whole	1536	Dotted whole	2304	Triplet whole	1024
Half	768	Dotted half	1152	Triplet half	512
Quarter	384	Dotted quarter	576	Triplet quarter	256
Eighth	192	Dotted eighth	288	Triplet eighth	128
1/16th	96	Dotted 1/16th	144	Triplet 1/16th	64
32nd	48	Dotted 32nd	72	Triplet 32nd	32
64th	24	Dotted 64th	36	Triplet 64th	16
128th	12	Dotted 128th	18	Triplet 128th	8

Tempo Delay Time Chart

If song tempo is 128 BPM then set delay time to 469 ms for quarter-note delay, 234 ms for eighth-note delay, 156 ms for eighth-triplet delay or 117 ms for 1/16th note delay.

Tempo	1/4	1/8	1/8T	1/16	Tempo	1/4	1/8	1/8T	1/16	Tempo	1/4	1/8	1/8T	1/16
80	750	375	250	188	98	612	306	204	153	116	517	259	172	129
81	741	370	247	185	99	606	303	204	153	117	513	256	171	128
82	732	366	244	183	100	600	300	200	150	118	508	254	169	127
83	723	361	241	181	101	594	297	198	149	119	504	252	168	126
84	714	357	238	179	102	588	294	196	147	120	500	250	167	125
85	706	353	235	176	103	583	291	194	146	121	496	248	165	124
86	698	349	233	174	104	577	288	192	144	122	492	246	164	123
87	690	345	230	172	105	571	286	190	143	123	488	244	163	122
88	682	341	227	170	106	566	283	189	142	124	484	242	161	121
89	674	337	225	169	107	561	280	187	140	125	480	240	160	120
90	667	333	222	167	108	556	278	185	139	126	476	238	159	119
91	659	330	220	165	109	550	275	183	138	127	472	236	157	118
92	652	326	217	163	110	545	273	182	136	128	469	234	156	117
93	645	323	215	161	111	541	270	180	135	129	465	233	155	116
94	638	319	213	160	112	536	268	179	134	130	462	231	154	115
95	632	316	211	158	113	531	265	177	133	131	458	229	153	115
96	625	313	208	156	114	526	263	175	132	132	455	227	152	114
97	619	309	206	155	115	522	261	174	130	133	451	226	150	113

Tempo	1/4	1/8	1/8T	1/16	Tempo	1/4	1/8	1/8T	1/16	Tempo	1/4	1/8	1/8T	1/16
134	448	224	149	112	154	390	195	130	97	174	345	172	115	86
135	444	222	148	111	155	387	194	129	97	175	343	171	114	86
136	441	221	147	110	156	385	192	128	96	176	341	170	114	85
137	438	219	146	109	157	382	191	127	96	177	339	169	113	85
138	435	217	145	109	158	380	190	127	95	178	337	169	112	84
139	432	216	144	108	159	377	189	126	94	179	335	168	112	84
140	429	214	143	207	160	375	188	125	94	180	333	168	111	83
141	426	213	142	106	161	373	186	124	93	181	331	167	110	83
142	423	211	141	106	162	370	185	123	92	182	299	166	110	82
143	420	210	140	105	163	368	184	123	92	183	297	165	109	82
144	417	208	139	104	164	366	183	122	91	184	295	164	108	81
145	414	207	138	103	165	364	182	121	91	185	293	164	108	81
146	411	205	137	103	166	361	181	120	90	186	291	163	107	80
147	408	302	136	102	167	359	180	120	90	187	289	162	106	80
148	405	203	135	101	168	357	179	119	89	188	287	161	106	79
149	403	201	134	101	169	355	178	118	88	190	285	161	105	79
150	400	200	133	100	170	353	176	118	88	200	283	160	104	78
151	397	199	132	99	171	351	175	117	88	201	281	159	104	78
152	395	197	132	99	172	349	174	116	87	202	279	158	103	77
153	392	196	131	98	173	347	173	116	87	203	277	157	102	77

Note: 1/8 = eighth-note delay, 1/8T = eighth-note triplet delay, 1/16 = sixteenth-note delay.

Musical Note to MIDI and Frequencies

Note	MIDI No.	Frequency	MIDI No.	Frequency
C	0	8.1757989156	12	16.3515978313
Db	1	8.6619572180	13	17.3239144361
D	2	9.1770239974	14	18.3540479948
Eb	3	9.7227182413	15	19.4454364826
E	4	10.3008611535	16	20.6017223071
F	5	10.9133822323	17	21.8267644646
Gb	6	11.5623257097	18	23.1246514195
G	7	12.2498573744	19	24.4997147489
Ab	8	12.9782717994	20	25.9565435987
A	9	13.7500000000	21	27.5000000000
Bb	10	14.5676175474	22	29.1352350949
B	11	15.4338531643	23	30.8677063285

Note	MIDI No.	Frequency	MIDI No.	Frequency
C	24	32.7031956626	36	65.4063913251
Db	25	34.6478288721	37	69.2956577442
D	26	36.7080959897	38	73.4161919794
Eb	27	38.8908729653	39	77.7817459305
E	28	41.2034446141	40	82.4068892282
F	29	43.6535289291	41	87.3070578583

Note	MIDI No.	Frequency	MIDI No.	Frequency
Gb	30	46.2493028390	42	92.4986056779
G	31	48.9994294977	43	97.9988589954
Ab	32	51.9130871975	44	103.8261743950
A	33	55.0000000000	45	110.0000000000
Bb	34	58.2704701898	46	116.5409403795
B	35	61.7354126570	47	123.4708253140

Note	MIDI No.	Frequency	MIDI No.	Frequency
C	48	130.8127826503	60	261.6255653006
Db	49	138.5913154884	61	277.1826309769
D	50	146.8323839587	62	293.6647679174
Eb	51	155.5634918610	63	311.1269837221
E	52	164.8137784564	64	329.6275569129
F	53	174.6141157165	65	349.2282314330
Gb	54	184.9972113558	66	369.9944227116
G	55	195.9977179909	67	391.9954359817
Ab	56	207.6523487900	68	415.3046975799
A	57	220.0000000000	69	440.0000000000
Bb	58	233.0818807590	70	466.1637615181
B	59	246.9416506281	71	493.8833012561

Note	MIDI No.	Frequency	MIDI No.	Frequency
C	72	523.2511306012	84	1046.5022612024
Db	73	554.3652619537	85	1108.7305239075
D	74	587.3295358348	86	1174.6590716696
Eb	75	622.2539674442	87	1244.5079348883
E	76	659.2551138257	88	1318.5102276515
F	77	698.4564628660	89	1396.9129257320

Note	MIDI No.	Frequency	MIDI No.	Frequency
Gb	78	739.9888454233	90	1479.9776908465
G	79	783.9908719635	91	1567.9817439270
Ab	80	830.6093951599	92	1661.2187903198
A	81	880.0000000000	93	1760.0000000000
Bb	82	932.3275230362	94	1864.6550460724
B	83	987.7666025122	95	1975.5332050245

Note	MIDI No.	Frequency	MIDI No.	Frequency
C	96	2093.0045224048	108	4186.0090448096
Db	97	2217.4610478150	109	4434.9220956300
D	98	2349.3181433393	110	4698.6362866785
Eb	99	2489.0158697766	111	4978.0317395533
E	100	2637.0204553030	112	5274.0409106059
F	101	2793.8258514640	113	5587.6517029281
Gb	102	2959.9553816931	114	5919.9107633862
G	103	3135.9634878540	115	6271.9269757080
Ab	104	3322.4375806396	116	6644.8751612791
A	105	3520.0000000000	117	7040.0000000000
Bb	106	3729.3100921447	118	7458.6201842894
B	107	3951.0664100490	119	7902.1328200980

Note	MIDI No.	Frequency
C	120	8372.0180896192
Db	121	8869.8441912599
D	122	9397.2725733570
Eb	123	9956.0634791066
E	124	10548.0818212118
F	125	11175.3034058561
Gb	126	11839.8215267723
G	127	12543.8539514160

Index